W0254378

About Island Press

Since 1984, the nonprofit organization Island Press has been stimulating, shaping, and communicating ideas that are essential for solving environmental problems worldwide. With more than 1,000 titles in print and some 30 new releases each year, we are the nation's leading publisher on environmental issues. We identify innovative thinkers and emerging trends in the environmental field. We work with world-renowned experts and authors to develop cross-disciplinary solutions to environmental challenges.

Island Press designs and executes educational campaigns, in conjunction with our authors, to communicate their critical messages in print, in person, and online using the latest technologies, innovative programs, and the media. Our goal is to reach targeted audiences—scientists, policy makers, environmental advocates, urban planners, the media, and concerned citizens—with information that can be used to create the framework for long-term ecological health and human well-being.

Island Press gratefully acknowledges major support from The Bobolink Foundation, Caldera Foundation, The Curtis and Edith Munson Foundation, The Forrest C. and Frances H. Lattner Foundation, The JPB Foundation, The Kresge Foundation, The Summit Charitable Foundation, Inc., and many other generous organizations and individuals.

Overtourism

Overtourism

Lessons for a Better Future

Edited by Martha Honey and
Kelsey Frenkiel

ISLANDPRESS | Washington | Covelo

Note: Chapters 1-6 were completed before the spread of the novel coronavirus. Chapters 7 explores how the solutions to overtourism can help create a responsible recovery.

Library of Congress Control number: 2020945054

All Island Press books are printed on environmentally responsible materials.

Manufactured in the United States of America
10 9 8 7 6 5 4 3 2 1

Keywords: Airbnb, cruise tourism, destination governance, historic cities, national parks, overtourism, resorts, traffic management, travel, United Nations World Tourism Organization (UNTWO), visitation caps/limits, World Heritage Sites, World Travel and Tourism Council (WTTC)

Contents

Foreword

By Elizabeth Becker

Overtourism: Lessons for a Better Future was written when the biggest problem in the world of travel was overtourism. Historic cities, rural hideaways, sun-drenched beaches, and forbidding mountains were under siege. Unrelenting crowds were destroying the very beauty, culture, and adventure that make these places so attractive. In May 2019, for example, the peak of Mount Everest was blocked by a deadly traffic jam of climbers.

Then overnight the problem became no tourism. The world shut down in early 2020 under a different siege: the pandemic caused by the novel coronavirus COVID-19. Planes, cruise ships, trains, and automobiles were grounded. We all stayed in place.

It sounds dismal, and it was. But within weeks, glimmers of light broke through. For the first time in recent memory, people from New York City to Agra, India, saw a blue sky with vistas stretching for miles. River water was running clean, and litter no longer piled on streets. Wildlife reappeared, freed from the humans and their noise, cars, and deadly pollution.

At the same time, businesses depending on tourism were going broke—many shuttering for good—revealing how much the global economy depends on the $8 trillion nonstop global travel industry. The economic devastation is expected to continue for years, which could well mean that fewer people will have the money to travel and that travel at all levels will take a greater slice of our disposable income.

Travel will return, albeit slowly. Much will depend on the discovery and availability of a vaccine to protect against COVID-19. Although this period of recovery is only a reprieve from overtourism, it is also an opportunity. Many destinations will push to fill hotels and resorts quickly, but others will take the time for a reset.

This illuminating book provides the perfect blueprint for a reappraisal by destinations wanting to avoid overtourism. It raises questions like how to quantify the invisible costs and burdens of unfettered tourism and how to protect what is essential while welcoming visitors. Essays describe how crowds have ruined the experience of wilderness parks and tropical beaches and how the climate crisis is threatening both. The goal now is to replace the unmanaged crowds with an approach known as sustainable travel: travel that enriches, not harms, an area and its local communities.

Each chapter in *Overtourism* stands alone. Arnie Weissmann explores overtourism from the vantage of the global travel industry, including how it has prospered and where it is looking for reforms. Andrea Sachs looks at beaches and coasts where the crush of overdevelopment has eroded the landscape, polluted the waters, and robbed localities of their way of life. She offers solutions, as does Martha Honey, who reveals the problems of winning World Heritage Site status and then becoming a hot spot of too much tourism. Jonathan B. Tourtellot challenges governments to manage and regulate tourism responsibly, enlisting every aspect of government—transportation, culture, environment. Many ask that officials see tourism not as a simple public relations issue measured by the number of visitors but as a critical, multidimensional planning issue addressed by a range of government agencies.

Finally, this book is meant for travelers. It is a twenty-first-century guidebook that will help us travel wisely and with a deeper respect for the places we visit. The overall message is to accept rather than fight the necessary solutions: limiting visitors and congestion, charging more to protect a habitat and society, requiring reservations, isolating fragile landscapes, and patronizing destinations that follow the new rules. The list is long, but the goal is the survival of the privilege of travel.

Chapter 1

A Growing Problem

By Arnie Weissmann

At its inception in the early 1990s, the World Travel and Tourism Council (WTTC), a leading trade association, had a singular focus: demonstrate to governments that it would be in their national interest to adopt policies that encouraged or facilitated cross-border visitation. In pursuit of this end, the organization began conducting research designed to show that if leaders eased entry requirements and fostered favorable business environments for travel industry corporations, their countries would collect more taxes, employ more citizens, and increase national prosperity.

The council, brainchild of then–American Express chief executive officer James Robinson, had its first official meeting in 1991, attended by thirty-two CEOs heading global travel-related companies.[1] That same year, approximately 450 million people crossed international borders.[2] In 2019, WTTC's Global Summit drew a record fifteen hundred attendees, among them more than one hundred travel industry CEOs, including the chiefs of Hilton and InterContinental Hotel Groups, cruise giants such as Carnival Corporation and Royal Caribbean Cruises, and tour operators TUI and Abercrombie & Kent. American Express was there, as was the chief executive officer of Expedia, the company that, in 2010, replaced American Express as the largest seller of travel products worldwide. Other companies operating at the intersection of travel and technology, including Airbnb and Uber, were represented, as were airline and airport executives, security experts, futurists, tourism ministers, and tourist boards. Heads of state, past and present, were on the program.[3] Also that year, an estimated 1.44 billion tourists would cross international borders.[4]

The parallel rise in the number of global travelers and expansion of WTTC membership from 1991 to 2019 isn't entirely coincidental—there's little question that WTTC has been an effective advocate for its private-sector membership. But it's also fair to wonder whether, at this point, its size and influence are more a mirror of the explosion of travel and tourism than its propagator. This rise has been fueled in part by the rising middle class, especially in China and India.[5] Consumers worldwide have come to view travel as both a right and a privilege, to be exercised as soon as free time and disposable income permit. The unintended consequence of putting the joys of travel within reach of a greater number of people is that, in some instances, the joy turns to a unique form of misery that has been labeled "overtourism."

One has to go back more than a decade to understand why, in some places, "popularity" has morphed from an asset to a liability. The word *overtourism* was coined in a 2008 academic article on coastal resource management in Vietnam and began showing up with noticeable frequency on Twitter in 2012. The timing of the mentions coincides with a number of developments, some directly travel related, some not.[6]

Causes of Overtourism

What makes overtourism a particularly vexing problem is that the forces behind it are myriad and involve a convergence of influences, some intertwining and others that are seemingly outliers. There have always been destinations with a reputation for being overcrowded; what's different now is that some are not merely overcrowded, but overwhelmed. An inundation of tourists is negatively affecting both once-remote and fragile environments and large cities that used to enjoy their status as popular tourist draws. Tourism is increasingly of the in-your-face variety. Not surprisingly, there is pushback.[7]

This volume examines different aspects of overtourism, and it's important to bear in mind that, currently, overtourism is a problem in a minority of the world's cities, villages, ports, beaches, national parks, World Heritage Sites, and other landscapes. Many of today's major resort areas, such as Cancun in Mexico and Punta Cana in the Dominican Republic, were designed specifically for large-scale tourism. Thought was given to provide sufficient supporting infrastructure, and the local population—those brought in to service the hotels and other businesses—knew what was coming and largely supported the efforts.

Despite headlines decrying overtourism, a goodly number of tourist boards continue to send out press releases crowing about their rising arrival numbers. For example, Gonzalo Robredo, president of the Buenos Aires Tourism Board, is actively seeking to increase the number of leisure arrivals into the city, but is also using technology to try to manage the concentration of visitors. Effort is made to capture both a tourist's cell phone number and interests, and by using intelligence about congestion in various parts of the city, the tourism board is able to send notifications and incentives in real-time to draw visitors away from congested areas and toward attractions that, at the moment, have the capacity and desire to handle more people.[8] Thus, some tourist boards are morphing from "destination marketing organizations" to "destination management organizations."

But few destinations have the sophistication or resources to develop data-intensive solutions. Overall, the problem of overtourism is intensifying and spreading at a rate that is worrisome to industry players. The 1.4 billion tourists who crossed borders in 2018 arrived two years earlier than the United Nations World Tourism Organization (UNTWO) had predicted, in 2010.[9] The unprecedented growth of international travel has been significantly amplified by technology, new business paradigms, and evolving social aspirations that, among other factors, are contributing to overtourism.

Soft Technology: Social Media, the Sharing Economy, and Bucket List Travel

The rise of social media and the sharing economy have affected the global travel industry, not unlike how warm Caribbean waters energize a hurricane. Looking at the curated, picture-perfect images posted by friends and influencers inspires people to travel to specific places, even if the present-day reality has been carefully edited out of the photo. For example, in 2009, only five hundred people visited Trolltunga, a tongue-like cliff in Norway that provides, at its tip, an amazing panorama of lakes and mountains. (See chapter 6.6.) It is Instagram-ready, but taking that "I'm alone at the top of the world" selfie now takes a bit more effort to make sure it appears that you're alone: in 2014, approximately forty thousand visitors were making the trek to the edge of this precipice.[10] So social media posts—coupled with the rise of "bucket list travel" to the iconic places tourists believe they must see in their lifetime—add to overcrowding. It's fine for guidebooks and travel sections to note that there are dozens of places within an hour of Venice that are very rewarding to visit (and that offer, by almost any objective measure, a much more pleasant experience than fighting the crowds for a table in St. Mark's Square), but will anyone who is that close to Venice give up the opportunity to say they've been there?

In parallel, the technology platforms of the sharing economy—Airbnb, Vrbo, Uber, Lyft, and others—have made areas of cities that were once almost exclusively residential more easily accessible to tourists by reducing the formerly formidable task of finding and booking homestays and by providing multilingual apps to minimize the impact of language barriers. Transactions are completed on an app that covers all details in the guests' language—they need not say a word to hosts to stay in their homes, nor to the drivers who bring them there. Professional property management companies have also jumped in to offer additional "homestay" inventory, kicking out residential renters as leases expire to make room for tourists and converting apartments buildings into, essentially, hotels. This practice subsequently reduces overall rental inventory, leading to rising rents and changes in the character of once-quiet neighborhoods that, previously, were unaccustomed to seeing tourists.[11]

Not surprisingly, it also contributes to the resentment of tourism by residents, which the WTTC's study identified as the first among the five leading challenges of overtourism. In one of the most extreme cases—Venice, Italy—locals are literally being replaced by tourists. On any given day, tourists outnumber residents, and the thickness of overtourism residue—rising rents, noise, displacement of local retail, change of the character of neighborhoods—has cut the resident population by about 30,000 since the early 1990s.[12] For those who remain, "resentment" likely doesn't come close to describing their feelings toward tourists.

The expressions of frustration aren't necessarily passive. In Barcelona, visitors may have passed signs, in English, reading "Refugees welcome. Tourists go home" and "Why call it tourist season if we can't shoot them?"[13]

Hard Technology: Advances in Aviation and Cruise Travel

Although "soft technology" such as social media and sharing-economy platforms contribute to the overtourism problem, "hard technology" and evolving business models play a role as well, particularly in aviation. Advances in engineering—such as the development of lighter and more fuel-efficient, carbon-fiber fuselages—coupled with relatively stable fuel prices have opened ultra-long-distance nonstop routes (Singapore–New York, London-Perth), which are also ultraconvenient for travelers.[14]

But of even greater significance than increasingly fuel-efficient hardware has been the rise of low-cost carriers, which have put international travel within the reach of more and more consumers. Among the operating expenses that airlines can't control, taxes and landing rights are some of the most vexing. A new breed of airlines, with a new set of assumptions about stimulating travel, made the bet that low prices could entice people to board planes bound for unfamiliar destinations or to land at airports located near, but not exactly conveniently near, popular destinations. EasyJet, RyanAir, and others approached Stansted and Luton (each less than an hour outside London by train), Charleroi (about 50 minutes by taxi from Brussels), and other secondary and tertiary city airports that had spare capacity and negotiated landing and gates fees that were significantly lower than what they would pay at London Heathrow or Brussels Airport. The airlines then passed on enough of the savings to passengers to offer what seemed to be amazingly low fares.

This move was complemented by a strategy to "unbundle" value that had always been assumed to be included in basic fares—checked bags, seat selection, meals, even printing boarding passes—and charge for them separately, as options. Because consumers tend to shop by comparing basic fares, some passengers didn't realize that, in the end, they might end up paying as much, or more, for a flight than they would have paid to a legacy carrier.

This approach has made low-cost carriers among the largest and most profitable airlines in the world and, as a side effect, has increased tourism to destinations that were often unprepared for an influx of budget-minded visitors. Among the cities served by budget airlines that saw a 40 percent lift in bookings on Expedia in 2018 were Vilnius, Lithuania; Seville, Spain; and Inverness, Scotland, all served by low-cost carriers.[15] And there is scant evidence that the introduction of low fares to "new" destinations—or, similarly, the introduction of "new" neighborhoods for lodging by Airbnb—has done anything to deflect visitation to legacy destinations and popular neighborhoods. Rather, they appear to have grown the pie of travelers.

Cruise lines are another sector of the travel industry that has drawn the ire of overtourism activists. Looking again to 1991 as a beginning reference point (the start of the WTTC), 4,168,000 people cruised that year. In 2020, more than 27,621,000 people were expected to cruise—an almost seven-fold increase in passengers in less than three decades.[16]

The three cities in Europe most frequently cited in stories about overtourism—Barcelona, Spain; Venice, Italy; and Dubrovnik, Croatia—all have active cruise ports. Of the more than twenty-seven million people who were expected to cruise in 2020, three million-plus were expected to disembark in Barcelona; more than 1,428,000 were expected to visit Venice; and more than 769,000 were expected to call at Dubrovnik.[17] Each is a multiple of the resident population.

Toward the end of the first decade of the twenty-first century, the optics associated with this fast-growing industry began to reflect its growing physical presence in ports as the first mega-cruise ships—the largest carrying fifty-four hundred passengers—began to be deployed. Several factors contributed to ships getting larger: Although cruising was gaining in popularity, the number of shipyards had dwindled, so if a cruise line could get onto a shipyard's schedule, it would make the most of it. There are certain fixed costs associated with any sailing, and if those costs could be defrayed over a larger number of passengers, yields would rise. And for many cruise lines, efforts had been made to position ships as destinations unto themselves, and the more options that could be made available onboard for dining, entertainment, recreation, and accommodation, the more attractive the ship as destination would become.[18]

Fixation on Continual Growth: The More, the Merrier

All this growth was occurring when most destinations still held the mindset that any increase in tourism was a plus. The more, the merrier, and why not? What the WTTC had promised turned out to be true: tourism brought in hotel bed taxes, cruise line head taxes, and rental car facility taxes that helped governments at the municipal, state, and federal levels avoid unpopular tax increases on residents. For instance, if one stays in a hotel in Chicago, one not only pays city and state taxes, but also a levy to Cook County, the Illinois Sports Authority, and the Metropolitan Pier and Exposition Authority (MPEA). Car rental prices include a 5 percent state automobile rental tax, a 6 percent MPEA charge, and a city fee of $2.75 per rental period, in addition to the city's 9 percent personal property lease transaction tax.[19]

Moreover, just as WTTC research had suggested, increased tourism has created business and job opportunities for local residents. Importantly, these are jobs that can't be offshored and businesses that can't be relocated. Tourism is also viewed as a clean industry, at least relative to manufacturing and mining, and is recorded as an export, which at the national level has helped governments with their balance of trade.

With the exceptions of the economic downturns after the September 11, 2001, terrorist attacks and the 2008 recession, travel has been on a roll in the twenty-first century. The travel industry, after an initial sputter, came roaring out of the Great Recession of 2008. As an industrial sector, it has outpaced global gross domestic product since 2010—and not by just a little.[20]

Popular Culture: Role of Film and Media

Even film and literature are doing their parts to lure people to destinations that are unaccustomed to heavy tourism presence. There has been a rise in the number of people seeking "authentic" or even "transformational" travel, inspired in no small way by books such as *Eat, Pray, Love*, in which an author tours the world to find meaning in life.[21] The *Lord of the Rings* series of films lifted tourism in its shooting locales in New Zealand, as well as other areas of that country.[22] The Singapore Tourism Board was quick to initiate campaigns in the United States, Australia, Japan, and Southeast Asia, in collaboration with Warner Brothers, timed with the release of *Crazy Rich Asians*.[23] Around the world, travel marketers have seized on the popularity of destinations featured in popular culture and created travel packages that, while touting "authenticity," often changed local realities. In some instances, destinations have been overwhelmed by the number of visitors anxious to visit a real-life movie set. Dubrovnik's role as the stand-in for "King's Landing" in HBO's *Game of Thrones* greatly exacerbated brewing overtourism-related issues.[24]

Many of the trends mentioned above have brought significant numbers of visitors to areas that were largely unprepared, and sometimes ill-equipped, to deal with a tourist invasion. A handful of these destinations, from Timbuktu to the salt flats of Bolivia, are the focus of the 2014 film *Gringo Trails*, directed by New York University anthropologist Pegi Vail and shot and coproduced by her cinematographer husband, Melvin Estrella.[25] The film presciently documents, over ten years, the effects of backpacker tourism on relatively minor destinations in Africa, Asia, and South America where tourism rose rapidly without appropriate planning or management. The results were, in many instances, tragic. The film serves as an overtourism microcosm, a collection of cautionary tales for what, writ large and propelled into prominence by international and social media, has become the overtourism crisis.

Rise in Nationalism and Nativism

If there is a "final factor" that has brought overtourism to the fore, it is the rise of nationalism and nativism. Those caught up in these movements will no longer be persuaded by the tourism industry's traditional economic arguments if part of the bargain is that their streets, restaurants, parks, buses, and bars are filled with people who look different, speak other languages, ignore traditional and familiar behaviors, and make life less resident-centric. Airbnb's senior vice president of global policy and communications, Chris Lehane, observed that nativism "tends to over-index in Europe, where the issue of overtourism also tends to over-index. Those cities are going through dramatic changes that are separate and apart from travel and tourism, [but] there is the perception that this industry is to blame for some of those issues."[26]

Travel Industry Response

For the most part, the WTTC and its members have adopted the position that they want to be part of the overtourism solution, not part of the overtourism problem. The travel industry, having enjoyed a reputation as providing a means of escape from the problems and pressure of daily life, is not keen to find itself in the role of a societal industrial villain, akin to Big Oil, Pharma, or the tobacco industry. On the whole, it has embraced dialogue with local and federal governmental agencies to work on solutions. At an overtourism roundtable, Airbnb's Lehane said, "We focus an awful lot of time on understanding what people's concerns are. . . . [Our] Office of Healthy Tourism is . . . designed to work with cities [and] sometimes at the national level with a series of specific tools."[27]

Given Airbnb's penchant for suing municipalities that attempt to restrict its spread or to collect taxes on homestays, the sincerity of its declared cooperative mindset has been questioned.[28] Airbnb resists some regulation on the grounds that the financial benefits it brings to its resident hosts provide, on balance, a net positive to cities. But that only underscores the perceived imbalance of tourism's benefits.

Although most people who directly benefit financially from tourism want more, for residents not involved in tourism, the benefit is not necessarily obvious. Their municipality may collect more tax money, which lessens their tax burden, but it's an ambient benefit. It's doubtful that many New Yorkers are aware that without the taxes collected from tourists, every household would pay, on average, about $2,000 more in taxes to receive the same city benefits.[29] It's invisible solace that seldom comes to mind when waiting in longer lines or being delayed in traffic or otherwise inconvenienced and annoyed by an overabundance of visitors.

In addition to home-sharing and aviation, the other sector that draws the most fire from overtourism activists is cruising. The bigger-is-better approach to building cruise ships has moderated—the average size of new builds in 2019 was 71,091 gross registered tons, down from 88,870 in 2018[30]—but the sheer size of existing large ships, particularly when juxtaposed against Renaissance or medieval skylines as they often are in photos illustrating articles about overtourism in Europe, has literally kept them a highly visible target for critics.

The cruise lines argue that they draw disproportionate attention, noting that, for instance, in Barcelona, they bring only 5 percent of the city's visitors. If every cruise line stopped calling at the port of Barcelona, Carnival Corp.'s David Dingle maintains, it wouldn't make a noticeable difference to the city's overcrowding problems.[31]

Further, cruise executives argue that they are simply responding to consumer and destination demands. At a 2018 industry roundtable discussing overtourism, Royal Caribbean Cruises vice chairman Adam Goldstein asserted, "The cruise industry goes to about 1,000 places. The vast majority want more tourists from us, not less."[32]

Darrell Wade, cofounder of tour operator Intrepid Travel, believes that one reason the cruise industry draws attention that may be disproportionate to the number of visitors it brings to a destination is rooted in economics. "One of the problems, or perceived problems, the cruise industry has is not only the visibility of a cruise ship coming into port, but how there's not a lot left behind from a cruise ship passenger," he said. "They're called 'ice cream tourists' because they get all their meals on the ship and then get off and buy only ice cream. That's the extent of economic impact left behind in the community."[33]

Keenly aware of these and other criticisms, the cruise line's trade group, Cruise Lines International Association (CLIA), signed a formal partnership with the city of Dubrovnik in mid-2019 to work together on destination stewardship, which could serve as a model for agreements with other cities concerned by the effects of cruise passenger traffic.[34] Two years earlier, the cruise lines had worked with the city to change the days they called at Dubrovnik to spread passengers out across the week rather than being concentrated in just three days.[35]

The global travel industry is well aware that it ignores this rising negative trend of overtourism at its own risk. Even among those wanting ever more travelers, there is a growing awareness of the dangers of overtourism. The WTTC hired the consulting firm McKinsey & Co. to define the roots and manifestations of overtourism, as well as formulate recommendations for policy to mitigate its effects.[36] But since the release of the report in 2017, the overtourism crisis has not abated. The genie, for the time being, appears to be out of the bottle.

Looking Ahead

What does all this portend for the future? Serious policies are being discussed, proposals are being implemented, and site-specific victories occur, some of which are explored in this book, but policy makers will, for the foreseeable future, continue to encounter headwinds. The myriad components that have arisen, independently, to foster overtourism can be tweaked and regulated, but other macro trends point to additional factors—let's call them innovations—that will enable continued growth in tourism going forward.

For example, even if the clash between tourists and residents or tourism and nature didn't currently exist, the number of border-crossing travelers is predicted to grow at a steady rate. As of February 2020, approximately 10 million people were flying each day on commercial aircraft; the International Air Transport Association, the airline industry's trade group, forecasts that traffic will grow to 21 million by 2035.[37] For airport authorities, whose concern is not overtourism but how, operationally, to handle the predicted rise in travelers, the situation is clear: they must either become more efficient or must build additional terminals and runways, which would be extremely costly and could involve expansion of their footprint, which often meets local resistance.[38]

As a result, airport authorities are turning to biometrics—primarily facial, fingerprint, iris, and hand-vein recognition. The advances have been largely motivated by the need to enhance security, but biometrics have also been embraced as a means to move more arriving and departing passengers quickly through an airport terminal. The technology, currently in relatively unsophisticated forms, has already been deployed in many international arrival halls, and airplanes have been boarded at record-breaking speed using facial recognition.[39]

Biometrics are but one example of what has also become the strategic goal of almost every sector within the travel industry: remove friction from the travel experience. Apps enable Hilton's guests to bypass the check-in desk. The Disney Magic Band moves people through gates and onto rides faster (and enables speedier purchases). Princess Cruises' wearable Ocean Medallion can unlock your stateroom door as you approach or summon a glass of champagne to be brought to you as you stand at a rail, watching the sunset. All are designed to ease travel's pain points and encourage more people to travel.[40]

Even the single biggest impediment to international travel—language barriers—may soon become trivial. Google Translate is currently the largest artificial intelligence project in the world. The Google Translate app already allows one to simply hold one's phone camera up to signage or menus and the translation will appear on the phone's screen.[41] The next step—also already here but, for now, in a less reliable form—is immediate translation of the spoken word. If two parties who don't speak a common language are both wearing Google Pixel earbuds when speaking to each another, the parties will hear each other translated into a language they understand.[42]

These significant steps in technology, coupled with booking apps that predict when fares and other travel prices will be lowest, will be abetted and enabled by advancements in artificial intelligence, 5G, blockchain, and augmented reality. In sum, as these and other efficiencies solve the industry's structural problems, they'll also likely stimulate travel and become additional contributing factors to overtourism.

Historically, traveling abroad has been a wonderful and rewarding experience. It can heighten cultural sensitivity and understanding, provide opportunities for much-needed relaxation and recreation, inspire creativity, and strengthen bonds in families who are otherwise focused on their jobs, studies, and phones. Although it's important to recognize that overtourism is not a problem in every destination right now, we also need to ask if there are any indications that tourism's long growth curve is nearing an end. The answer is likely to be, Friend, you ain't seen nothing yet.

Notes

1. World Travel and Tourism Council. (2020). https://www.wttc.org/about/organisation/history/.

2. World Tourism Organization. (August 2000). *Tourism Highlights 2000*. https://www.e-unwto.org/doi/pdf/10.18111/9789284403745.

3. World Travel and Tourism Council. (March 29, 2019). "2019 Seville WTTC Global Summit to Attract Record Delegates, World Travel & Tourism Council." https://www.wttc.org/about/media-centre/press-releases/press-releases/2019/2019-seville-wttc-global-summit-to-attract-record-delegates/.

4. World Tourism Organization. (January 21, 2019). "International Tourist Arrivals Reach 1.4 Billion Two Years Ahead of Forecasts." https://www2.unwto.org/press-release/2019-01-21/international-tourist-arrivals-reach-14-billion-two-years-ahead-forecasts.

5. Kristofer Hamel. (May 7, 2019). "Look East Instead of West for the Future Global Middle Class." *OECD Development Matters*. https://oecd-development-matters.org/2019/05/07/look-east-instead-of-west-for-the-future-global-middle-class/.

6. Nguyen Tac An, Nguyen Ky Phung, and Tran Bich Chau. (November 23, 2008). "Integrated Coastal Zone Management in Vietnam: Pattern and Perspectives." *Journal of Water Resources and Environmental Engineering*. http://citeseerx.ist.psu.edu/viewdoc/download?doi=10.1.1.585.6554&rep=rep1&type=pdf.

7. Lisa Abend. (July 26, 2018). "Europe Made Billions from Tourists. Now it's Turning Them Away." *TIME*. https://time.com/5349533/europe-against-tourists/.

8. Ibid.

9. World Tourism Organization. (January 21, 2019). "International Tourist Arrivals Reach 1.4 billion Two Years Ahead of Forecasts." https://www2.unwto.org/press-release/2019-01-21/international-tourist-arrivals-reach-14-billion-two-years-ahead-forecasts.

10. Carrie Miller. (January 26, 2017). "How Instagram is Changing Travel." *National Geographic*. https://www.nationalgeographic.com/travel/travel-interests/arts-and-culture/how-instagram-is-changing-travel/.

11. Christine Jelski. (Accessed March 2020). "Homesharing Shakeup." *Travel Weekly*. https://www.travelweekly.com/Travel-News/Hotel-News/Homesharing-shakeup.

12. Giorgio Ghiglione. (September 13, 2018). "Occupy Venice: We Are the Alternative to the Death of the City." *The Guardian*. https://www.theguardian.com/cities/2018/sep/13/occupy-venice-alternative-to-death-of-city-activists-tourism.

13. Graham Keeley. (May 12, 2017). "Barcelona Protesters Tell Tourists to Go Home." *The Times*. https://www.thetimes.co.uk/article/barcelona-protesters-tell-tourists-to-go-home-fmh9bz7q7?CMP=Sprkr-_-Editorial-_-TheTimesandTheSundayTimes-_-World-_-Imageandlink-_-Statement-_-Unspecified-_-FBPAGE&linkId=37533909.

14. Paul Lewis. (May 17, 2019). "Trends in International Travel Part 3: Aircraft, Polar Routes, and Flights to Asia." *Eno Transportation Weekly*. https://www.enotrans.org/article/trends-in-international-travel-part-3-aircraft-polar-routes-and-flights-to-asia/.

15. Agence France-Presse. (December 12, 2018). "Travelers Sought Out 'Secondary Cities' for 2018 in Wake of Overtourism." *The Jakarta Post*. https://www.thejakartapost.com/travel/2018/12/10/travelers-sought-out-secondary-cities-for-2018-in-wake-of-overtourism.html.

16. Cruise Market Watch. (2020). *Growth of the Cruise Line Industry.* https://cruisemarketwatch.com/growth/.

17. MedCruise. (2020). *Yearbook 2019–2020.* https://www.medcruise.com/tags/statistics.

18. Gene Sloan. (April 18, 2019). "The Limit Does Exist: Cruise Ships Keep Growing, but Can Only Get So Big." *The Points Guy.* https://thepointsguy.com/news/why-cruise-ships-keep-getting-bigger/.

19. Brendan Bakala. (December 20, 2017). "Planes, Trains and Automobiles: Chicago's High Travel Taxes." *Illinois Policy.* https://www.illinoispolicy.org/planes-trains-and-automobiles-chicagos-high-travel-taxes/.

20. World Travel and Tourism Council. (February 27, 2019). "Travel & Tourism Continues Strong Growth above Global GDP." https://www.wttc.org/about/media-centre/press-releases/press-releases/2019/travel-tourism-continues-strong-growth-above-global-gdp/.

21. Emma Weissmann. (Accessed March 2020). "Travel Milestones from the Last 50 Years." *Travel Age West.* https://www.travelagewest.com/Industry-Insight/Business-Features/Travel-Milestones-From-the-Last-50-Years.

22. Laura Del Rosso. (July 16, 1012). "New Zealand Anticipates Another Hollywood-Fueled Tourism Bounce." *Travel Weekly.* https://www.travelweekly.com/Travel-News/Travel-Agent-Issues/Insights/New-Zealand-anticipates-another-Hollywood-fueled-tourism-bounce.

23. Yeoh Siew Hoon. (September 24, 2018). "After 'Crazy Rich Asians,' Is Singapore Ready for Its Close-up?" *Travel Weekly.* https://www.travelweekly.com/Yeoh-Siew-Hoon/After-Crazy-Rich-Asians-Singapore-ready-for-its-closeup.

24. Jonnalyn Cortez. (April 16, 2019). "Dubrovnik's Overtourism: A Price to Pay with 'Game of Thrones' Fame." *Business Times.* https://en.businesstimes.cn/articles/110767/20190416/dubrovniks-overtourism-a-price-to-pay-with-game-of-thrones-fame.html.

25. Frank Scheck. (September 8, 2014). "Gringo Trails: Film Review." *Hollywood Reporter.* https://www.hollywoodreporter.com/review/gringo-trails-film-review-731088.

26. Arnie Weissmann. (Accessed March 2020). "Overtourism Roundtable." *Travel Weekly.* https://www.travelweekly.com/Travel-News/Government/Overtourism-Roundtable.

27. Ibid.

28. Paris Martineau. (March 20, 2019). "Inside Airbnb's 'Guerilla War' Against Local Governments." *Wired.* https://www.wired.com/story/inside-airbnbs-guerrilla-war-against-local-governments/.

29. NYC & Co. (2020). https://assets.simpleviewinc.com/simpleview/image/upload/v1/clients/newyorkcity/AM19_RackCard_8670cd0d-9ef7-4029-927a-c39dbeac2413.pdf.

30. Bill Hirsch. (Accessed March 2020). "New Cruise Ships Are Huge – Except When They're Not." *CruiseHabit.com.* https://www.cruisehabit.com/new-cruise-ships-are-huge-except-when-theyre-not.

31. Arnie Weissmann. (Accessed March 2020). "Overtourism Roundtable." *Travel Weekly.* https://www.travelweekly.com/Travel-News/Government/Overtourism-Roundtable.

32. Ibid.

33. Ibid.

34. Tom Stieghorst. (July 25, 2019). "CLIA and Dubrovnik in Tourist-Management Pact." *Travel Weekly.* https://www.travelweekly.com/Cruise-Travel/CLIA-and-Dubrovnik-in-tourist-management-pact.

35. Arnie Weissmann. (Accessed March 2020). "Overtourism Roundtable." *Travel Weekly.* https://www.travelweekly.com/Travel-News/Government/Overtourism-Roundtable.

36. McKinsey & Company and World Travel and Tourism Council. (December 2017). *Coping with Success: Managing Overcrowding in Tourism Destinations.* https://www.mckinsey.com/~/media/mckinsey/industries/travel%20transport%20and%20logistics/our%20insights/coping%20with%20success%20managing%20overcrowding%20in%20tourism%20destinations/coping-with-success-managing-overcrowding-in-tourism-destinations.ashx.

37. Arnie Weissmann. (Accessed March 2020). "Overtourism Roundtable." *Travel Weekly.* https://www.travelweekly.com/Travel-News/Government/Overtourism-Roundtable.

38. Staff Reporters. (May 31, 2019). "Extinction Rebellion Threatens 10-day Heathrow Protest If Government Doesn't Drop Airport Expansion Plans." *The Telegraph.* https://www.telegraph.co.uk/news/2019/05/31/extinction-rebellion-threatens-10-day-heathrow-protest-government/.

39. Arnie Weissmann. "Overtourism Roundtable." *Travel Weekly.* https://www.travelweekly.com/Travel-News/Government/Overtourism-Roundtable.

40. Arnie Weissmann. (January 22, 2019). "Carnival's Innovation: Rebate Time, Not Monday." *Travel Weekly.* https://www.travelweekly.com/Arnie-Weissmann/Carnival-innovation-Rebate-time-not-money.

41. Arnie Weissmann. "Context for the Future." *Travel Weekly.* https://www.travelweekly.com/Travel-News/Travel-Technology/Context-for-future-conversation-tech-blogger-Shelly-Palmer.

42. Ibid.

Chapter 2.1

Europe's Historic Cities

By Francesca Street

For many in Venice, it felt like the final straw. On June 2, 2019, along the Giudecca Canal, a bustling waterway that connects to the city's famous Basin of St. Marco, a colossal cruise ship smashed into the dock, hitting a tourist boat and sending passers-by fleeing.[1] It was as though someone had taken a heavy-handed metaphor for Venice's growing overtourism problem and conjured it to life: a monstrously large ship steamrolling into this fragile historic portside city, with seemingly little care for the consequences.

Further down the Mediterranean coast, the fortress city of Dubrovnik feels its ancient walls closing in. Visitors line up at the Pile Gate, the entry point to the Old Town, jostling against one another. As the day heats up, high on the city's ancient walls, streams of people morph into an insurmountable traffic jam, snapping selfies and straining to admire the views amid thousands of other tourists doing the same.

Meanwhile, in the canal-side city of Amsterdam, dawn breaks on another sleepless night filled with tourist revelry and loud music in the crowded city streets. Social media promotes Amsterdam as a one-stop destination for all-night partying, a city where anything goes. As one horde of visitors departs, another swarm arrives to party through the night, as has become de rigueur in the Dutch capital.

As destinations across Europe become increasingly submerged with tourists, concerns over the livability and endurance of historic cities such as Venice, Dubrovnik, and Amsterdam are growing ever louder. With protests intensifying and cities cracking down on visitor behavior, the region has become the proverbial canary in the coal mine for historic cities in the Americas, Asia, and elsewhere that are just beginning to experience the early symptoms of overtourism—and to experiment with potential reforms.

Causes of Overtourism in Historic Cities

The origins of overtourism in historic cities can be traced to a series of diverse but intertwined factors, and identifying and understanding these causes is central to crafting solutions. One fundamental reality is that tourism has been growing faster than even experts predicted. As noted in chapter 1.1, international tourist arrivals reached 1.4 billion in 2018, two years ahead of the 2020 date originally forecast by the United Nations World Tourism Organization.[2] And in Europe, there were 713 million international tourist arrivals over 2018, a 6 percent increase from 2017.[3]

Expansion of European Travel

So why are people traveling more? For one, travel today is easier and more affordable than ever. Budget flights are readily available, and there is a growing global middle class who can pay for and prioritize travel. According to the European Commission, the European Union's policy branch, 5.6 billion people will be designated "middle class" by 2030, two billion more people than at present, all with increased purchasing power and, by extension, power to travel.[4] The compact size of Europe, its concentration of historic cities, its relative affluence, the prevalence of budget flights, and good railway connections combine to make vacation travel in Europe increasingly easy and routine. As Tony Johnston, head of Tourism Studies at Ireland's Althone Institute of Technology, put it, "European cities are particularly affected [by overtourism] because of large urban populations, old city centers, budget airlines, geographic proximity and accessibility, and free movement of affluent [middle class] people."[5] The continent's historic highlights can be feasibly viewed in a fortnight: spend a few days touring Amsterdam and then head to the bright lights of Paris, southwest to Barcelona, westward to Florence, Rome, and Venice, and then east to Dubrovnik, ticking off the top sites in each historic urban center.

As travel becomes more commonplace for the middle class, exploring the world has become, somewhat, democratized, which has clear benefits: global travel opens our eyes to new cultures, old histories, and different ways of living. Travel experiences should be available to all. But as shorter, more frequent, cheaper so-called city breaks become the norm in Europe, the question is, at what cost? More people mean more congestion, more garbage, more noise, and more pollution. This tourism explosion threatens to ruin the experience of historic cities for both residents and visitors alike.

Growth of Cruise Tourism

For historic port cities, this threat is exacerbated by the exponential growth of cruise tourism. Several times a day, fresh boatloads of tourists pour into historic city centers. In 2019, there were an estimated 30 million cruise passengers across the world, up from 17.8 million just a decade earlier.[6] The Caribbean is the most popular cruising market, but cruise passenger arrivals in Europe grew by 3.3 percent in 2018 to 7.17 million.[7] The negative effects on portside cities include overcrowding at the specific times of the day when the cruise ships dock, pollution, noise, and less spending by cruise passengers than that generated by land-based stayover visitors.

Explosion of the Sharing Economy

In addition to cheaper flights, the mushrooming middle class, and the growth of cruise tourism, overtourism in historic cities has been fueled by the explosion of the "sharing economy." In just over a decade, this concept of home and car sharing has spawned megacorporations like Airbnb and Uber that have upended the urban accommodation and transport sectors. For the cost of an identical hostel or budget hotel, Airbnb promises an authentic opportunity not just to "go there," but "live there" as a 2016 commercial put it, in a hip house, quirky apartment, or other prime city center real estate.[8]

In some instances, Airbnb and its competitors have proved a win for both visitors and hosts. Vacation rental sites argue that they give power back to the residents, encouraging locals to profit from living in popular locations. Plus, in certain heritage city centers, short-term rentals have provided a welcome way to expand the accommodation pool. Historic cities usually have strict planning and zoning laws and may not have the space required to build new hotels or hostels. Eduardo Santander, executive director of the European Travel Commission, points to Valletta, the capital of Malta. "It was impossible," he explained, for Malta, "to grow organically through investment in more hotels and so on because the city center of Valletta is . . . already historically congested."[9] For this island nation, Airbnb was a lifesaver. Residents have quickly transformed their historic homes into vacation residencies for next to nothing. Between 2010 and 2018, tourist arrivals to Malta doubled, from 1.3 million to 2.6 million, thus moving this tiny European country squarely onto the global tourist map.[10]

Something similar occurred, albeit briefly, in Havana, Cuba. For decades, the US economic embargo, on the one hand, and Fidel Castro government's wariness of international tourism, on the other, worked to stifle the growth of hotels on the island. But, as tourism took off in the wake of Barack Obama–Raul Castro's historic December 2014 agreement to move toward a normalization of relations, Cuba's burgeoning home-based rental sector, known as "casas particulares," helped its historic capital, Havana, and other Cuban cities and towns meet the growing demand for tourist accommodation. By 2016, Cuba had become "the fastest-growing market in Airbnb history."[11] However with US President Donald Trump's hardline policy reversals beginning in 2017, Cuba's phenomenal Airbnb growth proved to be short lived.

Although Airbnb and its short-term rental competitors (including Vrbo, Booking.com, and Homeaway.com) have opened up accommodation options, in many cities the growth of this new market has limited long-term rental possibilities for locals, raised housing prices, and transformed neighborhoods into transient accommodation zones. What feels like a win while you are staying in a room with a view in Bruges, Belgium, becomes less appealing when you arrive back in your home city of Paris, France, to be woken up every morning to the sound of a new round of guests, fumbling with the keys in the lock box next door.

Social Media and the Instagram Effect

Yet another factor contributing to overtourism is the growth of social media and what's been dubbed the "Instagram effect." Travel photography has become ubiquitous on social media. Instagram is chockablock full of updates of friends abroad, snapped on a gondola in Venice's lagoon; captured laughing by San Francisco's Golden Gate Bridge; or posed, croissant in hand, on a picturesque Parisian street. The global Digital 2019 report stated that there are more than three *billion* social media users worldwide.[12] Social media both stimulates interest in already famous sites—the London Eye, New York's Empire State Building, the Eiffel Tower in Paris—and directs visitors to lesser-known destinations that become "Insta-famous"—picturesque residential streets, parks with panoramic city skylines, and rooftop bars on top of parking garages. "There's no more secrets out there to discover because someone posted about it on the Internet, on Instagram, on Facebook," said Santander.[13]

Impacts of Overtourism

These factors have coalesced to create overtourism in a number of Europe's historic cities, with a multitude of consequences. Especially at the social level, overcrowding has generated a series of negative impacts among visitors and residents that need to be urgently tackled.

Impact on Visitors

First, there's the impact on the visitor. When you imagine yourself wandering the streets of a historic city, you don't picture yourself fighting for room on the sidewalk amid thousands of fellow visitors or lining up for hours to visit landmark sites and possibly not even getting to see what attracted you to the city in the first place. Tour operators are acutely conscious of the moment when visitor pleasure morphs into visitor pain. "Sometimes it can be just overwhelming, hectic, and chaotic," said tour guide Marko Miloš, manager of Dubrovnik Local Guides in his home city. "For the tourists, this affects their experience of Dubrovnik. After all, they came here for vacation, to relax and unwind, not to elbow their way through people."[14]

Crowds elbowing for the perfect photos have, on occasion, degenerated into physical brawls. In August 2018, for instance, two women—one Dutch, one American—tried to simultaneously take selfies in the same spot at Rome's Trevi Fountain. Their jostling turned into fisticuffs as members of both families entered the fray, forcing police to intervene.[15] The Trevi Fountain, one of Rome's most popular attractions, has been beset with a variety of problems, from visitors wading in the fountain and sitting on its edges to people resisting police efforts to control pedestrian traffic. In November 2019, CNN reported that politician Andrea Coia suggested a "protective barrier" be installed to "prohibit people sitting on the edge of the fountain."[16] Meanwhile, sitting in Rome's public spaces can come with a hefty price tag: visitors to Rome's Spanish Steps can be fined if they are found loitering, with charges coming in upwards of €400 (US$448) if they are discovered to have damaged the iconic attraction.[17]

On occasion, extreme overcrowding in historic cities has also led to strikes and closures of popular attractions. With more than ten million tourists in 2018, the Louvre in Paris has earned the dubious distinction as the world's most visited museum.[18] In May 2019, the reception and security staff at the Louvre staged several short strikes to protest visitor overcrowding.[19] The workers union declared that "the situation is untenable" and "the Louvre is suffocating."[20]

Impact on Residents

Overtourism in historic cities also impacts residents' quality of life. Tour guide Miloš, who was born and bred in Dubrovnik, said that traveling around the city is much harder since the tourism boom. "Imagine walking down the main street, trying to get to work on time or going to an appointment, and then getting stuck in the crowds," he said. "It's frustrating."[21] One result is that locals move out. Miloš said that the street he lives on with his family has become "almost empty"

The crush of overtourism, the Trevi Fountain, Rome, Italy. Source: Julia (Flickr).

of permanent residents. He has seen his neighbors sell their homes and move out, as new owners snap up the prime real estate, transforming residential homes into vacation rentals.[22]

Barcelona and Venice have also seen their resident populations suffer in the wake of overtourism and its related problems. Take Barcelona's Ciutat Vella: from 2009–2017, this popular tourist district lost more than 11 percent of its residents.[23] In March 2017, Barcelona officials said that the city's Barri Gòtic neighborhood had lost 45 percent of its population in 2007, with Barcelona officials blaming "tourism-influenced urban development transformations on the district's housing and shops."[24] Something similar has happened in Venice: UNESCO, the United Nations Educational, Scientific and Cultural Organization, highlights that the Word Heritage Site has suffered a "significant decline in population."[25] Only one-third as many people live in Venice as lived there fifty years ago.[26]

In some cities, the rise of Airbnb and other short-term rental services has concentrated more tourists outside city centers. Although theoretically good for lessening overcrowding, this kind of concentration can lead to further frustration for locals as historically nontourist areas turn into short-term rental hotspots. "Some residents may feel that there is nowhere in the city where they can avoid tourists," said Santander.[27] That was the case for Ashleigh Gray, a former resident of Leith, a suburb of Edinburgh, Scotland's capital. "Living next to an Airbnb was like playing Russian roulette with the city's tourists, waking up every morning to catch a glimpse of the new arrivals, calculating how likely they look to disrupt yet another night's peace," said Gray.[28] Plagued by sleepless nights, Gray became increasingly resentful of her living situation, and, she said, the concept of short-term vacation rentals in general. There were other consequences, Gray explained, beyond just noise and disruption. "As a resident I want consistent neighbors that I can get to know. Neighbors who can take in my post, who can help with shared repairs, who can look out for me and my property," said Gray. "The only thing that pains me more than my shattered peace and quiet is the loss of a local community."[29] As cities become inundated with short-term rentals, residents feel the absence of a familiar neighborhood. "Communities are built on relationships between residents, and these relationships are becoming increasingly difficult to form in cities hollowed out by overtourism," Gray explained.[30]

Growing numbers of tourists in historic cities also generates a plethora of budget shops and restaurants selling cheap souvenirs and fast food that often displace stores catering to residents. "It's good for these new business owners and good for visitors, but not good for locals," said Geerte Udo, director of amsterdam&partners (formerly Amsterdam Marketing), Amsterdam's marketing organization. "It creates friction, and the locals say, 'Enough is enough.'"[31]

Civic Resistance

"Enough is enough" is the sentiment of a growing number of outspoken civic groups that have sprung up across Europe. In Venice, Gruppo 25 Aprile is actively campaigning for new regulations to control large cruise tourism, short-term rentals, and other overtourism issues. In Barcelona, citizens take to the streets to protest, plastering walls with anti–mass tourism graffiti and waving placards that proclaim, "Cruise ships kill Barcelona"[32] and "This isn't tourism: it's an invasion."[33] Ironically, this written resistance to overtourism has had the

Urban graffiti, Barcelona, Spain. Source: Amy (Flickr).

unintended consequences of spawning graffiti and urban art walking and biking tours of Barcelona.[34]

Arguably, historic cities are better suited to handle a sudden influx of tourists than, for instance, a glacier in Iceland or coral reefs on a small island in the Caribbean, but overwhelming numbers of people still take a toll on urban infrastructure. In Dubrovnik, for instance, tourists walking up the Stradun, the Old Town's main street leading from the Pile Gate entrance, are often unaware that the street's marble-like facade is misleading. The walkway is actually made of limestone, but the relentless footfall on the street each day has turned the limestone smooth and slippery, resembling marble. In response, the city has hired workers who intentionally rough up the stone to prevent it from being too slippery for pedestrians.[35] This act solves one particular issue, but when tourism grows exponentially, it is increasingly difficult for city planners to keep up. Plus, improving the infrastructure in historic cities is intrinsically challenging because there are often restrictions on repairs in protected historic centers.

Solutions

Amid the crush of visitors and the increasing protests and civil unrest, city officials and local organizations across Europe have started to craft solutions to overtourism. Some European municipalities are actively working to chart a new course, reevaluating the traditional marketing strategy of continual growth in tourist numbers as the main barometer of tourism success. Santander said that the European Travel Commission (ETC) is encouraging this type of reevaluation, describing the process as "reeducating demand."[36] Cities, including Amsterdam and Barcelona (see chapter 2.2), have made a conscious choice to flip from destination marketing to destination management, with a focus on increasing length of stay and spending per visitor rather than increasing visitor numbers. In addition, a variety of civic campaigns, often devised together with city governments, are aiming to change the public image and reputation of their cities and to improve visitor behavior. Although promising in the short-term, most of these initiatives are too new to assess if they will have long-term success in curbing overtourism.

Visitor Reeducation

In Amsterdam, the Enjoy and Respect campaign, spearheaded by Udo, is designed to shift the city's marketing strategy and reorient visitors' views of the Dutch capital.[37] Launched in May 2018, Enjoy and Respect targets Dutch and British tourists aged eighteen to thirty-four who visit Amsterdam for bar hopping and bachelor parties. Amsterdam has long been known as a socially tolerant city with a dynamic nightlife, said campaign director Udo, but problems including drunkenness, noise, litter, and public urination have escalated as tourist numbers have mushroomed.[38] Enjoy and Respect is working, for instance, with the British budget airline EasyJet to provide inflight information to travelers about how to enjoy their stay in Amsterdam while also respecting the city.[39] According to its website, the campaign "aims to inform the target group of the consequences of this kind of behavior, and raise awareness of what is allowed and—more importantly—what is not allowed in Amsterdam."[40] The hope, Udo said, is to gradually shift tourists' preconceptions, slowly transforming the city's reputation and gradually remapping visitor flows and activities to ensure long term sustainability.[41]

In Venice, there is another similarly named civic initiative: #EnjoyRespectVenezia, a visitor awareness campaign that emphasizes the Italian city's status as a UNESCO World Heritage Site.[42] Launched in 2017 as part of the United Nations' International Year for Sustainable Tourism and Development, the campaign is "designed to direct visitors towards the adoption of responsible and respectful behaviour towards the environment, landscape, artistic beauties, and identity of Venice and its inhabitants."[43] The objective, according to the #EnjoyRespectVenezia website, is "to raise awareness of tourist impact, with the belief that responsible travelling can contribute to sustainable development."[44]

Meanwhile in Vienna, Austria, another educational campaign, dubbed "Unhashtag Vienna," targets tourists' obsession with Instagram photos.[45] For example, "1.4 million people want to see *The Kiss* in Vienna every year" reads a web campaign that is accompanied by a video depicting tourists jostling for photos of Gustav Klimt's famous painting. "But," the site asks, "do they really see *The Kiss*?"[46]

Dispersal: Outside the City Center and into the Off-Season

Other initiatives aim to lessen congestion in historic city centers by spreading tourists more widely throughout the city or to nearby destinations. They also encourage more people to travel in the shoulder and off-peak seasons. In Amsterdam, for instance, tourists typically congregate around the Canal District and the Red Light District. Recently, city officials have focused on redirecting tourists from these bustling areas to less congested sections of the city.[47] The city's famous "I Amsterdam" letters, previously situated at Museumplein, the popular public square where three of the city's largest museums are located, were removed in December 2018 because of overcrowding concerns.[48] They have since been displayed in other neighborhoods, with the aim of drawing visitors to lesser-known areas of Amsterdam.

Another city initiative, "Unrating Vienna," seeks to encourage tourists to "discover Vienna on [their] own terms" by looking beyond the well-known online reviewing platforms and exploring lesser known locations.[49] The Unrating Vienna website highlights these "hidden" spots, with the aim of dispersing visitors more widely across the Austrian city.

Then there is the more comprehensive website, Avoid-Crowds.com, which uses tourism data to rate, on a scale from one to 100, how large crowds will be in key European cities at certain time intervals.[50] Enter a date and a destination, and Avoid-Crowds.com gives information on the size of crowds and offers alternative routes and itineraries. The website has recently moved beyond Europe to provide daily crowd forecasts for four popular US cities: New York City, Miami, Los Angeles, and Chicago.

In some cities, public and private sectors are working together to promote increased travel during the shoulder and off-peak seasons. At present, according to ETC executive Santander, more than 80 percent of tourism in Europe takes place in July and August, yet in many European destinations, warm weather lasts long into the fall.[51] Other factors, including school vacations, dictate why European travel largely takes place during the summer months. But it is also true that, for many, it is a learned behavior to go on vacation during the summer season. Offering more off-season flights and discounts and promoting attractions into the fall and winter can help disperse visitors more evenly.

Tour guide Miloš said that he would like to see more of this in Dubrovnik. "It's as if everything grinds to a halt on the 31st of October, and from forty flights per day in the peak of the summer, it goes down to five in winter," he said. "Why can't Dubrovnik have more connections over the winter? With better coordination and promotion, the flow of tourists could be managed better and dispersed equally throughout the twelve months."[52]

Regulation and Legislation

In addition to campaigns directed at tourist education, some cities are enacting new laws and urban planning to help address overtourism. In

Amsterdam, for instance, the city government proposed new measures to regulate tours of the Red Light District with the aim of controlling tourism numbers and improving working conditions for sex workers. These regulations set out "measures to reduce disruptions caused by overcrowding, create more space in the streets and provide for a decent working environment for sex workers."[53] The measures specify that guided tour groups to the Red Light District must have no more than 15 participants, must conclude by 10 p.m., must follow a permit system, and must pay an entertainment tax of €1.50 (US$1.75) per participant. In addition, the regulations state that tours "are not permitted to pass prostitution windows."[54]

In December 2018, in another move to regulate tourism and protect the city's traditional character, Amsterdam's highest court upheld a ban on tourist stores in the historical center.[55] Such measures aim to reorient the city's public image, change visitor behavior, better regulate tourism flows, and update infrastructure to ensure that Amsterdam is a livable and workable city for local residents, as well as a quality destination for visitors.

In Venice, legislation has also attempted to control visitor numbers in the most popular tourist areas of the city. Over the city's busy May Day weekend in 2018, mayor Luigi Brugnaro announced "urgent measures to guarantee public safety, security, and livability."[56] The temporary regulations, designed to manage pedestrian and water traffic and separate travelers from locals, included installing turnstiles, redirecting tourist flows away from the most popular sites, designating certain areas as accessible only to residents, providing regular visitors with a special Venezia Unica pass, relocating the disembarkment point for tourists arriving by water, and turning away motorists who had not reserved space in one of the city's parking lots. Visitors to Venice also face fines if they eat or drink in undesignated areas. Similar laws have been introduced in Florence and Rome.[57]

Reining in Short-Term Rentals

Several cities in Europe—and in the United States—have also introduced legislation to control the number and practices of short-term rentals. In Madrid, legislation was introduced to curb growth of vacation apartments in the city center, stipulating the need for separate access for guests and for full-time residents and effectively preventing many city center apartments from functioning as vacation rentals.[58] Another Spanish city, Valencia, limited private vacation rentals above the first floor to prevent tourists from nabbing the best ocean views.[59] In Barcelona, officials have devised a multipronged strategic plan to address overtourism (see chapter 2.2), which includes a zoning plan regulating areas of the city where new tourism accommodation facilities are permitted—both vacation rentals and hotels—and outlining measures to clamp down on illegal vacation rentals.

Although Europe's historic cities have captured much of the spotlight in the struggle against overtourism, a number of US cities have also seen their residential neighborhoods transformed into revolving accommodation districts and so have enacted legislation to regulate short-term rentals. The upscale Los Angeles beach suburb of Santa Monica has strict rules on short-term rentals, requiring homeowners to submit to stringent requirements, including that hosts must live on the property during a renter's stay, register for a business license, and post no more than two listings on any hosting platform.[60] In a statement, the government of Santa Monica described Santa Monica's Home-Sharing Ordinance as striking "a careful balance: it allows residents to invite visitors into their homes for profit when a resident is present in the home and possesses a City business license; but it otherwise prohibits property owners and online platforms like Airbnb, Inc., and Homeaway.com from booking residential properties for short-term vacation rentals."[61] Other US cities have enacted similar guidelines. In Charleston, South Carolina (see chapter 2.3), for example, strict rules govern short-term rental options in the city. In New Orleans, the city council imposed restrictions on vacation rentals in December 2019. The city's Short-Term Rental Handbook, available online, stipulates a ban on all short-term rentals in the city's historic Garden District, Riverfront Overlay, Bywater, and Marigny neighborhoods.[62]

Controlling Cruise Ship Tourism

Tackling the cruise ship issue has also been top of the agenda for several European cities, including Dubrovnik, Venice, and Barcelona, all

World Heritage Site cities. In 2015, UNESCO warned Dubrovnik against going above "the total sustainable carrying capacity of the city, which should not exceed 8,000 tourists a day" in order to preserve the city's World Heritage status and ensure its livability.[63] The Croatian city's dynamic mayor Mato Frankovic rose to the challenge, negotiating a deal with the Cruise Lines International Association (CLIA) to, beginning in 2019, permit a maximum of only two cruise ships per day with a total of no more than five thousand passengers.[64] The mayor had already implemented various other strategies to reduce overcrowding including reorganizing cruise schedules to stagger departure and arrival times, reducing the number of souvenirs stands by 80 percent, and cutting the number of restaurant tables and chairs by 20 percent—with another 10 percent cut beginning in January 2020.[65] *The Independent* reported that Frankovic explained his initiatives by saying, "We are ready to lose some money, but we will have a better quality of life for citizens and tourists."[66]

Venice's struggles with cruise tourism are even more long standing and complex. In 2014, the US-based World Monuments Fund (WMF)—which has been involved in conservation efforts in Venice for decades—put Venice on its watch list.[67] Then, in October 2015, a UNESCO mission report called on the Italian metropolis to develop "a sustainable tourism strategy, including alternative solutions to allow cruise tourists to enjoy and understand the value of Venice and also its fragility."[68] The following year, UNESCO threatened to place Venice on its World Heritage in Danger list if the city did not stop large cruise ships and tankers from entering the Venetian Lagoon.[69] So far, the municipal government has staved off the endangered designation by producing a series of studies and plans for addressing its cruise tourism issues. Still, cruise tourists continue to flock to the Italian city: AvoidCrowds.com pinpointed ten days in June 2019 where more than ten thousand passengers were due to arrive daily in the city.

Venice is also jeopardized by the climate crisis. In November 2019, following the city's worst flooding in recorded history, UNESCO described the city as "highly vulnerable," noting that "Venice faces several threats from overtourism, damage caused by a steady stream of cruise ships, and . . . the site is also highly vulnerable to the negative impacts of climate change."[70] UNESCO announced plans to send an advisory mission to Venice in early 2020.[71] While international pressure by the WMF and UNESCO has clearly brought global attention to Venice's plight, it has not, as yet, succeeded in compelling the government to implement comprehensive long-term solutions. And now, like some other historic coastal cities, including New Orleans, Louisiana and Charleston, South Carolina, in the United States, Venice is grappling with the complexities from the double assault of overtourism and climate change.

Tourism Taxes

Another policy change intended to help curb tourism numbers and finance reforms is the implementation of tourist taxes, a strategy that has been adopted in cities across Europe. Venice already taxes overnight visitors, but beginning in April 2021, it will charge day-trippers what it calls a "contributo di accesso" or access fee.[72] By 2021, day-trippers could pay up to €10 ($11) on the busiest days.[73] In the city of Edinburgh, Scotland, the council is currently looking into the possibility of adding a transient visitor levy of £2 ($2.40) per night to the price of any room—including Airbnb and other short-term rentals—for the first week of a stay (see chapter 2.4).[74] Although tourism taxes can be controversial, many officials argue these taxes help provide additional funds needed to address the problems linked to overtourism. For instance, according to the World Travel and Tourism Council's *Destination 2030* report, Barcelona's Tourism Commission raised €4.5 million (US$5.25 million) in 2017 and €9.6 million (US$11.2 million) in 2018 from the city's tourism tax.[75] These funds are earmarked to support infrastructure and accommodation regulations and improvements. As Santander stated, taxes "can improve residents' quality of life and visitors experience alike."[76]

Strengthening Sustainable Tourism

In addition to government-mandated reforms and visitor education campaigns are ground up civic initiatives designed to address overtourism through smaller-scale, sustainable tourism. Venezia Autentica, for example, describes itself as a "social business on a mission to make life and travel in Venice better."[77] Working together with local businesses and institutions, Venezia Autentica's online platform provides visitors a wide range of choices of "authentic Venetian experiences, tours and activities."[78] Another local nonprofit organization, BestVeniceGuides, which began in 2015, offers unique small group guided tours with 120 highly qualified local guides who showcase lesser known museums, monuments, and neighborhoods based on the principles of responsible tourism.[79]

Meanwhile a new home-sharing site, Fairbnb.coop, was launched in September 2019 to address some of the problems intrinsic to the vacation rental market. Describing itself as "a cooperative accommodation booking platform that promotes and funds local initiatives and projects," Fairbnb.coop permits only one home rental per host and donates a percentage of the booking fee to a local community project of the visitor's choice.[80] "We wanted to try to solve the issue not by protesting or only by working with local legislators to create better regulations, but also by trying to bring back the 'sharing' in the sharing economy, showing that is possible to combine business and social responsibility," said Venetian cofounder Emanuele dal Carlo.[81] To ensure that profits stay within the local community, Fairbnb.coop also advocates that only residents should be allowed to provide short-term rentals. Although the initiative is still in its infancy, Fairbnb.coop is aiming for an increasingly active presence, working with local authorities to challenge the status quo.

Effective Partnerships

Experts stress the importance of all sectors in a city working together to create effective, long-lasting overtourism solutions. "Destinations need to have management systems in place that prioritize long-term sustainable development, and these solutions will come from a collaboration between both [the] private and public sectors," said Kelley Louise, founder and executive director at the Impact Travel Alliance, a nonprofit promoting sustainable travel.[82] In Amsterdam, as outlined above, the tourism department is partnering with airport officials, airline personnel, online accommodation providers, and law enforcement officials on strategies to handle overtourism. Barcelona is often cited as

BestVeniceGuides guided tour, Venice, Italy. Source: BestVeniceGuides.

a leader in using good broad-based public and private-sector collaboration to tackle overtourism: the city has crafted a multifaceted strategic plan to address overtourism based on input from with a wide range of stakeholders involved in or affected by tourism (see chapter 2.2).

Santander stresses the importance of involving residents in the discussions. "There has to be a very honest and transparent dialogue with the local communities, because the reality of it is not everybody [is] on the same page," he said.[83] In Edinburgh, the city council consulted with more than 2,500 residents and local businesses to get their views about the proposed tourist tax (see chapter 2.4).[84] Such collaboration tends to produce positive results: the *Destination 2030* report, for instance, praised Paris and Sydney, Australia, for ensuring locals helped establish tourism goals.[85] When Sydney was developing its 2030 Tourism Action Plan, city officials embarked upon a large-scale community consultation involving residents, tourists, workers, industry associations, and community organizations, encouraging them to outline their "vision, goals and aspiration for the future of the city."[86] The *Destination 2030* report further explained that "the Parisian government regularly encourages the participation of the city's community in strategic decisions" and that nearly four hundred travel and tourism stakeholders worked together on Paris's *2022 Tourism Strategy*.[87]

While European cities are already experiencing the full force of the tourism boom, cities in Southeast Asia, the Americas, and beyond are beginning to be beset with similar challenges.[88] The *Destination 2030* report classifies urban centers such as New York and Sydney as "mature performers" that risk "future strains related to visitor volume, infrastructure or activity that is testing readiness for additional growth."[89] Delhi, India, and Mexico City, meanwhile, are two cities dubbed "emerging performers" that are just beginning to experience "pressures related to tourism growth."[90]

All evidence suggests that visitors will continue to flock to historic cities across the globe in search of authenticity, to experience other cultures, and to learn about bygone eras. Just as the causes of overtourism are multifaceted, so, too, must be the solutions. Led by local governments, reform strategies must also be informed by consultations with city residents, academics, tourism professionals, and other stakeholders, promoted by civic campaigns and social media, and designed to protect the historic character of the city, its ongoing livability, and its future development. Solutions must strike a balance between welcoming tourists and ensuring that the city continues to be a livable and affordable home for its residents. Some historic European cities are experimenting with suites of solutions, but it is still too soon to judge if these initiatives will be enough to curb and control overtourism effectively.

Notes

1. Valentina McKenzie and Sheena McKenzie. (June 3, 2019). "Cruise Ship Rams Tourist Boat in Busy Venice Canal, Four Hurt." CNN. https://www.cnn.com/2019/06/02/europe/cruise-ship-crashes-tourist-boat-venice-intl/index.html.

2. United Nations World Tourism Organization. (January 21, 2019). "International Tourist Arrivals Reach 1.4 Billion Two Years Ahead of Forecasts." https://www.unwto.org/global/press-release/2019-01-21/international-tourist-arrivals-reach-14-billion-two-years-ahead-forecasts.

3. Ibid.

4. European Commission. (2020). "Growing Consumerism—Knowledge for Policy European Commission." https://ec.europa.eu/knowledge4policy/foresight/topic/growing-consumerism_en.

5. Tony Johnston. (June 25, 2019). Athlone Institute of Technology. Personal communication with author.

6. Cruise Lines International Association. (2019). "2019 Cruise Trends & Industry Outlook." https://cruising.org/news-and-research//media/CLIA/Research/CLIA%202019%20State%20of%20the%20Industry.pdf.

7. Cruise Lines International Association. (2018). "2018 Europe Market Report." https://cruising.org/-/media/research-updates/research/final-market-report-europe-2018.pdf. p. 1.

8. Airbnb (April 19, 2016). "Don't Go There. Live There." YouTube. https://www.youtube.com/watch?v=bbdF8Muiwtk.

9. Eduardo Santander. (June 25, 2019). European Travel Commission. Personal communication with author.

10. Albert Galea. (February 5, 2019). "Number of Tourists Rises by 13% in 2018; 2.6 Million Tourists Visit Malta throughout the Year." *Malta Independent*. https://www.independent.com.mt/articles/2019-02-05/local-news/Tourism-rises-by-13-in-2018-2-6-million-tourists-visit-Malta-throughout-the-year-6736203216.

11. Airbnb. (March 20, 2016). "Airbnb Now Welcomes Guests from Around the World in Cuba." Airbnb. https://www.airbnb.com/press/news/airbnb-now-welcomes-guests-from-around-the-world-in-cuba.

12. Simon Kemp. (January 30, 2019). "Digital 2019: Global Internet Use Accelerates." We Are Social. https://wearesocial.com/blog/2019/01/digital-2019-global-internet-use-accelerates.

13. Eduardo Santander. (June 25, 2019). European Travel Commission. Personal communication with author.

14. Marko Miloš. (June 2019). Dubrovnik Local Guides. Personal communication with author.

15. Corpo di Polizia Locale di Roma Capitale. (August 9, 2018). "Fontana di Trevi #municipio1: rissa tra due famiglie di turisti." Facebook. https://www.facebook.com/PoliziaRomaCapitale/posts/1430928227050764.

16. Julia Buckley. (November 26, 2019). "Rome Considers Plans to Barricade Trevi Fountain from Tourists." CNN. https://edition.cnn.com/travel/article/trevi-fountain-barrier-overtourism/index.html.

17. Stacey Lastoe. (August 8, 2019). "Sitting on Rome's Famous Spanish Steps Can Now Lead to Hefty Fines." CNN. https://edition.cnn.com/travel/article/spanish-steps-rome-sitting-fine/index.html.

18. Themed Entertainment Association and AECOM. (2019). "The Global Attractions Attendance Report."

19. Musée du Louvre. (May 27, 2019). "Museum Closed Today." Twitter. https://twitter.com/MuseeLouvre/status/1132957844125487104?s=20.

20. Union Syndicale Solidaires. (May 27, 2019). "Le Louvre suffoque." Union syndicale Solidaires. https://solidaires.org/Le-Louvre-suffoque.

21. Marko Miloš. (June 2019). Dubrovnik Local Guides. Personal communication with author.

22. Ibid.

23. Direcció de Turisme. (March 2017). "Barcelona Tourism for 2020: A Collective Strategy for Sustainable Tourism." Ajuntament de Barcelona. https://ajuntament.barcelona.cat/turisme/sites/default/files/barcelona_tourism_for_2020.pdf.

24. Ibid.

25. World Heritage List. (2020). "Venice and Its Lagoon." UNESCO. https://whc.unesco.org/en/list/394/.

26. Chiara Albanese, Giovanni Salzano, and Federico Vespignani. (June 30, 2019). "The Long, Slow Death of Venice." Bloomberg. https://www.bloomberg.com/news/features/2019-06-30/venice-is-dying-a-long-slow-death.

27. Eduardo Santander. (June 25, 2019). European Travel Commission. Personal communication with author.

28. Ashleigh Gray. (December 2, 2019). Former Edinburgh resident. Personal communication with author.

29. Ibid.

30. Ibid.

31. Geerte Udo. (June 26, 2019). Amsterdam and Partners. Personal communication with author.

32. Victor Serri/SOPA Images/LightRocket via Getty Images. (April 7, 2018). "Protesters Seen with Banner during a Demonstration against . . ." Getty Images. https://www.gettyimages.co.uk/detail/news-photo/protesters-seen-with-banners-during-a-demonstration-against-news-photo/943526712?adppopup=true.

33. Luis Gene/AFP via Getty Images. (June 10, 2017). "Spain-Barcelona-Tourism-Demo." *Getty Images.* https://www.gettyimages.co.uk/detail/news-photo/protesters-carry-a-banner-that-reads-this-isnt-tourism-its-news-photo/694553898?adppopup=true.

34. Barcelona Street Style Tour. "Barcelona's Premier Street Art Tours & Activities." https://www.barcelonastreetstyletour.com/.

35. Mark Thomas. (February 2, 2016). "Dubrovnik Is Slippery When Wet." *The Dubrovnik Times.* https://www.thedubrovniktimes.com/news/dubrovnik/item/241-dubrovnik-is-slippery-when-wet.

36. Eduardo Santander. (June 26, 2019). European Travel Commission. Personal communication with author.

37. amsterdam&partners "Amsterdam Launches a Campaign to Stop Offensive Behaviour." (May 29, 2018). I amsterdam. https://www.iamsterdam.com/en/our-network/amsterdam-and-partners/news/2018/enjoy-and-respect.

38. Geerte Udo. (June 26, 2019). amsterdam&partners. Personal communication with author.

39. Ibid.

40. amsterdam&partners. "Amsterdam Launches a Campaign To Stop Offensive Behaviour." (May 29, 2018). I amsterdam. https://www.iamsterdam.com/en/our-network/amsterdam-and-partners/news/2018/enjoy-and-respect.

41. Geerte Udo. (June 26, 2019). amsterdam&partners. Personal communication with author.

42. Città di Venezia. (Accessed March 2020). "#EnjoyRespectVenezia." https://www.comune.venezia.it/en/content/enjoyrespectvenezia.

43. Ibid.

44. Ibid.

45. UnhashtagVienna. (Accessed March 2020). "Enjoy Vienna. Not #Vienna." https://unhashtag.vienna.info/en-us.

46. UnhashtagVienna. (Accessed March 2020). "See Klimt. Not #Klimt." https://unhashtag.vienna.info/en-us/article/see-klimt.-not-klimt.

47. Geerte Udo. (June 26, 2019) amsterdam&partners. Personal communication with author.

48. amsterdam&partners (Accessed March 2020). "I amsterdam Letters." I amsterdam. https://www.iamsterdam.com/en/about-amsterdam/overview/i-amsterdam-letters.

49. UnratingVienna. (Accessed March 2020). "So Who Decides What You Like?" https://unrating.wien.info/en-us.

50. Avoid-Crowds.com. (Accessed March 2020). "Check How Crowded It Is." https://avoid-crowds.com/.

51. Eduardo Santander. (June 25, 2019). European Travel Commission. Personal communication with author.

52. Marko Miloš. (June 2019). Dubrovnik Local Guides. Personal communication with author.

53. City of Amsterdam. (Accessed March 2020). "Rules and Permits for Guided Tours in the Red Light District." https://www.amsterdam.nl/en/business/rules-permit-tours/.

54. City of Amsterdam. "Entertainment Tax for Boats and Touring Cars." https://www.amsterdam.nl/en/municipal-taxes/entertainment-tax/.

55. AP News. (December 19, 2018). "Dutch Court Upholds Amsterdam's Ban on New Tourist Stores." https://www.apnews.com/eb9ac19952504a5ab81f327897106090.

56. Francesca Street. (April 28, 2018). "Venice to Separate Tourists and Locals over Busy May Day Weekend." CNN. https://www.cnn.com/travel/article/venice-separates-tourists-and-locals/index.html.

57. Comune.Roma.It. (November 14, 2018). "Presentato nuovo Regolamento di Polizia urbana." https://www.comune.roma.it/web/it/notizia.page?contentId=NWS196768; Dario Nardella. (September 3, 2019). "Quality Tourism." Facebook. https://www.facebook.com/darionardella/posts/1932165043537926.

58. Gloria Rodríguez-Pina. (March 27, 2019). "Madrid Adopts Rules That Will Shut Down Over 10,000 Holiday Apartments." ElPais.com. https://elpais.com/elpais/2019/03/27/inenglish/1553702152_849878.html.

59. Ignacio Zafra. (May 9, 2018). "Valencia Joins the Fight against Holiday Rentals." ElPais.com. https://elpais.com/elpais/2018/05/09/inenglish/1525851010_505130.html.

60. City of Santa Monica. (November 21, 2019). "City of Santa Monica Home-Sharing Ordinance Rules and Regulations." SMGov.net. https://www.smgov.net/uploadedFiles/Departments/PCD/Santa%20Monica%20HomeSharing%20Rules%20PDF%20July%202017.pdf.

61. City of Santa Monica. (August 19, 2019). "Santa Monica Home-Sharing Law Stands after Challenge by Airbnb, Inc. and Homeaway.com." SantaMonica.gov. https://www.santamonica.gov/press/2019/08/19/santa-monica-home-sharing-law-stands-after-challenge-by-airbnb-inc-and-homeaway-com.

62. City of New Orleans: Department of Safety and Permits. (December 3, 2019). "Short Term Rental Handbook." City of New Orleans. https://www.nola.gov/nola/media/311/STR-Handbook-version-20191203.pdf. p. 30.

63. UNESCO. (November 2015). "Report on the UNESCO-ICOMOS Reactive Monitoring Mission to Old City of Dubrovnik, Croatia, from 27 October to 1 November 2015." https://whc.unesco.org/en/documents/141053. p. 24.

64. Mark Thomas. (October 1, 2018). "From 2019 a Maximum of Two Cruise Ships a Day Allowed in Dubrovnik." *The Dubrovnik Times*. https://www.thedubrovniktimes.com/news/dubrovnik/item/5368-from-2019-a-maximum-of-two-cruise-ships-a-day-allowed-in-dubrovnik.

65. Julia Buckley. (November 5, 2019). "Croatian Port of Dubrovnik May Ban New Restaurants." CNN. https://cnn.com/travel/article/dubrovnik-restaurant-ban/index.html.

66. Helen Coffey. (October 2, 2018). "Dubrovnik to Cap the Number of Cruise Ships Allowed to Dock Each Day." *The Independent*. https://www.independent.co.uk/travel/news-and-advice/dubrovnik-cruise-ship-cap-croatia-overtourism-two-dock-a8565166.html.

67. World Monuments Fund. (Accessed March 2020). "Venice." https://www.wmf.org/project/venice.

68. UNESCO. (2016). "Mission Report: Venice and Its Lagoon (Italy C 394)." https://whc.unesco.org/en/list/394/documents/. p. 8.

69. UNESCO. (2016). "Decision: 40 COM 7B.52 Venice and Its Lagoon (Italy) (C 394)." https://whc.unesco.org/en/decisions/6717.

70. UNESCO. (2019). "UNESCO Closely Follows Tides and Flooding in Venice World Heritage Site." https://whc.unesco.org/en/news/2055/.

71. Ibid.

72. Città di Venezia. (April 4, 2020). "Contributo di accesso a Venezia: tutte le informazioni utili." https://live.comune.venezia.it/it/contributo-accesso-venezia-informazioni-utili.

73. Ibid.

74. Scottish Government. (September 2019). "Consultation on the Principles of a Local Discretionary Transient Visitor Levy or Tourist Tax." Gov.Scot. https://www.gov.scot/binaries/content/documents/govscot/publications/consultation-paper/2019/09/consultation-principles-local-discretionary-transient-visitor-levy-tourist-tax/documents/consultation-principles-local-discretionary-transient-visitor-levy-tourist-tax/consultation-principles-local-discretionary-transient-visitor-levy-tourist-tax/govscot%3Adocument/consultation-principles-local-discretionary-transient-visitor-levy-tourist-tax.pdf.

75. Rochelle Turner, Tiffany Misrahi, Jonathan Mitcham, Nejc Jus, Dan Fenton, Lauro Ferroni, and Eva Chan. (June 2019). "Destination 2030: Global Cities' Readiness for Tourism Growth." World Travel and Tourism Council and JLL. https://www.wttc.org/publications/2019/destination-2030. p. 23.

76. Eduardo Santander. (June 26, 2019). Personal communication with author.

77. Venezia Autentica. (Accessed March 2020). "Venezia Autentica." https://veneziaautentica.com/.

78. Ibid.

79. BestVeniceGuides. (Accessed March 2020). www.bestveniceguides.it.

80. Fairbnb.coop. (Accessed March 2020). https://fairbnb.coop/.

81. Emanuele dal Carlo. (June 28, 2019). Fairbnb.coop. Personal communication with author.

82. Kelley Louise. (June 28, 2019). Impact Travel Alliance. Personal communication with author.

83. Eduardo Santander. (June 26, 2019). European Travel Commission. Personal communication with author.

84. City of Edinburgh Council Consultation Hub. (2020). "Transient Visitor Levy: Summary of the Response to the City of Edinburgh Council's Draft Proposal for a Transient Visitor Levy." https://consultationhub.edinburgh.gov.uk/ce/tvl/user_uploads/tvl-consultation-report.pdf. p. 1.

85. Rochelle Turner, Tiffany Misrahi, Jonathan Mitcham, Nejc Jus, Dan Fenton, Lauro Ferroni, and Eva Chan. (June 2019). "Destination 2030: Global Cities' Readiness for Tourism Growth." World Travel and Tourism Council and JLL. https://www.wttc.org/publications/2019/destination-2030.

86. Ibid.

87. Ibid.

88. Ibid.

89. Ibid.

90. Ibid.

Chapter 2.2

Barcelona, Spain

By Albert Arias-Sans, Aina Pedret, and
Natalia Sánchez Castro

In 2015, Barcelona elected its first female mayor, Ada Colau, who also became Europe's first high-profile politician to campaign on a tame tourism ticket. Colau's election reflected the rising tide of resentment among Barcelona residents that out-of-control tourism was making their beloved city unlivable. Tourism had become a public issue, encompassing a range of problems that needed to be addressed in an organized and inclusive way. Indeed, as evidence of tourism's central importance, one of the Colau government's first documents was titled "Actions Required for a Participatory and Sustainable Tourism Strategy."[1] It laid out a collective approach to develop the tools required to manage tourism in a sustainable and responsible way. Barcelona's new policies tried to overcome the traditional sectorial approach to tourism and create an innovative and holistic framework that goes beyond the narrative of "success-by-growing." Despite the challenges and unforeseen twists and turns, Barcelona stands out as one of the first destinations that has worked creatively and thoughtfully to implement a multidimensional strategy for dealing with overtourism.

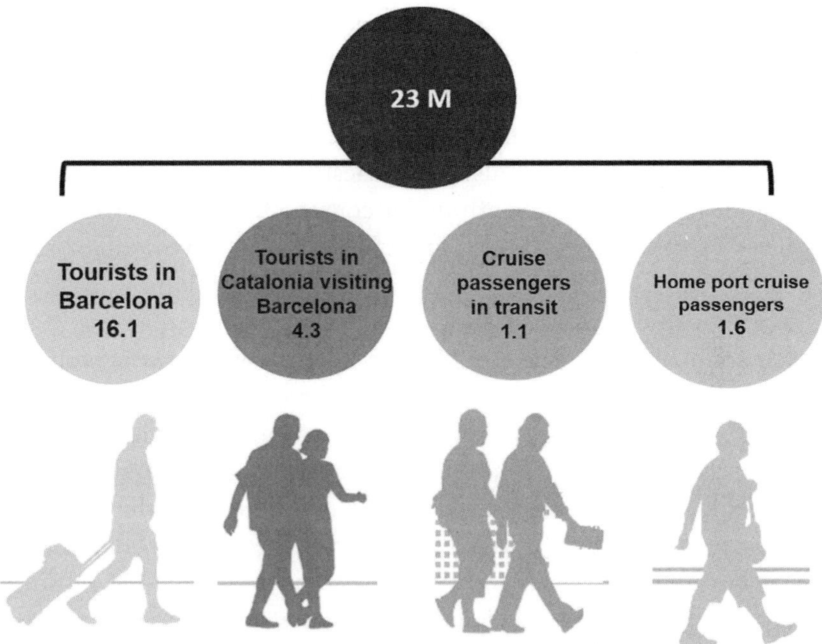

Types of visitors to Barcelona, Spain. Source: Barcelona City Council.

Causes and Impacts of Overtourism

Barcelona, one of Europe's most densely populated and cosmopolitan cities, has also become a leading tourist destination in a very short time. In 2016, Barcelona had almost 23 million visitors, of whom some 1.1 million were cruise passengers in transit, 1.6 million were home port cruise passengers, 16.1 million stayed overnight in all types of tourism accommodations, and 4.1 million were day visitors who arrived mainly by tour bus. For comparison, the number of tourists who stayed in city hotels grew from 1.7 million in 1990 to 8.8 million in 2016.[2]

This steady increase in visitors didn't happen by accident. Beginning with Barcelona's hosting of the Olympic Games in 1992, the city made an economic and political commitment to position itself as one of the most visited urban destinations in Europe. Barcelona city, together with the port and airport authorities, set out to enlarge its infrastructure and visitor capacity. Doing so involved doubling the number of passengers transiting the international airport between 2008 and 2018, remodeling and upgrading hotels, expanding the supply of legal vacation rentals, investing heavily in cruise port infrastructure, and creating specialized visitor areas, all of which have drastically changed the city. Today Barcelona is the most popular cruise destination in Europe, with six cruise terminals and a seventh under construction. Cruise passengers visiting Barcelona (both in transit and home port) have doubled since 2010, totaling in recent years between 2.5 and 2.7 million.[3]

Barcelona's commitment to tourism as a leading economic sector was redoubled during the economic crisis that began in 2007; tourism and real estate development, both closely linked, were the two sectors that offered the greatest resistance to the economic downturn. These sectors, combined with several other factors—the appearance of the city-break phenomenon or short holidays to Barcelona, the explosion of new technologies and social media, and the emergence of digital platforms and new forms of tourist accommodations—have spread images of Barcelona and widened the commercial appeal. The number of tourist flats, for instance, increased dramatically, growing from 2,300 to more than 9,600 rental licenses between 2011 and 2014.[4]

By 2017, Barcelona had 145,901 registered tourist accommodation beds, including 70,129 in hotels and 58,129 in hostels, guesthouses, and other legal short-term rentals, compared with only about 70,000 registered accommodation beds in 2010.[5] On average, Barcelona hosts 150,000 tourists a day who congregate in specific areas, use city services, and spend on average €79.4 ($88.38) per night, which is more than residents spend per day.[6]

Despite Barcelona's significant earnings from tourism, critical voices began to emerge warning that the growth in visitor numbers was overwhelming public spaces, degrading residents' quality of life, creating tensions between residents and visitors, and raising housing and real estate prices. For the first time, in 2016, polls found that the number of residents who thought that Barcelona had reached the limits of its ability to host more visitors were higher than those who thought that Barcelona could keep attracting more tourists. In 2017, residents said tourism was "the worst" problem facing the city. And that same year, 50 percent of visitors also said that there were too many people in Barcelona for them to enjoy their visit.[7]

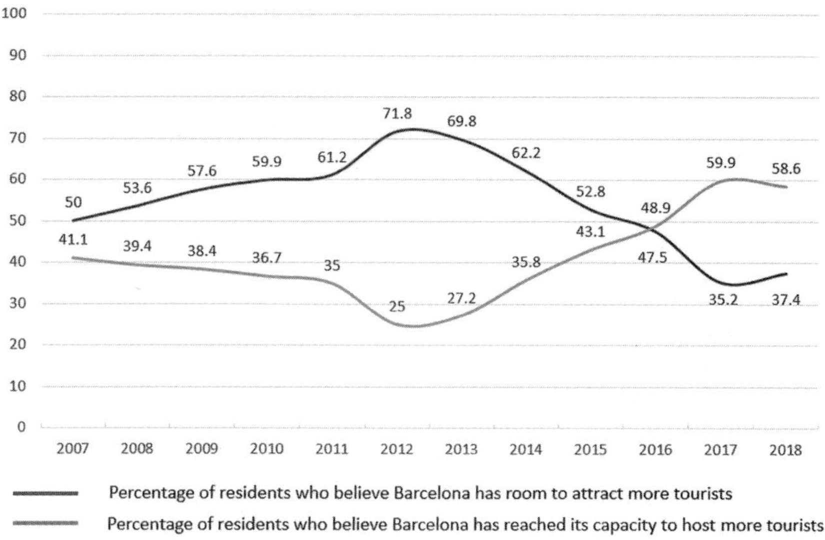

How tourism is perceived by residents in Barcelona. Source: Barcelona City Council.

Solutions

As tourism in Barcelona has turned into a source of conflict and unrest, overtourism has been identified as one of the key challenges facing the city.[8] The municipal government elected in 2015 quickly began assessing overtourism issues and enacting a number of reforms including setting regulations to suspend permits for new tourist accommodations prior to the drafting of the Special Urban Development Plan for Tourist Accommodation (PEUAT, described below), drafting a new Strategic Plan for Tourism 2020, and creating a new institution, known as the Tourism and City Council. Barcelona thereby made a strong political commitment to address overtourism as a public policy issue. Barcelona's Strategic Plan for Tourism 2020 seeks to shift the paradigm from promoting continual growth of tourism numbers to seeking the maximum social return for the city and its residents.[9]

The plan recognizes that the challenge is not simply managing tourism in isolation, but managing a city dominated by tourism. It views tourism not as a separate phenomenon unconnected with the city but rather as an integral part of the urban environment. In practice, that means taking into account the needs of the tourism industry and tourists in areas such as urban design and mobility planning and management. The plan therefore seeks to balance tourism with the many other diverse activities in Barcelona, breaking down barriers that have traditionally separated tourism from the rest of the city.

To accomplish that, the plan defines a governance model and set of planning tools that prioritize coordination. It is intentionally intradepartmental so as to integrate tourism in each of the municipal planning and management departments. Likewise, to involve residents in the tourism policy debate, the city government created the Tourism and City Council as an open, diverse, and participatory civic body set up to discuss what kind of city is desired.

In addition, the Strategic Plan for Tourism 2020 relies on analytical tools, Big Data, and real-time information, which have proved essential to effectively managing overcrowding. By monitoring pedestrian flows to find behavior patterns of visitors, Barcelona is able to acquire very useful information for city planners.[10] This information is helping optimize mobility and plan better services within the prime tourism zones. Another key tool, the Tourism Observatory, is designed to generate quantitative and qualitative data on tourist activities in coordination with other public and private institutions and to share the data with citizens to help with decision-making.[11] Finally, the Strategic Plan for Tourism 2020 launches five specific strategies—territorial, tourism mobility, accommodation, economic development, and marketing—to address the city's main tourism challenges.

Territorial Strategy

The plan takes into account the four geographic levels affecting tourism in Barcelona—the overcrowded spaces, the ten districts, the city as a whole, and the entire metropolitan area—and includes 261 measures with three main objectives: (1) to maximize the social return from tourism activities for local neighborhoods, (2) to improve the integrated management of tourism, and (3) to build different strategies across the territory using a bottom-up process. The Territorial Strategy focuses on the need to manage the seven most overcrowded spaces of the city. Those are the places that host the largest transient populations,

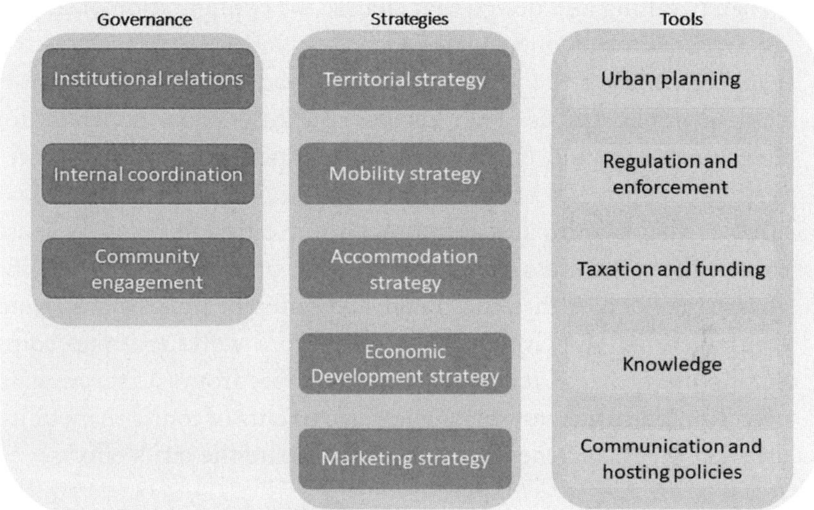

Integral destination management system. Source: Barcelona City Council.

including two in the Old City (Ciutat Vella and Les Rambles) and five areas surroundings the tourism hotspots (Sagrada Familia, Park Güell, Montjuïc, the coast and beaches, and Turó de la Rovira). These "must see" areas experience large volumes of traffic (both vehicle and pedestrian) and an overcrowding of their public spaces. Although these popular spaces have long been used for diverse purposes, the growth of tourism has added extra complexity, intensifying the pressures and degrading daily life for many residents.

Some of the measures being carried out in these prime tourism areas include:

- Closing certain places to guarantee they are reserved for local children's recreation.
- Reforms in visitor flows designed to reduce the pressure on public spaces. Among the steps taken are better management of queues and waiting times, establishment of visitor entrance quotas per hour, promotion of advance bookings, and spreading visitor arrivals throughout the day, thus flattening out the peak periods.

- Urban planning and design that change the configuration of streets, sidewalks, terraces, and squares.
- Tools to improve the planning, control, and supervision of events held in public spaces. They include tracking social networks for illegal activities and encouraging community-based events to reverse the dominance of tourism-driven activities. This requires governance that coordinates key actors from different areas. Because each of the seven most overcrowded spaces of Barcelona has specific governance bodies that coordinate the different policies and avoid contradictions, the city council has set up a working group comprising political and technical representatives from the seven main overcrowded spaces, as well as the departments of tourism, mobility, urban strategy, and the districts, to coordinate the city's efforts.

Tourism Mobility Strategy

To better manage the flows of both residents and visitors, the Tourism Mobility Strategy has put forth twelve action proposals. They include developing specific mobility plans for overcrowded spaces; prioritizing the movement of visitors (and residents) on foot; designing specific solutions for the public transport network's congestion problems and reviewing its fare policy; regulating tourist use of bicycles, scooters, or Segways; and strengthening the quality of the taxi service catering to tourists. The collection of data, counting of people, and monitoring of uses and patterns within key tourism spaces are fundamental to making a diagnosis, verifying the impacts of the new measures, and carrying out better decision-making.

The city council, together with leading tour operators, has also developed a Code of Good Practice for guiding tour groups on the streets of Barcelona. Among other measures intended to reduce the impact of tourism on the daily lives of residents, the sixteen good practices include keeping tourist groups small and adjusting their size to the spaces visited, use of audio guides or whispers to reduce noise pollution, and respecting children's privacy and safety (for example, not taking photos or interfering with games).[12] Furthermore, the council has created a crew of unarmed civic agents to patrol overcrowded tourism areas, communicate with visitors, and redirect flows to less crowded areas.

Accommodation Strategy

A central focus of the Strategic Plan is the management and regulation of the city's supply of tourism accommodations. For this purpose, the Accommodation Strategy addresses four large and interconnected blocks of actions:

1. Special Tourist Accommodation Plan (PEUAT)
Due to Barcelona's rapid growth of tourist accommodations, a zoning plan now regulates any new facilities (hotels, tourist apartments, youth hostels, multioccupancy residences, and so forth). The intent is to make them compatible with a sustainable urban model that guarantees fundamental rights of city residents. The plan defines four specific geographic areas with different regulations for growth depending upon current levels.

2. Inspection, Enforcement, and Coexistence
Parallel to the deployment of the PEUAT, the city council has invested significant human and economic resources in ending illegal tourist accommodations. They include an emergency plan with a network of viewers and inspectors who help detect unlicensed short-term rentals. As a result, from 2016 to 2018, more than 4,900 illegal tourist flats were identified and closed in Barcelona. Similarly, an online "flat detector" tool has been created that allows local community members and tourists to enter the address of an advertised rental property to find out if it is legally licensed or not.

The city council has also focused on improving relations between tourists and residents in those communities where there are legal tourist flats. The council has set up a special mediation service that encourages locals and visitors to reach agreements on how to handle problems such as excessive noise and trash in common areas.

3. Communications Campaign
The actions listed above are supported by an intense communications campaign to raise public awareness about the problem of illegal tourist accommodations in the city. One of the most iconic symbols of the campaign has been the art installations in main tourism areas that feature king-size beds. The aim is to warn residents not to rent out their

apartments illegally, as well as encourage tourists to only use private short-term rentals that are properly licensed.

4. Collaboration and Networking
In addition, Barcelona is connecting with other European cities that have similar problems with short-term rentals. New digital platforms encourage communities to share experiences and increase the transfer of data, information, and strategies. In 2019, for instance, Barcelona was one of ten European cities that asked the European Union for assistance in coping with Airbnb and similar short-term housing platforms.[13]

Economic Development Strategy

As part of its economic development strategy, Barcelona is working to foster use of local resources, promote quality employment, and facilitate investments in innovative and responsible tourism businesses and projects. Two of the initiatives designed to boost the multiplier effect of tourism are the following:

1. Impulsem el que fas! ("We promote what you do!")
This program, funded by a tourism tax, is designed to provide financial subsidies to socio-economically innovative tourism projects and businesses. Two examples of projects that were financed in 2018 and 2019 are gastronomic workshops held in a hostel and run by refugees for both tourists and local residents and a project to connect tourism accommodations with sustainable sources of food, cleaning products, and office supplies.

2. Training for Tourism Companies
The city has developed a project to encourage tourism companies to adopt environmentally and socially sustainable practices, with an emphasis on working conditions (gender equality, work-life balance, salary improvements based on labor agreements, and so forth). The city council provides training sessions and advisors that offer personalized help to companies.

Marketing Strategy

The Tourism Marketing Strategy for Barcelona is a central component of the city's Strategic Plan for Tourism 2020. Its main goals are to guarantee sustainability of the destination, ensure that Barcelona's tourism activities are competitive so as to generate the greatest social return, boost the multiplier effect of tourism, and promote the integrated management of Barcelona. The strategy focuses on the introduction of responsible practices rather than on increased tourist numbers. The marketing strategy targets not only the city but the region as well.

Although Barcelona is addressing many issues related to overtourism, several have been particularly difficult to resolve. Finding the right tools to ensure good wages and working conditions, including affordable housing near workplaces for hotel and restaurant employees, has proved challenging. Another issue is control of the airport and cruise port. In an effort to control day-trip visitors, Barcelona's mayor has vowed to curb the number of cruise ships docking and to put the brakes on expansion of the international airport. As is the case in many other cities, however, the port and airport are overseen by separate governing bodies, not the city of Barcelona.[14]

What Can We Learn from Barcelona?

Although still a work in progress, the Barcelona city council's experience in setting an agenda and creating a five-year strategic plan to tackle overtourism is providing insights for other destinations in Europe. In addition, it is creating the foundation on which future overtourism policies and strategies for Barcelona will be developed. There are three important takeaways—both lessons learned, and questions raised—from the Strategic Plan for Tourism 2020.

First, Barcelona has demonstrated that sustainable solutions for overtourism are not a matter of simply roping off an area and deciding the carrying capacity. Unlike national parks, museums, or heritage sites, Barcelona, like other cities, is an open system, and it is not possible to set finite limits on the total number of visitors. Furthermore, overtourism is not just a matter of volume, but also of the power dynamic among different urban activities that overlap in the same space and time. The first question is how to set up the proper framework, instruments, and tools to deal permanently, not just seasonally, with such an intense, fluid, and powerful phenomenon as overtourism.

Second, overtourism and its related issues cannot be addressed as separate from other urban issues. In fact, overtourism just amplifies and feeds off existing urban problems, including rising rents, low-paid service jobs, and lack of sufficient public services. Overtourism is not only a side effect of global tourism flows; it is also a symptom of inequalities within cities. So, the second question is how to promote the greatest social return from tourism as a way to confront overtourism.

The third question is how to turn tourism into a public issue rather than just an economic one. There are no magic solutions, and most of the time there are also no clear technical solutions. Instead, answers require political leadership. Consequently, it is critical to implement good governance practices that include as many voices as possible beyond the economic sector of tourism or the tourism ministry. This approach is much more complex and harder to manage, but it is also more democratic. It is the path Barcelona must pursue to ensure that it continues to be a livable city.

Notes

1. Barcelona City Council. (July 20, 2015). *Mesura de govern d'impuls del procés participatiu sobre el model de turisme de Barcelona.* https://ajuntament.barcelona.cat/turisme/sites/default/files/documents/150720_mesura_de_govern_-_impuls_proces_participatiu.pdf.

2. Albert Arias, Manel Valdés, and Manel Villalante. (November 2017). "Estratègia de mobilitat turística de Barcelona [Tourism Mobility Strategy]." *Barcelona City Council.* https://ajuntament.barcelona.cat/turisme/sites/default/files/memoria_emt_20171204.pdf.

3. Ibid.

4. Barcelona City Council. (2017). *Special Tourist Accommodation Plan.* http://ajuntament.barcelona.cat/pla-allotjaments-turistics/en/.

5. Observatory of Tourism in Barcelona: City and Region. (2017). *Barcelona Tourism Activity Report 2017.* https://ajuntament.barcelona.cat/turisme/sites/default/files/informe_act_tu_2017_complet_1.pdf.

6. Ibid.

7. Barcelona City Council. (2017). *Perception of Tourism in Barcelona.* https://ajuntament.barcelona.cat/turisme/sites/default/files/percepcio_del_turisme_2017_informe.pdf.

8. Antonio Paolo Russo and Alessandro Scarnato. (2018). "Barcelona in Common: A new Urban Regime for the 21st-Century Tourist City?" *Journal of Urban Affairs.* https://www.tandfonline.com/doi/abs/10.1080/07352166.2017.1373023.

9. Barcelona City Council. (2017). *Strategic Plan for Tourism 2020 Barcelona.* https://ajuntament.barcelona.cat/turisme/sites/default/files/barcelona_tourism_for_2020.pdf.

10. EurecatEG (November 17, 2016). *Barcelona Studies Visitor Flows around the Sagrada Familia to Improve Its Management.* https://eurecat.org/es/barcelona-flujos-visitantes/.

11. Barcelona Home Guide. (2019). "Barcelona Tourism. They Launch an Observatory to Manage It!" https://barcelona-home.com/events-and-guide/event/tourism-barcelona-launch-an-observatory/.

12. .Ship the Line Barcelona. (November 8, 2019). "Code of Good Practice for Guiding Tour Groups on the Streets of Barcelona." https://www.skipthelinebarcelona.com/code-of-good-practice-for-guiding-tour-groups-on-the-streets-of-barcelona/.

13. Tom Stainer. (July 12, 2019). "Barcelona Is Cracking Down on Cruise Ships and Airport Expansion." Lonely Planet. https://www.lonelyplanet.com/news/2019/07/12/barcelona-cruise-ships/.

14. Ibid.

Chapter 2.3

Charleston, South Carolina

By Dan Riccio

Charleston, South Carolina, is a charming coastal city with important historic, cultural, and environmental value. Situated on a peninsula framed by the Ashley and Cooper Rivers, the historic district dates from the 1600s. Its walkability, compact size, abundance of locally owned shops, world-class gastronomy, theater and arts scenes, immaculately preserved antebellum homes and buildings, and rich history make Charleston a wonderful city to live in and visit. Details of the Revolutionary and Civil Wars, stories of enslaved peoples, plantation life, and the current-day Gullah-Geechie culture create a fabric of living history to discover. A number of beaches are a short drive away, and dolphins and manatees can be seen going about life in the important coastal ecosystem. Since 1998, Charleston has been ranked as the Best US City for several years by readers of *Travel + Leisure* magazine in the annual World's Best Awards.[1] "Charleston keeps winning because it doesn't just tout history—it's always coming up with something new," according to *Travel + Leisure* editor in chief Nathan Lump.[2]

Causes and Impacts of Overtourism

Charleston's popularity has taken its toll: in 2017, 6.9 million visitors came to the Greater Charleston Area, a 26 percent increase from 2016 (compared to little more than 136,000 residents in 2018).[3] Although businesses thrive on these numbers (the estimated economic impact from tourism in 2017 was $7.37 billion; estimated labor earnings were $2.7 billion), the profit comes with downsides.[4] Residents raised concerns with mobility and transportation, visitor orientation, enforcement, and residential quality of life as far back as 1978, when Charleston's mayor, Joseph P. Riley, implemented the city's original Tourism Management Plan.

As downtown historic neighborhoods have become increasingly popular, they have experienced unprecedented congestion primarily caused by motor coach buses and smaller tour buses, horse-drawn carriages, walking tour groups, pedicabs, and cruise ship activity. Cruise tourism in Charleston began in April 1973.[5] Following the establishment of Charleston as Carnival's home port in 2010, the city was placed on the National Trust for Historic Preservation's watch list in 2011 and the 2012 World Monuments Fund's watch because of threats to the city's historic and cultural preservation.[6] Almost 225,000 cruise ship passengers came to Charleston in 2017, a nearly 20 percent increase from 2015.[7] In 2018, there were more than one hundred cruise ship visits to the Charleston pier, and legal battles have ensued over the proposed expansion of the terminal. Aside from environmental impacts of the expansion, residents are most concerned about traffic, congestion, and emissions (soot and smoke) from cruise ships.[8] Additionally, although the South Carolina Ports Authority receives $75 per cruise passenger and $15 daily for each of the five hundred cars parked in the port terminal, the City of Charleston does not receive any direct revenue and does not have authority over the port.[9]

The rapid growth of short-term rentals has also put pressure on residential neighborhoods in the city, raising concerns about noise, litter, congestion, and transient visitors. Residents also worry about rentals diminishing housing stock and raising the price of homes. The city did not anticipate this phenomenon when it adopted its Tourism Management Plan Update in 2015, but the following year joined other

Cruise ship docked at Charleston's port. Source: Ron Cogswell (Flickr).

cities in the United States and Europe in adopting measures to combat the problem, mainly through compliance software designed to identify and monitor unpermitted short-term rental properties.

Community members and hospitality leaders are concerned that there is an overconcentration of hotels, which may displace office, retail, or residential spaces.[10] The Charleston Planning Department projects that anticipated hotel developments will leave the city with twenty hotel rooms per one hundred residents, reducing the potential for a mixed-use downtown area.[11] Although most developments on the peninsula are restricted to fifty rooms (a typical size for boutique hotels) and developers must follow a multilevel approval process, residents still worry that their community will turn into a "hotel district."[12]

Solutions

Charleston's current Tourism Management Plan Update, prepared by the 2014 Tourism Advisory Committee (a group of residents, industry representatives, and city staff), was completed in part to address the

70 percent increase in tourist arrivals since 1998.[13] It provides recommendations to address five principal areas: Tourism Management and Enforcement, Visitor Orientation, Quality of Life, Special Events, and Mobility/Transportation.[14] The main goal of the 2015 plan is to maintain "Charleston's identity as a place to work and live," as this identity is critical to tourism industry and quality of life.[15] The plan includes "the creation of a centralized, coordinated approach to tourism management and special events" and an annual public review of tourism management, as well as some specific recommendations:

> Improve visitor orientation through signage for destinations and parking. . . .
>
> Conduct a comprehensive traffic and parking study. . . .
>
> Monitor and review tourism activities during peak months (for example, impact of future hotel room increases). . . .
>
> Continue dialogue about and review areas of opportunities (regarding cruise ship policies in the historic district). . . .
>
> Study the possibility of a passenger head tax (to mitigate negative effects of cruise tourism). . . .
>
> Create transportation apps and integrate with "tourism-related orientation resources. . . .
>
> Mitigate traffic congestion by improving public transportation, bicycle-sharing programs, and sidewalk safety.[16]

To date, the city has implemented many of the recommendations of the 2015 Tourism Management Plan, including creating the Department of Livability and Tourism. The department's mission is to carry out the plan's recommendations, but also treat the plan as a living document that can be revised as necessary. Formed from two existing city divisions (previously Livability and Tourism separately), the department is in charge of tourism management and oversees officers responsible for commercial, residential, tourism, and short-term rental code enforcement. Although the department does not have all the answers, Charleston has developed three innovative strategies that may serve as examples for other destinations grappling with the challenges of success: enforcement of policy, mobility and transportation, and short-term rentals.

Tourism Ordinances

At the onset of the planning process, members of the 2014 Tourism Advisory Committee identified enforcement as a key issue. Many people believed that tour operators and businesses could violate tourism ordinances with impunity, leading to the perception that industry was in control and the city was powerless. In response, the city hired and trained three tourism enforcement officers for the Livability Division, sending the message that it was committed to implementation of policy.

Tourism enforcement officers, who are often the first to interact with tourists, also serve as ambassadors for the city and have proven instrumental in effectively regulating tourism activities.[17] As of early 2020, six officers patrolled the downtown peninsula every day, primarily by bicycle. Their responsibilities include monitoring carriage tour zones, coach buses, tour guides, and rickshaw drivers to ensure that they are following regulations. Each day one officer answers the "tourism hotline" cell phone, responding to questions and complaints from citizens and investigating situations as required.

Tour operators who violate tourism ordinances can receive a maximum fine of $1,087, which has been an effective deterrent for repeat offenders. In general, the presence of enforcement officers is widely seen as a success and has been well received by both residents and tourists.

Carriage Tours

Downtown residents were particularly concerned that horse carriage tours were causing gridlock in their neighborhoods and on main streets throughout the historic district. In 2018 alone, 41,011 carriage tours were conducted through downtown neighborhoods.[18] To address this issue, the city convened a committee of carriage industry owners, representatives from historic neighborhoods directly affected by carriage tours, and City Livability and Tourism staff to examine the existing carriage tour routes and offer recommendations for improvement. As a result, two newly regulated zones were added, creating five regulated zones; several new entrance points to each zone were created to allow carriages to spread out and reduce congestion; and "A and B" routes

Carriage tours in Charleston's historic district. Source: Ann Marie Brasco. (Flickr.)

going in opposite directions were implemented for all zones, allowing for more carriage separation.

Additionally, the city began using global positioning system (GPS) technology to efficiently track carriages while on tour. Not only do the GPS devices track the carriages to ensure that the drivers remain in their assigned zones, but they also allow drivers to alert tourism enforcement officers when horses relieve themselves on city streets. The officers and a waste removal company are assigned computer tablets to monitor the spill locations and make sure they are cleaned promptly. All information is recorded to ensure accountability of both the carriage driver, who is required by ordinance to report waste spills, and of the waste management employee charged with cleaning the spill.[19]

Coach Buses

In 2018, Charleston issued 5,156 coach bus transportation permits and 3,470 coach bus parking permits to bus operators.[20] Prior policy required coach bus operators to obtain permits upon arriving to Charleston, thus creating delays at the city's permit center. To make the process more efficient, bus operators are now required to apply for permits well in advance of arrival to Charleston. They also must provide an itinerary outlining dates, times, destinations, and loading/unloading locations so that staff can plan ahead, with the goal of improving mobility throughout the city.

The city requires coach bus operators to use a perimeter route around the Charleston peninsula, prohibiting the large vehicles from traveling on narrow streets and allowing for separation of the buses.[21] Bus operators receive a map that outlines the perimeter route, which also denotes authorized loading zones and parking areas. Bus operators and tour guides often complained that the city had not updated the map in many years and that it was difficult to navigate between the northern and southern portions of the city. They also said that there were not enough loading zones, which lead to congestion and gridlock. Frustrated with these obstacles, tour guides had been instructing drivers to remain in the more popular tour areas in the Lower Peninsula, contributing to a higher concentration of coach buses in the historic district.

In response, the city made changes to the perimeter route map that allowed for easier connectivity to the northern portion of the city. In addition, an agreement with the local transit authority added seven loading zones in the historic district, spreading the coach buses throughout the entire peninsula and creating more separation. As an added measure, the city now offers an interactive perimeter route map accessible by the operator's cell phone, allowing access at any time. The interactive map includes GPS tracking and current real-time street and lane closures to assist the operator while visiting the city.[22]

Short-Term Rentals

In 2017, Charleston mayor John Tecklenburg appointed a task force comprising residents and city staff to explore comprehensive recommendations to legalize short-term rentals.[23] In April 2018, the city council passed an ordinance to legalize short-term rentals, but with strict guidelines. A permit and business license are required, and the property must be owner-occupied with a tax assessment of 4 percent. The operator, who must be present on the property while renting short term, is required to include the issued registration number on all online listings; simply advertising without a permit is a violation of the ordinance. Whole-house rentals are prohibited, and the city was divided into three separate categories, each with its own requirements.[24] To enforce the new regulations, the Department of Livability and Tourism hired three code enforcement officers to focus full time on short-term rental compliance.

The city has purchased newly developed software designed to identify properties that are advertised on vacation rental websites but do not have a permit. The software is capable of extracting information from websites that conceal the address, name of the property owner, and dates that the property had been rented.[25] Officers can then locate the property, take photographs, and attempt to make contact with the renters for additional evidence.

If found guilty of violating the short-term rental ordinance, the operator can receive a maximum fine of $1,087. When enforcement began in August 2018, approximately twenty-five hundred unpermitted short-term rental properties were being advertised in the incorporated areas of Charleston. Based on Livability and Tourism statistics to date, the city has experienced a 51 percent reduction in unpermitted advertisements on vacation rental websites, more than two hundred court summonses have been issued to violators as a direct result of enforcement, more than ninety-nine properties have returned to full-term rentals, and sixty-six homes have been sold and returned to the housing stock.

Hotel Development

Since 1987, Charleston has regulated the growth of hotels through zoning requirements. In 2019, the city updated those requirements to create the most stringent set of regulations yet.[26] Proposed new hotels must meet conditions which regulate parking, traffic, number of rooms, and any potential negative impact on surrounding neighborhoods.

Success to Date

Centralizing key functions of tourism management and quality of life under one city department has proven successful to adequately address the equilibrium between Charleston residents, visitors, and the tourism industry. Successes include resourcing tourism officers to monitor and enforce tourism industry activities; implementing sophisticated software to identify, track, and enforce unpermitted short-term rentals; resourcing code enforcement officers to regulate and enforce residential and commercial property cleanliness; managing and regulating all special events; maintaining a good working relationship with the South Carolina Ports Authority to efficiently address concerns with cruise ship activity; and serving as a model for other cities by implementing in 2019 stringent zoning requirement for all new proposed hotel projects.

Experience in Charleston has shown that tourism and livability go hand in hand. Adopting management approaches that are realistic and enforceable requires an integrated approach that involves all stakeholders. The strategies undertaken in Charleston have not only improved the city's tourism, but have also restored confidence that the local government is committed to its residents' quality of life.

Notes

1. Emily Williams. (July 10, 2019). "Travel + Leisure Readers Name Charleston No. 1 U.S. City for 7th Year in a Row." *Post and Courier*. https://www.postandcourier.com/business/travel-leisure-readers-name-charleston-no-u-s-city-for/article_0866afde-a315-11e9-ac3f-137105f15384.html.

2. Dave Munday. (July 10, 2018). "Charleston Ranked No. 1 US City by Travel Leisure Readers for 6th Year in a Row." *Post and Courier*. https://www.postandcourier.com/business/charleston-ranked-no-us-city-by-travel-leisure-readers-for/article_f1ead056-83ac-11e8-9176-cbf9af4b2c16.html.

3. CHS Today. (May 7, 2018). "A $7.37 Billion Economic Impact." https://chstoday.6amcity.com/the-economic-impact-of-tourism-in-the-greater-charleston-s-c-area/; United States Census Bureau. (July 1, 2018). "Charleston City, South Carolina." https://www.census.gov/quickfacts/fact/table/charlestoncitysouthcarolina/PST045218.

4. CHS Today. (May 7, 2018). "A $7.37 Billion Economic Impact." https://chstoday.6amcity.com/the-economic-impact-of-tourism-in-the-greater-charleston-s-c-area/.

5. World Monuments Fund. (February 2013). *Harboring Tourism Cruise Ships in Historic Port Communities*. https://www.wmf.org/sites/default/files/article/pdfs/Charleston-Report.pdf. p. 106.

6. World Monuments Fund. "Charleston Historic District." (Accessed March 2020). https://www.wmf.org/project/charleston-historic-district; Mikaela Porter.

(June 10, 2019). "Charleston Groups Head to Supreme Court over Controversial Cruise Ship Development." *Post and Courier*. https://www.postandcourier.com/news/charleston-groups-head-to-supreme-court-over-controversial-cruise-ship/article_00e8071c-87a9-11e9-b223-4380385b6031.html.

7. David Wren. (December 16, 2017). "Big Pleasure Ships Gaining Popularity among Charleston Tourists but Growth Is Limited." https://www.postandcourier.com/business/big-pleasure-ships-gaining-popularity-among-charleston-tourists-but-growth/article_38f2702c-e0fa-11e7-b6a6-6f64ecd8dcda.html.

8. Office of Tourism Management. (2015). "Tourism Management Plan Update 2015." https://www.charleston-sc.gov/DocumentCenter/View/10419/Tourism-Management-Plan-2015?bidId=; World Monuments Fund. (February 2013). *Harboring Tourism Cruise Ships in Historic Port Communities*. https://www.wmf.org/sites/default/files/article/pdfs/Charleston-Report.pdf. p. 121.

9. Ibid.

10. Understand SC. (August 13, 2019). "Episode 16: Controlling Hotel Development." Post Courier Podcasts. https://www.postandcourier.com/understandsc/understand-sc-what-s-driving-charleston-s-rapid-hotel-boom/article_707840d8-bd19-11e9-8b87-3f6e006fe653.html.

11. Patrick Hoff. (May 27, 2019). "Task Force Exploring Potential Changes to Charleston Hotel Regulations." *Charleston Regional Business Journal*. https://charlestonbusiness.com/news/hospitality-and-tourism/76503/.

12. Understand SC. (August 13, 2019). "Episode 16: Controlling Hotel Development." Post Courier Podcasts. https://www.postandcourier.com/understandsc/understand-sc-what-s-driving-charleston-s-rapid-hotel-boom/article_707840d8-bd19-11e9-8b87-3f6e006fe653.html.

13. Official Website for Charleston, SC. (2015). Tourism Management Plan 2015 Update. https://www.charleston-sc.gov/DocumentCenter/View/10419.

14. Ibid.

15. Ibid. p. 3.

16. Ibid.

17. *Post and Courier*. (April 3, 2015). "Tourism Plan Brings Balance." https://www.postandcourier.com/opinion/tourism-plan-brings-balance/article_770447bc-fbc2-54c7-aa4f-ce01a71fe0a1.html.

18. Department of Livability and Tourism. Official Website for Charleston, SC. (Accessed March 2020). https://www.charleston-sc.gov/index.aspx?nid=1380.

19. Emily Williams. (January 24, 2019). "Charleston Starting to Track Carriage Horse Pee with GPS Technology." *Post and Courier*. https://www.postandcourier.com/business/charleston-starting-to-track-carriage-horse-pee-with-gps-technology/article_8db095a2-1f52-11e9-b8e3-17d68894460d.html.

20. Department of Livability and Tourism. Official Website for Charleston, SC. (Accessed March 2020).

21. Ibid.

22. "Touring Historic Charleston, SC by Motor Coach." (Accessed March 2020). Official Website for Charleston, SC. https://gis.charleston-sc.gov/interactive/motorcoach/.

23. Angie Jackson. (September 25, 2017). "Charleston Short-Term Rental Taskforce Votes to Send Proposal to City Planning Commission." *Post and Courier*. https://

www.postandcourier.com/news/charleston-short-term-rental-task-force-votes-to-send-proposal/article_5248de42-a24f-11e7-b0db-9b4638832d0b.html.

24. Abigail Darlington. (April 10, 2018). "Charleston Finally Lifts Ban on Short-Term Rentals, but the New Rules Are Strict." *Post and Courier*. https://www.postandcourier.com/news/charleston-finally-lifts-ban-on-short-term-rentals-but-the/article_51d57ee2-3d00-11e8-aa86-23d87515 80dc.html.

25. David Slade. (January 10, 2019). "Charleston Now Uses Software to Find Short-Term Rental Violations on Apps Like Airbnb." *Post and Courier. https://www.postandcourier.com/news/charleston-now-uses-software-to-find-short-term-rental-violations/article_6ad8bb20-1428-11e9-be47-83cf1c4bff30.html.*

26. Jacob Lindsey. (January 6, 2020). City of Charleston Planning Department. Personal communication with author.

Chapter 2.4

Edinburgh, Scotland

By Aileen Lamb

Edinburgh, Scotland's capital city, acts as the home to world-class academic institutions and is the center for Scottish political life. It is also a hub for finance, data science, and technology. Recently ranked as the "most livable city in the world" by Arcadis, residents enjoy a high quality of life—international students, travelers, and workers swell the city's population, creating a vibrant mix of cultures and ideas.[1] Edinburgh's attractiveness was confirmed when the capital claimed the top spot in 2019 Lonely Planet Ultimate UK Travel list.[2]

As a tourist destination, Edinburgh has been welcoming travelers for hundreds of years. Today, Edinburgh is the United Kingdom's second leading tourist destination for international visitors, just after London. Generating 5 percent of the GDP and supporting 207,000 jobs, tourism in Scotland is recognized by the government as vital to the economic and social structure of the country.[3] Edinburgh has seen notable growth in its tourism sector between 2010 and 2017. Visits increased by 30 percent to 4.26 million, and day visitors have grown by almost half a million each year.[4] Over the same period, visitor expenditure has increased by 46 percent to £1.478 million (US$1.902 million), and jobs supported have risen from 28,000 to nearly 34,000.

However, alongside the accelerating appeal of this historic city and tourism's central role in the economy, Edinburgh is experiencing the dark sides of rocketing tourism demand. Recently, Edinburgh was named as "one of the world's most serious 'overtourism hotspots'" by *CNN Travel*.[5] With total annual visitor arrivals nudging close to ten times the number of residents, this small city of around 513,000 is currently facing the challenge of finding a balance for its people.[6]

Causes of Overtourism

Although Edinburgh has been welcoming visitors for centuries, 1995 proved to be a landmark year for the development of modern-day tourism—and eventually overtourism at certain times—in the city. During this year, Edinburgh secured UNESCO World Heritage Site status for its city center, opened the first purpose-built conference space (Edinburgh International Conference Centre), saw the launch of the low-cost flight revolution (Edinburgh to London was the second "budget flight" to depart Scotland after Glasgow to London), and hosted the premiere of the hugely successful *Braveheart* movie, which is set in Edinburgh.[7] A comprehensive program of city center regeneration brought the old town back to life for residents, while also improving its quality for visitors. The city prioritized investments in hotels, attractions such as Dynamic Earth (a visitor attraction that aims to tell the story of the planet), and "pedestrianization" that facilitated viewing of the city's art and architecture as well as human interactions.

These beneficial initiatives have also brought unforeseen complexities, however. As seen in many other places, UNESCO designations have now contributed to a "honeypot effect," that is, increasing visitor numbers many times over and countering the original conservation goals.[8] Although Edinburgh has welcomed this influx of customers who bring cash and drive economic opportunity, it simultaneously feels the pressure of such rapid development.

This growth was later compounded in 2012, when the Edinburgh Tourism Action Group (ETAG) ramped up its collaborative efforts, publishing the Edinburgh Tourism 2020 strategy.[9] The document articulated how stakeholders from industry and public bodies would work together to reduce seasonality, undertake collaborative innovation, and focus on delivering new overnight visitors from international markets.

The sector saw success from this strategy leading up to 2015. Hotel investment and occupancy were buoyant. Arrivals and new routes into the city's airport exceeded previous records. Entrepreneurs built new types of dynamic businesses to service experience-hungry travelers, and efforts began to improve visitor flow out from the city. The capital's longstanding short-term rental marketplace saw Airbnb enter the mix.

Edinburgh was delivering a significant injection of jobs and cash into Scotland's economy. A midterm review revealed that strategy targets—£1.5 billion (US$ 1.929 billion) in tourism revenue annually by 2020—were likely to be achieved early, and so it was agreed to increase goals, now aiming for £1.6 billion (US$2.05 billion) in tourism revenue by 2020. Critics of this strategy have seen it as a clear indication of openly encouraging overtourism.

Reputation as the "Festival City"

Since 1947—when cultural events were first used to help bring life to a population worn down by World War II—Edinburgh has been seen as an arts destination. The city now annually hosts more than three thousand events with more than twenty-five hundred performers and more than four-and-a-half million attendances from seventy countries worldwide.[10] Home to the world's largest arts gathering, the Edinburgh Festival Fringe, the city has been dubbed the "Festival City."

During August, the heart of the city effectively becomes a stage, and Edinburgh begins to feel the distinct challenges of being a tourist destination while also being a living, working place. The most in-demand festival locations exist within a few square miles, drawing massive crowds. Within Edinburgh Festival Fringe, the founding principle of inclusion asserts that no one who can secure a performance venue will be turned away. So, as travelers and artists soak up the unique atmosphere of events with a medieval castle as a backdrop, locals, who often make up more than one-fourth of attendees, regularly report their perception that the city is "being taken over."

By 2015, this criticism was extending from the peak August festival season to include concerns about the rest of the year, and effective visitor management was becoming a significant consideration. The development of Edinburgh's festivals has created benefits and challenges, supporting growth and driving investment in hotels and attractions, but also generating criticism about the extent to which the city is "being overrun" with visitors at peak times.

Rising International Recognition

Edinburgh has built a position as a powerhouse for the Scottish visitor economy, and the city's tourism industry has partnered with public bodies to guide Edinburgh's tourism sector. ETAG led a comprehensive program of industry development coupled with global consumer engagement, which has seen Edinburgh's profile on Weibo, a Chinese microblogging site often compared to Twitter, ranked number eleven in the world in the Overseas Destination Weibo Influence Ranking—only marginally below cities including New York and nations including Thailand and Germany.[11]

Like other iconic places, however, a significant amount of tourism in Edinburgh has also been driven by factors and global trends out of the destination's control. The weak pound, the city's attractiveness to Instagrammers and as a film location, the rise of low-cost airlines, and the popularity of the city to emerging markets able to produce huge numbers of new travelers have driven increased visits.

Impacts of Overtourism

By 2017, the press was suggesting that Edinburgh faced the same scale of overtourism as seen in Venice, Barcelona, and Amsterdam. The challenges have continued to grow, fueled by the rapid growth of short-term rental platforms such as Airbnb and a growing sense that the city prioritizes visitors' needs over those of the residents. Prominent Edinburgh resident and author Val McDiarmid believes that authorities have not acted quickly enough. She contends that "the Scottish capital had become over-run due to the number of properties being used for short-term holiday lets" and "the mass influx of tourists to Edinburgh is at risk of tearing up the fabric of the city."[12]

Furthermore, property advisor Knight Frank has called for the introduction of a "protection plan" to mitigate the impacts of the "unsustainable" rate at which city center office space is being redesignated for apartment and hotel accommodation. One advisor at the firm stated, "Tourism has proved a double-edged sword for Edinburgh. Visitors coming to the city and spending money is great news for hotels and retailers, but it's had unintended consequences for other sectors." They believe that the "perilously low" supply of office space and rising rent could dent the city's position as a top location for business.[13]

Solutions

Despite the positive backdrop of strong demand from travelers and investors, the city is alert to the need to address tourism challenges. City leaders recognize that they must listen to citizen views, but doing so may raise tougher decisions about how to manage tourism's future growth. One main concern is losing revenue if tourism is reduced. As Roddy Smith, chief executive officer of Essential Edinburgh, noted, "Edinburgh is the headline act for Scottish tourism, delivering over 25 percent of the country's visitor spending and attracting 63 percent of all international visitors. That means decisions made in this city will have very real impacts on the rest of the country."[14]

In response to this paramount need to balance tourism with livability, ETAG has expanded its focus. It continues to influence strategic and policy decisions and articulate the importance of the city's tourism sector.[15] In tandem, significant work, including Scotland's largest stakeholder engagement exercise around tourism, has also been undertaken to ensure that citizens are heard and their views incorporated into future strategy.[16] The result is that ETAG has shifted its position from driving growth to managing growth.

Research and Development

City partners believe that all policy must be based on data and address resident housing and employment challenges such that the implemented policy will enable the market to operate effectively for participants. Unlike other industries that can relocate with relative ease, tourism delivers jobs rooted locally to maximize the assets travelers enjoy—scenery, nature, culture, and heritage. Given the importance of this employment, policymakers are keen to see the industry improve job quality and make better use of worker skills to increase productivity and profitability.

Evidence of data's role in shaping policy decisions can be seen with a recent Scottish Government study, which looked at the impact of short-term rentals on communities in Scotland.[17] Findings revealed that almost 17 percent of dwellings in Edinburgh's city center are currently promoted via short-term rental platforms, thereby providing the evidence needed to back citizen calls for change. The city now welcomes an upcoming Scottish Government policy that will introduce statutory registration for all short-term property rentals. The policy aims to improve oversight and therefore management of accommodation stock.

The local government is also encouraging collaboration between Scotland's technology, business, academic, and tourism industry with investments, including a £1.3 billion (US$1.67 billion) city deal for the Edinburgh region. The investment includes a significant focus on data-driven innovations for tourism, festivals, and culture.[18] A ten-year program will include the creation of a Scottish tourism data hub and tourism observatory to help businesses, policy makers, and citizens understand and manage tourism effectively using data. Projects using sensors to help manage visitor flow, research to forecast future tourism demand, and collaborations between transport and events organizations to manage burdens on residents are all being undertaken to help mitigate overtourism.[19]

Transient Visitor Levy

Edinburgh is likely to be the first UK city to introduce a transient visitor levy.[20] Although the Scottish Government has already led a public consultation, the required legislation is ongoing. The City of Edinburgh Council has also announced plans for a levy as soon as legislation allows. The authorities plan to invest revenues into initiatives to improve the visitor and resident experiences in the city and on tourism management; many stakeholders, however, are opposed to what they perceive to simply be another cost on the traveler and a burden on accommodation providers (who are likely to be the collection point).

Visitor Dispersal

Recently, the city has taken steps to encourage visitors to discover unexplored areas of the city. The City of Edinburgh Council and ETAG have collectively worked to improve wayfinding systems (prescribed cycle, bus, and walking routes), implemented a new bicycle hire system, and introduced proactive marketing for peripheral attractions away from the city center.[21] Edinburgh has recognized that these strategies would spread the wealth and deliver economic benefits to more communities, while providing travelers who want to "live like a local" the opportunity to do so.

Destinations Leaders Program

Simultaneously, a unique initiative to develop tourism industry leadership—the Destination Leaders Programme—has been launched in Edinburgh.[22] Currently, and after several years since its launch in 2014, it focuses on creating an ongoing supply of people from across the sector with a deep and practical understanding of what makes a modern destination work. Participants are taught about strategic tourism planning, sustainable business practices, and how to work with and influence stakeholders. The city now has approximately fifty alumni who take active roles in Edinburgh's tourism development and management.

A Sustainable Future Ahead

Driven by growing tension between citizens, tourists, and other city businesses, along with the need for inclusive growth and fair work, Edinburgh is determined to be an exemplar of how to create a genuinely sustainable tourism strategy. The Edinburgh 2030 Tourism strategy, launched in January 2020, clearly states, "It's time to adapt our approach by working to make tourism work better for the city."[23]

The strategy has the residents of the city at the heart of its policy and seeks to balance their needs with those of businesses and visitors. It focuses on five priorities, which were determined following consultation with stakeholders: place, people, environment, partnerships, and reputation.[24] ETAG is developing action plans that lay out what needs to be done, including projects, citizen participation, public and private investments, stakeholder management plans, and environmental management.[25]

Local council leader Adam McVey highlighted the significant difference between this strategy and those which have preceded it. "This is a fundamental shift in policy from generating growth to managing the continued success of our tourism industry for our people, environment, and sense of place," he said.[26]

The city's strengths provide an opportunity for a bright future. Edinburgh's tourism, data science, and academic expertise offer the chance to create modern solutions in this historic city, such as the Scottish tourism observatory to monitor and provide tourism management information.

The key for Edinburgh is to embed the 2030 tourism strategy with wider city development strategy. As Peter Jordan, tourism and place expert and author of the Edinburgh 2030 strategy, noted, Edinburgh 2030 is a "strategy for the city" that "provides a guide for all on where they should focus their time and resources." Following this strategy means focusing on broad priorities, including implementing low-carbon targets, improving the quality of jobs in the sector, and ensuring that the tourism sector supports other sectors, such as food and drink.

Edinburgh will have to make some tough decisions and radical changes going forward. Doing so effectively will require even more effective partnerships and tourism management. The Edinburgh 2030 tourism strategy provides the road map needed to guide this change. City councilor Scott Arthur voiced the ambitions of many when he recently said, "I am proud that tourists want to visit Edinburgh, but ultimately our capital must be a livable city."[27]

Notes

1. Arcadis. (2018). *Sustainable Cities Index 2018: Citizen Centric Cities*. https://www.arcadis.com/en/united-kingdom/our-perspectives/sustainable-cities-index-2018/united-kingdom/.

2. Lonely Planet. (August 12, 2019). "Lonely Planet Reveals the 10 Best Travel Experiences in the UK." https://www.lonelyplanet.com/articles/lonely-planet-uk-best-experiences.

3. Scottish Government. (April 24, 2018). *Tourism in Scotland: The Economic Contribution of the Sector.* https://www.gov.scot/publications/tourism-scotland-economic-contribution-sector/pages/5/.

4. Brian Ferguson. (July 3, 2019). "Edinburgh Named One of the World's Most Serious 'Overtourism Hotspots.'" *The Scotsman.* https://www.scotsman.com/heritage/edinburgh-named-one-of-the-world-s-most-serious-overtourism-hotspots-1-4958194.

5. Joe Minihane. (July 2, 2019) "Destination Trouble: Can Overtourism Be Stopped in Its Tracks?" CNN. https://edition.cnn.com/t.ravel/article/how-to-stop-overtourism/index.html.

6. Brian Ferguson. (July 3, 2019). "Edinburgh Named One of the World's Most Serious 'Overtourism Hotspots.'" *The Scotsman.* https://www.scotsman.com/heritage/edinburgh-named-one-of-the-world-s-most-serious-overtourism-hotspots-1-4958194.

7. Edinburgh World Heritage Website. (2020). https://ewh.org.uk/.

8. Rosie Spinks. (July 11, 2019). "UNESCO Could Have Helped Save Venice from Overtourism: Why Didn't It?" *Skift.* https://skift.com/2019/07/11/unesco-could-have-helped-saved-venice-from-overtourism-why-didnt-it/.

9. Edinburgh Tourism Action Group. (2020). https://www.etag.org.uk/about-us/.

10. Edinburgh Festival City. (2020). *History of Festivals.* https://www.edinburghfestivalcity.com/the-city/history-of-the-festivals.

11. Alice He. (2019). *Our Weibo Is NO.11 in the World in the Overseas Destination Weibo Influence Ranking in July!!* Tweet, September 16, viewed March 4, 2020. https://twitter.com/AliceHe_/status/1173539389164720129.

12. Brian Ferguson. (October 9, 2018). "Tourism 'Tearing Up Fabric' of Edinburgh, Warns Val Mcdermid." *The Scotsman.* https://www.scotsman.com/lifestyle-2-15039/tourism-tearing-up-fabric-of-edinburgh-warns-val-mcdermid-1-4811871.

13. Scott Reid. (August 19, 2019). "Over-Tourism and Population Growth Threaten Edinburgh's Office Market." *Edinburgh News.* https://www.edinburghnews.scotsman.com/business/over-tourism-and-population-growth-threaten-edinburgh-s-office-market-1-4986514.

14. Roddy Smith. (September 10, 2019). "Over-Tourism Debate Is a £1.4 Billion Question." *The Scotsman.* https://www.scotsman.com/news/opinion/columnists/over-tourism-debate-is-a-1-4-billion-question-roddy-smith-1-5000802.

15. Edinburgh Tourism Action Group. (2020). https://www.etag.org.uk/about-us/.

16. Edinburgh Tourism Action Group. (2020). *Developing Edinburgh's Nest Tourism Strategy Phase 3 Consultations.* https://www.etag.org.uk/edinburgh-2020/developing-the-edinburgh-2030-strategy/developing-edinburghs-next-tourism-strategy-phase-3/.

17. Scottish Government. (October 28, 2019). *Short-Term Lets—Impact on Communities: Research.* https://www.gov.scot/publications/research-impact-short-term-lets-communities-scotland/pages/1/.

18. Accelerating Growth: Edinburgh and South East Scotland City Regional Deal. (2020). http://www.acceleratinggrowth.org.uk/.

19. Data Driven Innovation. (2019). *Expert Blogs: Joshua Ryan-Saha, Tourism and Festivals* https://ddi.ac.uk/expert-blogs/joshua-ryan-saha-tourism-and-festivals/.

20. BBC News. (January 9, 2019). “Bid to Introduce a 2 Euro Tourist Tax in Edinburgh Wins Support.” https://www.bbc.co.uk/news/uk-scotland-edinburgh-east-fife-46797446.

21. United Nations World Tourism Organization. (January 2019). *‘Overtourism’? Understanding and Managing Urban Tourism Growth beyond Perceptions Volume 2: Case Studies*. https://www.e-unwto.org/doi/book/10.18111/9789284420629.

22. Edinburgh Napier University. (2020). https://www.napier.ac.uk/courses/browse-interests/tourism/destination-leaders-programme.

23. Edinburgh 2030 Tourism Strategy. (January 2020). *Final Draft Edinburgh 2020 Tourism Strategy*. https://www.etag.org.uk/wp-content/uploads/2014/01/Final-Draft-Edinburghs-Tourism-Strategy-2030.pdf.

24. Edinburgh Tourism Action Group. (2014). *Industry & Stakeholder Consultations Findings*. https://www.etag.org.uk/wp-content/uploads/2014/01/Edinburgh-Tourism-Strategy-2030-Phase-2-Industry-Stakeholder-Consultation-Findings.pdf.

25. Peter Jordan. (January 1, 2020). “Edinburgh 2030 Tourism Strategy Presentation to ETAG Conference.” *Toposophy*. https://www.etag.org.uk/wp-content/uploads/2014/01/Ed-2030-Tourism-Strategy-PJordan-v2-1.pdf.

26. City of Edinburgh Council. (January 22, 2020). “New People ‘First Strategy’ Welcomed” https://www.edinburgh.gov.uk/news/article/12724/new-people-first-tourism-strategy-welcomed.

27. C. S. Arthur. (April 20, 2019). Twitter. https://twitter.com/CllrScottArthur/status/1119486719718559744.

Chapter 3.1

US National Parks: The Unfatigued Allure of Wild Places

By Richard Bangs with Martha Honey*

John Wesley Powell, the one-armed brevet colonel, made his water-breaking trip down the Colorado River through the Grand Canyon in 1869. Fifty years later, on February 26, 1919, the sublime and tremendous ditch became a national park, the fifteenth in the system. But although thousands flocked to view the deep cut in the skin of the continent from the rims, few had dared float the river below. In 1934, Bus Hatch, a carpenter by trade, captained the first paying participants down the Colorado River through the Grand Canyon in a wooden boat he had handcrafted, but that trip did not initially lead to a booming tourism business. By 1949, only one hundred people had navigated the 277-mile steep-walled corridor. By 1961, the one thousandth person had made the trip.[1]

Then, in 1963, the diversion tunnels for the newly constructed Glen Canyon Dam, just upstream of the Grand Canyon, were closed, and the river flows downstream were now managed, making safe passage down the Colorado less dependent on the unpredictable periods between spate or drought. Surplus stores around the West began to sell

* Martha Honey contributed to the research and writing of the sections outside the Colorado River and NPS history.

inflatable pontoon bridges, left over from the Korean War, which, with an outboard hung off the stern, proved ideal as passenger-carrying rafts.

Overtourism on the Colorado River

The seismic event for river tourism came in 1967 when Bobby Kennedy took forty-one of his closest friends and family on a four-day raft trip guided by Hatch River Expeditions down the Colorado through the Grand Canyon. The flotilla, dubbed "Ship of Fools" by the media, included humorist Art Buchwald, journalist George Plimpton, singer Andy Williams, and some two dozen children.[2] The story and photos were picked up in media around the world, and suddenly the trip became not just known, but something very cool to do.[3] And the number of vacation rafters began to rise: more than three thousand in 1968, six thousand in 1969, and sixteen thousand in 1972.[4]

In 1969, at the age of eighteen, I sent a letter to Ted Hatch, son of Bus Hatch, then the head of Hatch River Expeditions, saying that I would love to be a river guide on the Colorado, although I had never left the East Coast and had only canoed down waterways in the Shenandoah, Appalachian, and Blue Ridge Mountains. Sight unseen, he hired me. The company needed guides to meet the growing demand. My first trip was a charter by the Four Corners Geological Society, with 120 members packed into eleven pontoon rafts. It was a Wild, Wild West experience. We camped at whatever broad beach we came to in the late afternoon. We made our cookfires with driftwood spread upon the fine-grained sand. We peed and pooped behind rocks. We buried the toilet paper in the sand and washed our dishes in the river.

Whatever our habits in this cathedral, I felt joyfully robbed of any conventional grip on time; human history seemed a momentary blink against the immense scroll of eternity embedded in the rock. When we ran the enormous rapids, I felt a seam open onto a void whose terrible content was the inevitability of finality, and it raced my blood and evoked feelings of being more alive. It was an experience of enthusiastic terror, of shivering pleasure, of a thrilling, delightful horror, and the axis of my identity spun to a new North. I was a River Guide.

But over the next few seasons, as the numbers of rafters grew, things began to change. The once pristine beaches were now often blackened from previous campfires. Scraps of toilet paper glanced through the sand behind almost every accessible rock. The campsites began to stink and were buzzing with flies. Driftwood for fires was harder to find, so we guides would cut down some of the scarce trees that had found purchase up the side canyons. There was now competition between the mounting number of rafting companies for the best beaches, and we would often send one boat ahead to claim a prime spot, only to be beaten by another who had launched a bit earlier.

And then there was the noise. I recall my first trips as passing through uncontaminated wilderness, we as pioneers on our planet's first morning. But, by 1971, there were no quiet camps in summer, when more than ten thousand rafters were jockeying down the river.[5] The din of twenty-horsepower engines, powering ever larger rafts with more passengers, echoed across the canyon until twilight. Where once rafting through the cavernous architecture that contained the Colorado was a transcendental experience, profound in its purity, it was becoming more like a crowded amusement park ride, but with unpleasant smells and biting bugs.

Early Overtourism Regulations

River guides, park administrators, and visiting rafters became concerned that, without management and intervention, use would continue to surge and would progressively damage ecological and experiential values. As a result, Grand Canyon National Park managers froze tourism rafting numbers at 1972 levels to give time to evaluate how much use the river could withstand, and use levels have been controlled ever since. A series of requirements for rafters were implemented, including a rule that portable toilets must accompany each trip and that all waste, human and otherwise, be carried out. Cooking had to take place on gas stoves, and fires were only allowed on fire pans. And the rafting visitor numbers were capped at some sixteen thousand commercial passengers, with no additional outfitters permitted. "This marked the first time access to a national park was managed in an allotment way," explained Donald Leadbetter, Tourism Program manager for the National Park Service (NPS). "It worked, and it's still in place today. It's been modified and improved over the years. But it's provided a higher quality visitor experience and preserved resources at the bottom of the canyon."[6] Today under the allotment system for rafting through the Grand Canyon, there is a mandated quota of twenty-two thousand "people floats"[7] each year.[8]

Growth of Tourism in US National Parks

The Colorado River–Grand Canyon story reflects to a degree the rising popularity of national parks throughout the United States. In 1969, when I started working in the Grand Canyon, there were about 50 million visitors to all the US national parks. In 2018, there were just over 318 million recreation visits[9] to the 418 national park sites, covering more than 85 million acres in every state, the District of Columbia, American Samoa, Guam, Puerto Rico, and the Virgin Islands.[10] Although visitation varies considerably depending on the park, the US national parks saw big increases between 2015 and 2017 and then a modest decline to just over 318 million in 2018.[11] One reason for the drop is that international visitors, about one-third of whom visit a national park, flattened out during the Trump administration.[12] Through the end of August 2019, visitation to parks increased less than 1 percent, noted Leadbetter. "There is a real slowdown emerging system wide," he said.[13]

Roots of Overtourism in National Parks

The roots of this rising popularity go back to the nineteenth century and are linked, most fundamentally, to a change in mindset of the national psyche, as well as to the two dueling mandates within the mission of the NPS. To these have been added in recent years a cacophony of new factors that have coalesced to further propel upward visitation to our parks.

Change in the National Mindset

Few foresaw the bloating visitor demand for national park experiences when Yellowstone National Park was established in 1872, making it the nation's first national park. Through the middle nineteenth century, most Americans found grand nature perilous, irregular, and dislikable.

The wild areas were places to avoid or, if one were a merchant, soldier, or pilgrim, to go around.

So, how did this mindset change? Some of the shift had to do with the palpable deterioration of US cities, especially smoggy, coal-blackened New York, with its rising fatigue and social atomization. As more people flocked from farming to the economic opportunities that cities promised, sewage got worse, crime increased, and urban blight spread. In this claustrophobic urban culvert, influenza, measles, scarlet fever, and diphtheria became epidemic. By the middle of the nineteenth century, tuberculosis, which was spread by the dense population, the damp climate, and the concentration of smoke in Northeast cities, accounted for one death in six. But the urban trend was irreversible. By 1920, more people lived in cities than in the countryside for the first time in US history, and more were employed in industry than in agriculture.[14]

In the wake of this corrosive wave there evolved the romantic notion of the virtues of the simple, clean, and healthy life in nature. It was, however, the Industrial Revolution that led to vastly improved means of transport and inspired travel to places such as the Northeast's high cliffs and steep valleys and west of Kansas, where the mountains set the minds wandering—and a traveler perhaps got rid of that nasty cough.

Poets and artists called the feelings they felt when experiencing wilderness and mountains "sublime," and the public became inspired. The American wilderness became a literary topography as much as a geographical designation. Writers and painters, such as Thomas Cole, famed as the founder of the Hudson River School, provided a vocabulary of these landscapes, and the seekers came. The flocking of Americans to the remote reaches of the continent was not about growing crops, grazing animals, or extracting minerals. It was about observing, feeling, and articulating a personal response.

National Parks Mission: A Management Paradox

The seeds for explosive growth of national parks is also embedded within the conflicting goals of the NPS Mission Statement, which says in part, "The National Park Service preserves unimpaired the natural and cultural resources . . . for the enjoyment, education, and inspiration of this and future generations." These two central concepts, preservation and enjoyment, are in opposition, which creates an ongoing tension in how the parks should be managed, explained Leadbetter.[15] Jeffrey Marion, a recreation ecologist at Virginia Tech, concurred: "Parks have been commanded by the Congress to accommodate people and to preserve and protect these places. It's a huge management paradox."[16]

Striking the right balance between protection and openness is complicated as well by the park service's historic commitment to equal and free (or low-cost) access for all visitors. As Leadbetter explained, "There's a well-established tradition of egalitarian access to our parks. These places belong to the American people and the global citizenry so we must keep them accessible and affordable for everyone."[17]

Initially, the national parks were all free to visitors and were financed by private hotels that paid concessionaire fees to the parks. But the growth of automobile travel led to a significant change in consumer preferences. Many visitors now day-tripped to the parks by car, rather than by train or stagecoach, which were relatively expensive and prompted visitors to stay for several days. Therefore, visitor fees were gradually enacted as a substitute.[18] In 1908, Mount Rainier in Washington became the first national park to charge for automobile admissions.[19] Today, 111 of the 419 parks with the national park system currently charge entrance or visitor fees of some sort, per vehicle, person, motorcycle, or a park-specific annual pass. (Unlike in many countries around the world, US national parks do not charge higher entrance or other fees for foreign visitors.) Fees range from a modest $5 for vehicles entering the Chattahoochee River National Recreation Area to $80 for an annual pass to many of the most popular parks, including Zion, Yosemite, Yellowstone, Rocky Mountain, Grand Teton, and Grand Canyon and Glacier.[20] Of the fees collected, 80 percent stays in the park, and the other 20 percent is used to benefit parks that do not collect fees.[21] Park fees along with concessions actually contribute only a small slice of the total NPS budget; the vast majority of the budget, some 90 percent, comes from Congress.

A Perfect Storm of Recent Factors

These early factors that help build popularity for the US national parks have been accelerated by a perfect storm of more recent trends that are contributing to overtourism in the United States, as well as in national parks in other countries around the globe. Nobody predicted the effects of selfies and social media as hyperincentivizing visitation to national parks. "People come to the park, take a crazy selfie with a beautiful backdrop, put it on social media, and then their 100 friends want to go there and do the same thing," said John Marciano, a spokesman for Zion National Park. "It cascades."[22] That, compounded with baby boomers retiring in record numbers—seven thousand to ten thousand Americans turn sixty-five each day—and taking to the road, the DIY (do-it-yourself) movement that shuns packaged tours, the rise in domestic vacations out of fear of overseas terrorism or contagions, and the hare-footed rising middle class in Asia seeking sanctified wilderness experiences, all converge in a tumultuous crush in the gazetted national parks.

From his vantage point as the NPS chief of tourism, Leadbetter ticks off what he sees as the reasons for the increased visitation: The United States has been in its longest period of economic expansion ever, global tourism arrivals are increasing every year, and in 2016 the NPS celebrated its one hundredth anniversary, accompanied by a huge marketing campaign. The increase in demand may also reflect a larger trend involving personal priorities. "People are spending more on experiences and less on things," said Brandon Lake, the chief marketing officer of Western River Expeditions, which runs rafting trips on the Colorado. "We're seeing an increase in multi-generational trips with families traveling together."[23]

And then there's Instagram, which has shone a spotlight on certain national parks. Take, for instance, Horseshoe Bend, located in Arizona on a spectacular bend above the Colorado River, which has become "a social media superstar," according to the NPS website.[24] By 2018, the #horseshoebend hashtag had been viewed more than 425,000 times, and park officials say that is a big reason why Glen Canyon attracted 4.2 million visitors in 2018—nearly double the 2.1 million visitors it saw in 2010, the year Instagram was founded.[25]

In reflecting on overtourism and its causes, Leadbetter took umbrage with the terminology. "We call it 'too much love for our national parks.' This is an okay problem to have. We'd rather have this challenge than the opposite. And," he added, "We call it 'congestion,' not overtourism." Leadbetter said that "congestion" is particularly apparent in "your bucket list parks," those perennial vacation spots. Included here are marquee parks such as Acadia, Yosemite, Zion, Glacier, Yellowstone, Grand Canyon, and Rocky Mountain, parks where "veteran park administrators are aghast at the 'greenlock'—gridlock in natural surroundings," wrote Jim Robbins in Yale University's School of Forestry online environmental magazine.[26] Even smaller sites are receiving increases in visitors. For instance, Muir Woods National Monument, north of San Francisco, saw a 35 percent jump, from 830,000 in 2010 to 1.1 million in 2016. Although the reasons for the steep increases to particular parks are not always clear, in the case of Muir Woods, Leadbetter's educated guess is that it's due to its proximity to San Francisco, rising Asian tourism, and social media. He said that such visitor spikes are "portending a new stage and the need to find solutions for this new level of congestion."[27]

Impacts of Overtourism

Just as with my vernal experiences in the Grand Canyon, the visitor interactions in many parks today often find nature compromised with too many like-minded human visitors, and one can't see the forest for the elbows and knees. With thousands, and in some cases tens of thousands, of visitors a day, parks ignore the issues at their own peril. Just leaving it alone leads to accidents and injuries, environmental degradation, and very chaotic visitor experiences, with overflowing parking lots and people leaving their cars haphazardly on the edges of the park. According to Robbins, "The visitor crush is creating two main problems—a steep decline in the quality of visitor experience that a national park is supposed to provide, and damaging impacts on the ecology of these intact natural places."[28]

For many, Zion National Park in southwestern Utah is top of the list for dramatic scenery; it's also become the poster child for overcrowding. With less than 150,000 acres, this relatively small park is now getting more than 4.3 million visitors a year—more than Yellowstone, which is fifteen times larger. "In the last few years, this huge uptick in visitation has overwhelmed our infrastructure facilities, our trails, our backcountry, it goes on and on and on," said Marciano.[29]

A shared "wild" experience, Zion National Park, 2017.
Source: Edna Winti (Flickr).

As Robbins explained:

The park now swarms with tens of thousands of people each day, and the season lasts almost all year, instead of 8 or 9 months as previously. Flocks of tour buses pour in from Los Angeles and Las Vegas. Instead of coming to get a sense of nature transcendent, people wait an hour or two in traffic just to get through the park gates,

> and day hikers jostle with hundreds of other people on one-lane trails eroded by overuse. Trash bins can't be emptied fast enough and overflow onto the ground. . . . Amusement park-style lines form to get on the shuttle and into restrooms.[30]

If that weren't enough, Robbins's expose continued, that "in addition to a decline in visitor experience, there is an impact on park ecology. . . . Now hundreds of people a day splash and wade their way up the [Virgin River Narrows] riverbed into the half-light of the canyon. All of these feet trample vegetation, aquatic insects, and fish habitat."[31]

In Yellowstone, visitation has gone from two million in 1980 to more than four million every year beginning in 2015. Not only is the visitor experience being degraded, but the park's physical environment is as well. "A growing number of visitors are walking off boardwalks, making their own trails, throwing stuff into hot springs, or driving off roads and trampling fragile natural areas," Robbins said, "and in many areas of the park, the roar of traffic drowns out the subtle sounds of nature."[32]

"These are irreplaceable resources," said Joan Anzelmo, a retired NPS superintendent. "We have to protect them by putting some strategic limits on numbers, or there won't be anything left. Nobody will want to visit them. Everyone I know who lives, works, and is involved in these issues says something has to be done. It can't go on like this anymore."[33]

Solutions

"The park service has dramatically changed its tune during the past year and acknowledged that overtourism is something it must address," wrote Dan Peltier in a 2017 *Skift* article.[34] Zion's Marciano concurred: "We can't sit on our hands anymore. We have to come up with some kind of management plan to be able to preserve resources and to make sure our visitors have a good and safe experience."[35]

Solutions for controlling overtourism in national parks must be different from those of historic cities and other alluring environments because the park service's mission is to preserve unimpaired and for future generations the natural and cultural assets. Therefore, as Robbins wrote, "Wider roads and more hotels and campgrounds [in the parks]

Cars lined up at the entry, Yellowstone National Park. Source: NPS/Jacob W. Frank (Flickr).

could create sprawl, diminish the experience of nature, and encourage yet more people to come."[36] On the other hand, parks cannot ignore their gateway communities, the built environments that often surround the protected areas and houses the tourism infrastructure, providing essential jobs and income. As Leadbetter explained:

> From the point of view of managing visitor experience and protection of resources, a decrease in visitation of 20 percent to a park would be great. But the truth is we can't take that approach because gateway communities rely on visitor numbers. A vibrant and strong community is our best advocate and supporter for the parks. So, we have to be responsive to concerns and interests in gateway communities and to find solutions that win support across the park's entire landscape, including surrounding communities.[37]

Both the NPS headquarters and the individual parks play a role in crafting visitor policies and management. The NPS has a centralized planning office with experts who conduct projects and planning for the entire park system, working in collaboration with the US Forest Service, Interior Department, and other services that make up the interagency Visitor Use Management Council. Each park has a set of foundation documents describing its purpose, significance, core resources and values, and interpretive themes. These documents serve as the touchstone for reforms and policy changes developed by individual parks that are then sent up the chain of command for review and to assess compatibility with other initiatives. Nothing happens very quickly: "We are cautious, iterative, and adaptive when it comes to planning, and we need to understand phases and trends," explained Leadbetter. "We have a fleet with the best individual ships—the parks—but not the best management overall."[38]

In seeking solutions to overtourism and congestion, the NPS uses a planning framework known as visitor use management (VUM)[39] that brings a structured approach to the challenge. VUM offers a more multidimensional and effective set of tools and strategies for managing visitor access to and use of national parks than the older carrying (or visitor) capacity concept.[40] Although too much visitation can certainly degrade resources and visitor experiences, setting a fixed visitor capacity number is not a panacea. Instead, "managing use levels to a visitor capacity is one of many strategies for dealing with visitor use management issues. Changing visitor behavior, modifying where and when use occurs, or building facilities that can accommodate heavy use are [all] critical strategies for protecting desired conditions," explains an analysis by the government's Visitor Use Management Council.[41] As the NPS's Leadbetter succinctly put it, "We don't use visitor carrying capacity. It's too simplistic." Rather, he explained, the VUM tool enables park managers to (1) proactively protect resources, (2) protect visitor safety, and (3) enhance access and improve visitor experiences. Leadbetter also explained that the NPS is using three broad categories of solutions in addressing congestion/overtourism: education, engineering, and enforcement.[42]

Education Solutions

Education solutions include a wide range of programs and tools to connect visitors to resources and provide travelers with trip-planning information. They include websites, park newspapers, signage, and digital platforms that show real-time conditions such as how many spots are left in a parking lot. Educational tools and programs are developed by each park, tailored to its needs and conditions. The goal is to provide more information to help people make decisions about when, where, and how to visit parks and help ensure that they arrive with informed expectations.

A growing number of parks are using their official websites to put up information about weather conditions, crowd levels, and what activities are available for each month the park is open. For instance, the Experience Yellowstone website tells visitors, "Yellowstone is seasonal. Plan Your Visit by learning about current conditions, seasons, road conditions, services, activities, and more." Its "Plan Your Visit" page lists ten "Top Things to Know" about visiting Yellowstone National Park.[43] Montana's Glacier National Park has a no-nonsense "Crowds," which includes "Where and When to Expect Crowds," along with a pristine Instagram photo side by side with a "reality" photo of an overlook filled with visitors. The website also details "Three Tips for Dealing with Glacier's Crowds": (1) adjust your expectations, (2) have a backup plan or two, and (3) consider one of Glacier's neighbors. This last tip contains a link to nearby attractions. Other pages and charts on Glacier's website contain real-time updates for parking, campground availability, current weather conditions, and area closures.[44]

Finally, educational efforts by the NPS, travel websites, and the media are promoting lists of less- or least-visited national parks as a way of spreading out the crowds. One website, Reservations.com, carries a list of the twenty least-visited "hidden gems" among the US national parks. Although the list includes some geographically far-flung parks—in Alaska, the Arctic, Samoa, and the Virgin Islands—it also contains a selection within the contiguous states. Similar lists have been put forth by travel magazines, including *Travel + Leisure*, *Outdoor*, *E Magazine*, *USA Today*, and *National Geographic*.[45] As Talia Avakian wrote in *Travel + Leisure*, the NPS tracks the total number of visits made to each of the parks each year, revealing the most and least visited. Although the country's least-visited parks can take a bit more planning to reach, they offer incredible experiences to all those who make the trek: watch synchronous fireflies, hike among the world's oldest trees, take in views of the northern lights, or enjoy wildflower blooms at these lesser known national treasures."[46]

Engineering Solutions

Engineering solutions include new, improved, or redesigned roads, visitor centers, trails, parking lots, park entrances, and shuttle services. One engineering solution aims to reduce the bottleneck at Yellowstone's entrances by adding a fast lane, à la E-ZPass, for those with park passes, with entry passes available for purchase in the gateway communities. Another solution is to direct vehicles to less-crowded entrances. As Leadbetter explained, "Many popular parks still have capacity, but it's clustered. Most people could be entering through one gate, but other entrances of a park may have much more capacity."[47]

Acadia National Park, along the central coast in Maine, has been developing since 2015 a robust transportation plan, based on extensive preparation, collection of visitor data, broad-based public engagement, and an environmental impact analysis. According to Acadia's website, the park's visitation hit 3.5 million in 2017, a 59 percent increase over the preceding decade, and "crowds in the summer and, increasingly, into the fall and spring, are making it difficult for people to experience the best of Acadia." The roads and parking lots are unable to handle the traffic and "crowded conditions in the park also cause safety hazards," according to the website.[48]

In May 2019, the NPS officially issued the final Transportation Plan for Acadia, which began to be implemented in 2020. Designed to improve visitor experiences and reduce congestion, its three main components are to:

1. Provide visitors with transportation options, which will eventually include an expanded network of commercial tours, Island Explorer buses, and on-demand taxis.
2. Construct the Acadia Gateway Center and a new visitor center to function as a park-and-ride site and significantly improve accessibility for visitors.
3. Institute an advance parking reservation system for private vehicles to access the three most popular destinations, Ocean Drive, Cadillac

Mountain, and Jordan Pond. Private vehicles will be allowed in all other destinations in the park without parking reservations.[49]

Park shuttles, first introduced some twenty years ago to reduce car traffic and offer guided tours, have become popular solutions in Zion, Yosemite, Glacier, Acadia, and Great Smoky Mountain, among other parks. Access to the popular Mariposa Grove of Giant Sequoias, located in the southern end of Yosemite, for instance, has been renovated and the parking lot pushed back. After being closed three years for restoration, the Mariposa Grove reopened in 2018. According to the official website, "Crews improved habitat for sequoias by removing parking lots and roads and restoring the natural flow of water to the trees. Parking was relocated two miles away from the grove and is connected by shuttle buses. The restoration also added accessible trails and improved bathrooms. This was the largest restoration project in the history of the park."[50] Leadbetter said that this solution offers a much better experience: the shuttles provide interpretation, and, in the grove, there is improved signage and more guides on site.[51]

Horseshoe Bend, the Instagram heavyweight, is today employing a multitude of measures. Just Google "Horseshoe Bend Guide to Planning Your Visit" to find the NPS web page and scroll down to see what's been done: a new parking lot, new trails, and overlooks with rails. And now when the parking lot is full, visitors may need to use the shuttle.[52]

The Great Smoky Mountains National Park, with the largest swath of protected upland forests east of the Mississippi river, is the most visited national park in the United States. Between 2015 and 2019, the Park experienced five consecutive years of record-breaking visitation, reaching more than 12.5 million visitors in 2019.[53] This visitor spike led to increasingly crowded trailheads, packed shuttle services, and car traffic, as well as increased water consumption, air pollution, litter, and waste products.[54] In addition, cuts to the National Park Service's annual budget have limited park staff from undertaking maintenance on much-needed roads and other infrastructure.[55]

Despite its popularity, the park has moved cautiously in introducing new congestion control measures. To date the park's primary congestion mitigation effort is its shuttle bus service, first introduced in 2008 to decrease car traffic in the park and provide visitors with high-quality educational tours. The shuttle fleets have recently been upgraded with start-stop engine technology that limits noise and air pollution, and local businesses and the NPS are marketing the shuttle bus tours as unique, educational experiences.[56]

But Leadbetter acknowledged that shuttle bus services and parking modifications cannot solve congestion programs in every park. "Some parks are insanely crowded, almost like a theme park," Leadbetter told *Skift* in 2017, adding that "in some parks, shuttle buses are beyond capacity and are dumping huge numbers of people onto trails all at once. Parking is also an issue in some parks."[57] Many park administrators are left looking for the next phase of solutions, which often involves some form of compulsory reservation system that caps visitation at predetermined limits.

Enforcement Solutions

Enforcement solutions are today in place or under development at a number of parks. They typically involve tools such as reservation systems, managed access, permitting, and visitor caps, all designed to curb congestion. Since at least 2016, the NPS has been considering a reservation system for entrance into the Great Smoky Mountains, Zion, Arches, Acadia, and a handful of other popular parks, but implementation has been delayed in part out of persistent concerns about the accuracy of its visitation data collection.[58] Another issue is that, if recent trends hold, visitation numbers appear to be leveling off, yet the time visitors spend in parks has increased—and a reservation system that limits visitors but doesn't account for time spent in the parks won't solve the problem.[59] In addition, although the NPS has hoped that a reservation system will help visitors consider coming during off-peak periods, the reality is that work and school schedules often dictate when people can visit. Given this and other constraints, Leadbetter has predicted that reservation systems for entrance into parks are a couple of years away.[60]

A number of parks do, however, have reservation systems and visitor caps for specific trails, activities, and areas. For instance, Utah's Zion National Park, the fifth most visited national park, requires wilderness permits for many activities, including backpacking, canyoneering, river trips, some rock climbing, and a few day hikes.[61] In 2017, Zion became the first park to consider a broad reservation solution, for the entire main corridor of Zion Canyon, that would require all visitors to preapply for permits prior to entrance and capping off at a predetermined and seasonally appropriate limit.[62] Although it is hoped that reservations will curb overcrowding, park officials also recognize its daunting challenges. As Zion's Marciano explained, "Many people may not realize how complex an undertaking this is."[63]

Muir Woods National Monument just north of San Francisco saw visitor numbers jump 35 percent from 2010 and 2016—to more than one million. The park responded by introducing parking and shuttle bus reservation systems, both of which require advanced bookings.[64] By 2018, Leadbetter said, "Visitation subsided to a more manageable 950,000. Today, with better management, the park is offering a higher quality experience."[65]

Half Dome, rising nearly 5,000 feet above Yosemite Valley, offers an iconic hike that has become so popular that the park introduced a permit system and daily cap on hikers in 2008. During the hiking season when the cables are open, normally May to October, a maximum of 300 hikers—approximately 225 day hikers and 75 backpackers—are allowed per day.[66] Prospective hikers apply for permits through a lottery system, which costs $10 for the application and another $10 when the permit is issued. By all accounts, the permit system is working well in controlling congestion. According to the Sierra Club, before 2008, "nearly 1,200 visitors per day made it to the cables section of Half Dome."[67] Now with the permit system capping this number at 300, "up to 66 percent fewer hikers [have] summited each day."[68]

The NPS divides its overtourism tools into the three categories of Education, Engineering, and Enforcement, but most national parks opt for a mix of solutions. As Leadbetter explained, the NPS—unlike many other destination authorities dealing with overtourism—does have a process and a structured approach that engages both national level and local park officials, involves public consultation, and is grounded in science and social science. "It's a very nuanced and challenging conversation. But we are developing a toolkit that has lots of different tools within it." And Leadbetter believes that in the United

States, the national parks "are leading the way in finding ways to address overtourism."[69]

Back to the Future

We continue to seek the mystery, power, and resonance of national parks as hunting grounds for our imaginations and better selves. But, as my early experiences on the Colorado River demonstrated, unmanaged growth leads to overtourism, which can collapse entire systems and ultimately blacken the *élan vital* that makes our parks—America's greatest idea—so special and so necessary. But because of wise management and well-prescribed environmental regulations, the Colorado River today is just as cleanhanded and magnificent as when I first traversed it in 1969. And, it seems, most visitors today feel the same about the national parks they visit. In a 2019 Visitor Survey Card Data Report of 336 units in the National Park System, visitors rated their experiences as very high even as visitation topped 300 million for the fifth consecutive year.[70] The survey asked visitors to rate their overall satisfaction with the quality of facilities, services, and recreational opportunities available at parks. Some 98 percent of the nearly 60,000 respondents ranked the quality of their experience as "very good" or "good," whereas only 2 percent said it was "average." No respondents rated their experience as "poor" or "very poor."[71]

As human knowledge grows, the space we share is becoming smaller. We have to think our way to solutions. Wild places, ones that offer the tonic of solitude, can be gazetted, but they also need enlightened management. The NPS began developing tools for abating congestion early on and today invokes systems that address and help alleviate the destructive love that comes from overtourism. These fixes include capping visitation numbers, lotteries, dispersal to alternative sites, shuttle buses, and education. They may seem like Band-Aids when viewed with the larger issue of global population escalation, and they may be just the top drawer in the red toolbox of remedies. But knowing how wild places tremble the spirit with wonder, how the exquisite shock of silent joy in a wood or desert soars the soul, how the sweet communion with Nature fills the cup of validation, the exercise, if ever challenging and evolving, is worth it.

Notes

1. Larry Stevens. (1996). *The Colorado River in Grand Canyon: A Comprehensive Guide to Its Natural and Human History*. Red Lake Books.

2. Elaine Raines. (November 19, 2008). "Bobby Kennedy Would Be 83 Years Old." *Arizona Daily Star.* https://tucson.com/bobby-kennedy-would-be-years-old/article_4b57b55e-fa31-56fe-ad08-afe56eae7bcc.html.

3. Hatch River Expeditions. "Hatch River Expeditions: Our History." https://www.hatchriverexpeditions.com/about-us/our-history.

4. Kim Crumbo. (1992). "The Canyon by River." In *The Grand Canyon, Intimate Views*. Eds. Robert Euler and Frank Tikalsky. University of Arizona Press. p. 86.

5. William Calvin. (1986). *The River That Flows Uphill: A Journey from the Big Bang to the Big Brain*. Macmillan.

6. Donald Leadbetter. (October 28, 2019). National Park Service. Personal communication with editors.

7. A "people float" is one person on three-day or more commercial float trip down the Colorado River through the Grand Canyon.

8. National Park Service. (Accessed March 2020). "3 to 18 Day Commercial River Trips: River Concessioners. https://www.nps.gov/grca/planyourvisit/river-concessioners.htm

9. This is the total number of parks visits, not park visitors per year. One person could make multiple visits.

10. National Park Service. (Accessed March 2020). "Frequently Asked Questions." https://www.nps.gov/aboutus/faqs.htm.

11. National Park Service. (Accessed March 2020). "Annual Visitation Highlights." https://www.nps.gov/subjects/socialscience/annual-visitation-highlights.htm.

12. Donald Leadbetter. (October 28, 2019). National Park Service. Personal communication with editors.

13. Ibid.

14. Khan Academy. (Accessed March 2020). "America Moves to the City." https://www.khanacademy.org/humanities/us-history/the-gilded-age/gilded-age/a/america-moves-to-the-city.

15. Donald Leadbetter. (October 28, 2019). National Park Service. Personal communication with editors.

16. Jim Robbins. (July 31, 2017). "How a Surge in Visitors Is Overwhelming America's National Parks." *YaleEnvironment360*. https://e360.yale.edu/features/greenlock-a-visitor-crush-is-overwhelming-americas-national-parks.

17. Donald Leadbetter. (October 28, 2019). National Park Service. Personal communication with editors.

18. Adam Summers. (April 2005). *Funding The National Park System: Improving Services and Accountability with User Fees.* Reason Foundation. https://reason.org/wp content/uploads/2005/04/0b6c6302bfcc621638fcdbdebec8b63a.pdf.

19. Barry Mackintosh. (1983). Visitor Fees in the National Park System: A Legislative and Administrative History. https://www.nps.gov/parkhistory/online_books/mackintosh3/index.htm.

20. National Park Service. (Accessed March 2020). "Entrance Fees by Park." https://www.nps.gov/aboutus/entrance-fee-prices.htm.

21. National Park Service. (Accessed March 2020). "Your Fee Dollars at Work." https://www.nps.gov/aboutus/fees-at-work.htm

22. Jim Robbins. (July 31, 2017). "How A Surge in Visitors Is Overwhelming America's National Parks." *YaleEnvironment360.* https://e360.yale.edu/features/greenlock-a-visitor-crush-is-overwhelming-americas-national-parks.

23. Jessica Colley Clarke. (December 19, 2017). "If You Want to Raft the Grand Canyon, Prepare to Sign Up Far in Advance." *New York Times.* https://www.nytimes.com/2017/12/19/travel/plan-in-advance-grand-canyon-rafting.html.

24. National Park Service. (Accessed March 2020). "Horseshoe Bend." https://www.nps.gov/glca/planyourvisit/horseshoe-bend.htm.

25. Matt Wastradowski. (April 5, 2019). "What's Being Done to Save Wild Spaces from Instagram." *Outside.* https://www.outsideonline.com/2393025/whats-being-done-save-wild-spaces-instagram#close.

26. Jim Robbins. (July 31, 2017). "How a Surge in Visitors Is Overwhelming America's National Parks." *YaleEnvironment360.* https://e360.yale.edu/features/greenlock-a-visitor-crush-is-overwhelming-americas-national-parks.

27. Donald Leadbetter. (October 28, 2019). National Park Service. Personal communication with editors.

28. Ibid.

29. Ibid.

30. Ibid.

31. Ibid.

32. Ibid.

33. Ibid.

34. Dan Peltier. (March 2, 2018). "U.S. National Parks Still Aren't Sure How to Deal with Overtourism." *Skift.* https://skift.com/2018/03/02/u-s-national-parks-arent-sure-how-to-deal-with-overtourism/.

35. Jim Robbins. (July 31, 2017). "How a Surge in Visitors Is Overwhelming America's National Parks." *YaleEnvironment360.* https://e360.yale.edu/features/greenlock-a-visitor-crush-is-overwhelming-americas-national-parks.

36. Ibid.

37. Donald Leadbetter. (October 28, 2019). National Park Service. Personal communication with editors.

38. Ibid.

39. VUM is the latest iteration of National Park Service management tools that began in the 1930s with the concept of carrying (or visitor) capacity and have included more multidimensional concepts such as Limits of Acceptable Change, Carrying Capacity Assessment Process, Visitor Impact Management, and Visitor Experience and Resource Protection.

40. InterAgency Visitor Use Management Council. (July 2016). *Visitor Use Management Framework A Guide to Providing Sustainable Outdoor Recreation.* Edition One. p. 118. https://visitorusemanagement.nps.gov/Content/documents/highres_VUM%20Framework_Edition%201_IVUMC.pdf.

41. InterAgency Visitor Use Management Council. (July 2016). *Visitor Use Management Framework A Guide to Providing Sustainable Outdoor Recreation.* Edition One. pp. 83–94. https://visitorusemanagement.nps.gov/Content/documents/highres_VUM%20Framework_Edition%201_IVUMC.pdf.

42. Donald Leadbetter. (October 28, 2019). National Park Service. Personal communication with editors.

43. National Park Service. (Accessed March 2020). "Yellowstone: Plan Your Visit." https://www.nps.gov/yell/planyourvisit/index.htm.

44. National Park Service. (June 13, 2019). "Glacier National Park Website Provides Real-Time Updates." https://www.nps.gov/glac/learn/news/19-30.htm.

45. Talia Avakian. (March 14, 2019). "The 15 Least-Visited National Parks Have All the Beauty, and None of the Crowds." *Travel + Leisure.* https://www.travelandleisure.com/trip-ideas/national-parks/least-visited-national-parks; Devon O'Neil. "The 10 Least Popular National Parks. *Outdoor.* https://www.outsideonline.com/2082041/10-least-popular-national-parks; Nicole Villegas. (November 6, 2019). "Overtourism Antidote: Indulge at Least Visited National Parks." *E Magazine.* https://emagazine.com/how-overtourism-impacts-our-environment/; USA Today. (April 26, 2019). "Avoid the Crowds: The Least-Visited National Parks." *USA Today.* https://www.usatoday.com/picture-gallery/travel/destinations/2019/03/14/national-parks-least-visited-20-ranked/3160486002/; Christine Bednarz. (August 23, 2019). "Go Wild at the 10 Least-Visited U.S. National Parks." *National Geographic.* https://www.nationalgeographic.com/travel/destinations/north-america/united-states/national-parks/least-visited/.

46. Talia Avakian. (March 14, 2019). "The 15 Least-Visited National Parks Have All the Beauty, and None of the Crowds." *Travel + Leisure.* https://www.travelandleisure.com/trip-ideas/national-parks/least-visited-national-parks.

47. Donald Leadbetter. (October 28, 2019). National Park Service. Personal communication with editors.

48. "Acadia National Park, Maine." (Accessed March 2020). National Park Service. https://www.nps.gov/acad/learn/news/national-park-service-finalizes-transportation-plan-at-acadia-national-park.htm.

49. Christie Anastasia. (May 9, 2019)."National Park Service Finalizes Transportation Plan at Acadia National Park." National Park Service. https://www.nps.gov/acad/learn/news/national-park-service-finalizes-transportation-plan-at-acadia-national-park.htm.

50. National Park Service. (Accessed March 2020). "Restoration of the Mariposa Grove of Giant Sequoias." https://www.nps.gov/yose/planyourvisit/mariposagrove.htm.

51. Donald Leadbetter. (October 28, 2019). National Park Service. Personal communication with editors.

52. National Park Service. (Accessed March 2020). "Horseshoe Bend." https://www.nps.gov/glca/planyourvisit/horseshoe-bend.htm.

53. Emily Wolff. (January 27, 2020). "Great Smoky Mountains National Park Breaks Record Number of Visitors (Again!)." Visit My Smokies. https://www.visitmysmokies.com/blog/smoky-mountains/great-smoky-mountains-national-park-visitors-record/.

54. Nicole Villegas. (November 6, 2019). "Overtourism Antidote: Indulge at Least Visited National Parks." *E Magazine.* https://emagazine.com/how-overtourism-impacts-our-environment/.

55. Dan Peltier. (March 2, 2018). "U.S. National Parks Still Aren't Sure How to Deal with Overtourism." *Skift.* https://skift.com/2018/03/02/u-s-national-parks-arent-sure-how-to-deal-with-overtourism/.

56. Amy Leinbach Marquis. (2008). "Moving Stories." *National Parks,* 82(3): 11. https://search-proquest-com.proxyau.wrlc.org/docview/220213899?accountid=8285.

57. Dan Peltier. (March 2, 2018). "U.S. National Parks Still Aren't Sure How to Deal with Overtourism." *Skift.* https://skift.com/2018/03/02/u-s-national-parks-arent-sure-how-to-deal-with-overtourism/.

58. Ibid.

59. Ibid.

60. Ibid.

61. National Park Service. (Accessed March 2020). "Zion National Park: Permits & Reservations." https://www.nps.gov/zion/planyourvisit/permitsandreservations.htm.

62. Lindsay Whitehurst. (July 21, 2017). "Overtourism Leads Zion National Park to Consider New Reservation System." *Skift.* https://skift.com/2017/07/21/overtourism-leads-zion-national-park-to-consider-new-reservation-system/.

63. Reuben Wadsworth. (July 18, 2017). "This Is What Zion National Park Might Do to Solve Overcrowding Issues; How to Comment." *St. George News.* https://www.stgeorgeutah.com/news/archive/2017/07/18/raw-this-is-what-zion-national-park-might-do-to-solve-overcrowding-issues-how-to-comment/#.XkHjZDFKiLc.

64. National Park Service. (Accessed March 2020). "Muir Woods: Enter the Redwood Forest." https://www.nps.gov/muwo/index.htm.

65. Donald Leadbetter. (October 28, 2019). National Park Service. Personal communication with editors.

66. National Park Service. (Accessed March 2020). "Yosemite: Half Dome Day Hike." https://www.nps.gov/yose/planyourvisit/halfdome.htm.

67. Amy Graff. (May 7, 2019). "Study Finds Half Dome Permits Didn't Make Yosemite Safer." *SF Gate.* https://www.sfgate.com/bayarea/article/Yosemite-Half-Dome-permit-lottery-study-safety-13825841.php.

68. Kevin Stark. (May 21, 2019). "The Half Dome Lottery System Was Supposed to Make the Hike Safer. This Study Says It Hasn't." KQED. https://www.kqed.org/science/1941829/the-half-dome-lottery-system-was-supposed-to-make-the-hike-safer-this-study-says-it-hasn't.

69. Donald Leadbetter. (October 28, 2019). National Park Service. Personal communication with editors.

70. Neal Herbert. (February 27, 2020). "National Park Visitation Tops 327 Million in 2019." Press release. National Park Service. https://www.nps.gov/orgs/1207/2019-visitation-numbers.htm.

71. Pacific Consulting Group and National Park Service. (2019). "National Park Service 2019 Visitor Survey Card Data Report." https://irma.nps.gov/DataStore/DownloadFile/632882.

Chapter 3.2

Brazil's National Parks

By James R. Barborak, Juarez Michelotti,
Thiago do Val Beraldo-Souza, and
Paulo Eduardo Pereira-Faria

Brazil is a massive country. It is the fifth largest and sixth most populated nation in the world. Brazil's federal protected area system, managed by the Chico Mendes Institute for Biodiversity Conservation (ICMBio), is one of the largest and most biologically diverse on the planet, conserving coral reefs, coastal and freshwater wetlands, rainforests, and savannas. Brazilian federal conservation units have features in high demand by tourists, including tropical rivers, sun-drenched beaches, caves, waterfalls, archaeological sites, wildlife spectacles, and indigenous and Afro-Brazilian cultures. The federal protected area system, begun approximately seventy-five years ago with the creation of Itatiaia, Brazil's first national park, now includes more than 330 protected areas, ranging from small, highly visited parks near major cities to remote parks of millions of hectares in the Amazon. Brazilian and international visitation to the protected area system is rapidly increasing, often by more than 10 percent a year.[1]

Recreational and outdoor activities practiced by visitors to Brazil's federal parks and reserves include diving, swimming, wildlife watching, biking, hiking, sport fishing, adventure sports, and simple contemplation of the wonders of nature and associated cultural resources. Total combined visitation to Brazil's federal protected areas—around

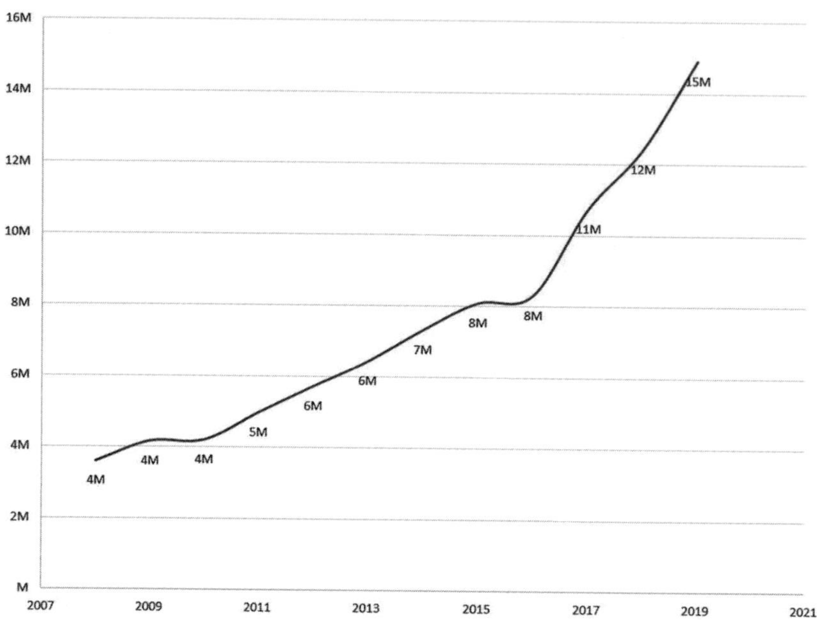

Visits, in millions, to Brazilian federal protected areas, 2008–2019. Source: Prepared by Paulo Eduardo Pereira-Faria and CREST with data from ICMBio.

15 million individuals in 2019—is still low comparatively; visitation to just one US park in 2019, the Great Smoky Mountains, was 12.5 million.[2] However, two of Brazil's most iconic national parks in the Atlantic Forest biome are among the most visited in Latin America: Iguaçu National Park, located along the Paraná River, which separates Brazil and Argentina and, along with another national park, Iguazú on the Argentinian side, is best known for its waterfalls; and Tijuca National Park, in Rio de Janeiro, site of the Christ the Redeemer statue. Together, these parks receive around 5 million visitors a year, or 40 percent of total visitation to Brazil's entire federal protected area system.[3] Managing extremely high and still growing levels of tourism at those two parks is a challenge. Brazil is taking steps to ensure that staff, infrastructure, and services at other protected areas do not become overwhelmed and natural and cultural resources are not overly impacted by tourism.

Causes and Impacts of Overtourism in Brazil

Iguaçu and Tijuca rank among the most visited protected areas in Latin America. These two parks are the poster children for overtourism in Brazil. Their location in Brazil's densely populated southeast, their iconic cultural and natural sites and global reputation, their well-developed infrastructure, and their proximity to other attractions mean that they are traditional destinations for national and international tourists.[4]

Location

Iguaçu is in the prosperous southern Brazilian state of Paraná, near several major metropolitan areas like Curitiba and Foz do Iguaçu, and within a day's drive of many millions of Brazilians. Tijuca is in the heart of Rio de Janeiro, Brazil's capital for much of its modern history until the creation of Brasilia, and is close to other heavily populated Brazilian states like São Paulo, Minas Gerais, and Paraná. Rio, with a population of over six million, is an extremely important national tourism destination; its two major airports combined have the most air traffic of any Brazilian city after São Paulo.[5] With nearly three million visitors in 2019 and a yearly increase around 10 percent, Tijuca, located in the heart of Rio de Janeiro, is Brazil's most visited national park. Its location has led to parking issues and traffic congestion. There is a high concentration of visitors to a limited number of easily accessible and well-known park destinations, particularly the Christ the Redeemer statue on Mount Corcovado, which gets about two-thirds of all park visitation. Limited public transportation options to the heart of the park have also led to high levels of congestion.[6]

Political and Economic Landscape

Brazil recently liberalized its tourism visa policy, reducing the costs and hassles of travel to Brazil from major international markets such as the United States and European nations.[7] Also, a devaluation of the Brazilian currency in recent years created economic disincentives for Brazilians to travel abroad and boosted domestic tourism while making Brazil, traditionally quite expensive for many potential foreign visitors, a more attractive destination.[8]

Marketing and Reputation

Although global events including the Summer Olympics and the World Cup failed to create major lasting boosts in international tourism to Brazil, other factors are increasing tourism to these parks.[9] Iguaçu and Tijuca are on the UNESCO World Heritage List, and both were selected for the "Seven Modern Wonders" lists, Tijuca for the Christ the Redeemer statue and Iguaçu for its waterfalls.[10]

Access and Infrastructure

In 2019, visitation to Iguaçu topped two million visitors for the first time.[11] Like many global tourism destinations battling overtourism, much of this flow is compressed into a few days, weeks, or months during "high season," corresponding to major holiday and vacation periods. During peak times in 2018 and 2019, such as Carnival, Holy Week, Independence Day, and Christmas, visitation was well over ten thousand persons a day and reached sixteen thousand one day, three times the average daily flow.[12] The existing infrastructure was designed decades ago for a peak of eight thousand visitors a day, and the main trail to the waterfalls has engineering, access, and capacity issues. In addition, visitation is concentrated in just one trail route, with other attractions receiving a small percent of total visitors, which has led to severe crowding and taxed the staff, facilities, and concessionaire partners.

Since almost all visitation is along one corridor to the waterfalls, one problem at Iguaçu common to many parks is that some park neighbors and community leaders along secondary access points are clamoring for more, not less, tourism to the park. In recent participatory workshops to update the park management plan, many local residents, entrepreneurs, and leaders said that they would like to see more economic benefits to communities not on the main access route. Similar issues—not a clamor to reduce tourism, but rather, to democratize its impact—also affect Tijuca.

Solutions to Overtourism at Brazil's Flagship Parks

Fortunately, the Brazilian federal government—with support from local and state governments, nongovernmental organizations, the business sector, national and international universities, the US Agency for International Development (USAID), the US Department of Agriculture Forest Service (USFS), the US National Park Service (NPS), and other donors and partners—has taken many steps to better manage peak visitation at highly visited parks like Tijuca and Iguaçu, while simultaneously preparing for anticipated, and desired, increased visitation to other lesser known and visited parks and reserves. Many of these are scalable, replicable, and worth review by other developing nations also working to manage peak visitation to well-known destinations while promoting greater visitation to undervisited sites.

Improved Planning Frameworks

Until recently, Brazil had a problem faced by many other nations: the unwieldy methodologies used to create management plans for parks led to massive documents and unrealistic expectations that did little to improve protected area management. Plans were very expensive and took years to produce, resulting in planning backlogs that stymied efforts to deal with issues like booming but uncontrolled tourism at some parks. At the same time, frustration grew among local governments, the business sector, and communities waiting for plans to be finalized to expand park tourism. Overall standardized and modernized policies and regulations were not in place across the protected area system on several types of concessions and public-private partnerships, such as guiding services, land- and water-based transportation services, food service, and equipment rental, which led to a hodgepodge of inconsistent rules.

In recent years and with the support of USAID and USFS, ICMBio has adapted an approach developed by the NPS to deal with its own planning backlog and created a streamlined planning methodology that is already being applied at highly visited parks like Iguaçu.[13] ICMBio also developed a new framework for recreation and public use planning for their parks and adapted the Recreational Opportunity Spectrum approach to visitor management developed by the USFS. This approach uses zoning and the development of indicators and standards for recreational activities in protected areas to reduce use conflicts and environmental impacts while improving visitor satisfaction.

Policy-Making

ICMBio has recognized the need to update its policies to supplement its limited staff and budget. For example, it has created improved concessions policies to expand the use of public-private partnerships, leading to a considerable increase in the number of active tourism concessions, particularly in more visited parks. Such changes help ICMBio better manage visitor flows and provide improved visitor services. In addition, new policies enacted since 2016 expanded the scope of the ICMBio volunteer program to provide more opportunities for citizen involvement in parks, including trail construction and maintenance and provision of visitor services.[14]

Monitoring and Data Collection

Since 2014, ICMBio, working in partnership with USAID, USFS, and the state universities of Ponta Grossa (Brazil) and West Virginia (US), has development improved systems for counting or estimating visitation numbers and determining visitor characteristics and preferences. Several studies have also been done to improve estimation of the economic impact of tourism and other environmental services provided by protected areas, demonstrating to the government the high rate of return on federal investments in tourism programs and facilities in conservation areas.[15] To have a more accurate accounting of the total visitation and visitor characteristics, preferences, and economic impact of national and international tourism in Brazilian federal protected areas, ICMBio has improved its record keeping on visitor numbers and expanded visitor monitoring to many additional parks and reserves.[16]

Training Staff and Partners

ICMBio has trained hundreds of its own staff, volunteers, and partners in public use planning, managing volunteer programs, trail construction and management, visitor counting, concessions planning and management, and management and interpretation planning. Education has been provided through courses for staff and partners given at ICMBio's national training academy in São Paulo state, ACADEBio, through field courses throughout Brazil and through intensive short courses and seminars given by the University of Montana and Colorado State in the United States and in Brazil in partnership with USFS and NPS.

Research Initiatives

Until recently, few Brazilian academics and practitioners were devoting considerable effort to studying tourism in Brazilian protected areas. Lately, though, a vibrant research community on protected area tourism and public use has emerged. Much of the effort of this community has been devoted to study protected area visitor perceptions and motivations and the environmental and economic impacts of ecotourism, all of which are helping ICMBio better plan and manage tourism.[17] This community of practice has grown to encompass ICMBio staff and researchers at universities across Brazil, often working with colleagues from US colleges, such as West Virginia, North Carolina, Montana, and Colorado State Universities. This community of practice now meets annually to share research results and discuss information gaps and research needs related to tourism and public use in Brazilian protected areas. Several publications highlighting the economic importance of tourism to federal protected areas have also been produced by international organizations such as Conservation International and national organizations such as the Semeia Institute and Instituto de Pesquisas Ecológicas (IPÊ), both of São Paulo.[18] Semeia has played a key role in promoting public-private partnerships to expand support from Brazilian civil society for protected areas, and IPÊ has worked with ICMBio to synthesize good practices in protected area management, included tourism management and involvement of local communities in tourism.[19]

Emphasis on Interpretation and Expanded Trail Systems

Until recently, very few staff of ICMBio or guides from tour companies, park concessionaires, or local community-based ecotourism initiatives had any formal training in interpretation. The number of interpretive trails, exhibits, and guided hikes in protected areas was limited. This lack of capacity building for guides and orientation for visitors contributed to widespread problems of littering, vandalism, and visitors going off the marked trails. With substantial support from USAID and USFS, which also enlisted the help of Colorado State University and the National Association for Interpretation, a core team of ICMBio staff were trained and certified as interpretive hosts and guides. They were tasked with providing guidance on the development of model interpretive trails, exhibits, and visitor centers at protected areas in both highly visited southern Brazil and in more remote Amazonian parks. To grow the interpretive profession in Brazil, this team rapidly replicated the skills they obtained through a series of training events for communities, other ICMBio staff, and other partners. Several Amazonian parks have served as pilot and demonstration sites for development of tourism facilities and interpretive programs and the training of local guides.[20] At the same time, USFS and Colorado State University collaborated with ICMBio on organizing trail training events to build ICMBio capacity to design, build, and maintain trail networks in protected areas. This effort has also led to revitalized development of a long-distance trail network in Brazil that connects protected areas across the country. There is now a rapidly growing cadre of trail enthusiasts in Brazil involved in this effort.[21]

Applications in Tijuca National Park

ICMBio's team has moved to tackle head-on the challenges of what many consider South America's most visited national park. At Tijuca, an abandoned hotel was repurposed into a state-of-the-art environmental education and visitor center, gift shop, and restaurant. The concessions contracts with private partners at Tijuca have also been renegotiated to ensure that there is a better fit between what ICMBio needs to complement its limited budget, staff, and facilities and what concessionaires and other partners can offer to improve infrastructure and visitor services. A planned new parking facility near the visitor center will further resolve the parking issue on peak days. New attractions, such as access to a previously inaccessible waterfall, have diversified tourism offerings. The famous rack railway, more than a century old, was overhauled with new trains nearly twice as fast that ferry more than six hundred thousand visitors a year to the Christ statue. In addition, the concessionaire now offers a shuttle bus service from three different Rio locations to complement the traditional railway access to the park. Volunteer programs have been expanded as well. A co-management agreement with the municipal government and increased law enforcement have improved visitor security despite Rio's much publicized crime issues. An active program of restoration of extirpated native fauna is also underway.[22] Finally, to link Tijuca National Park and a number of other protected areas and promote dispersed outdoor recreation, ICMBio and many local and state government agencies and recreation and environmental groups are collaborating on the development of the Transcarioca Trail, which forms part of a larger long-distance trail system throughout Brazil.[23]

Applications in Iguaçu National Park

Key to the response to increased tourism in Iguaçu has been updating the long-term management planning for the park through a participatory process. This process is an adaptation of the approach the US National Park Service is using to produce briefer, more executive "foundation documents" that are then complemented by much more detailed plans for specific management programs or development sites. The new Iguaçu management plan was prepared with active participation of the park's management council, a technical team, and other partners, including a five-day intensive planning workshop with scientists, concessionaires, nongovernmental organizations, and key representatives of the local community. The plan will be valid for six years (2017–2023), and a more detailed public use plan is currently in the final stages of approval.[24] Four key strategies are being used to address visitation issues: (1) creation of other visitor attractions and infrastructure to spread out public use over a wider area; (2) use of studies of visitor satisfaction to determine a peak overall capacity; (3) analysis of a differential fee structure and a change in operating hours to encourage more visitation on less visited days and hours and reduce peak visitation; and (4) improvement in overall infrastructure and services to be able to accommodate more visitors in peak visitation periods without overloading facilities, such as elevators, overlooks, the visitor center, boardwalks to the falls, and the internal transportation network.

Conclusion

Although it shares challenges faced by many conservation agencies, such as limited funding and staff and increased but still inadequate levels of public and decision-maker support, in recent years ICMBio has taken important steps to improve management of peak tourism flows to its most iconic parks. From those experiences, ICMBio staff have learned to proactively prepare for expected expansion of tourism numbers, facilities, and programs at many other parks and reserves that still have low levels of visitation, to reduce potential negative environmental impacts, and to maximize potential economic benefits at these sites. The steps that ICMBio has taken—building capacity, expanding partnerships and interpretive programs, improving planning and policies, and promoting research and monitoring on protected area tourism—have helped manage expanding protected area tourism at its flagship parks while working to build public and decision-maker support for effective overall management of one of the world's largest and most diverse protected area systems.

Notes

1. Instituto Chico Mendes de Conservação da Biodiversidade. (February 14, 2019). "Visitação em Parques Nacionais bate novo recorde em 2018." https://www.icmbio.gov.br/portal/ultimas-noticias/20-geral/10216-visitacao-em-parques-nacionais-bate-novo-recorde-em-2018.

2. Karen Chávez. (February 13, 2020). "Great Smoky Mountains National Park Hits Record Number of Visitors, Boosts Local Economy." MSN. https://www.msn.com/en-us/news/us/great-smoky-mountains-national-park-hits-record-number-of-visitors-boosts-local-economy/ar-BBZXdso.

3. Alana Gandra. (June 2, 2020.) "Parque Nacional da Tijuca recebe quase 3 milhões de turistas em 2019"; New 7 Wonders of the World. (2020). "Christ the Redeemer." https://world.new7wonders.com/wonders/christ-redeemer-1931-rio-de-janeiro-brazil/.

4. New 7 Wonders of Nature. (2020). "Iguazú Falls." https://nature.new7wonders.com/wonders/Iguazú-falls-argentina-and-brazil/.

5. Acritica. (February 6, 2020). "Parque Nacional da Tijuca recebe quase 3 milhões de turistas em 2019." https://www.acritica.net/editorias/geral/parque-nacional-da-tijuca-recebe-quase-3-milhoes-d-20004d9aa06431b47e7/433090/.

6. Ibid.

7. RPC Foz do *Iguaçu.* (December 30, 2019). "Parque Nacional Do Iguaçu Bate Recorde Anual Com Mais de 2 Milhões de Visitantes." *Oeste E Sudoeste.* https://g1.globo.com/pr/oeste-sudoeste/noticia/2019/12/30/parque-nacional-do-Iguaçu-bate-recorde-anual-com-mais-de-2-milhoes-de-visitantes.ghtml.

8. Wilson Abrahão Rabahy. (December 15, 2019). "Análise e perspectivas do turismo no Brasil." *Revista Brasileira de Pesquisa em Turismo,* 14(1): 1–13.

9. Roberto Meurer and Hoyêdo Nunes Lins. (December 6, 2017). "The Effects of the 2014 World Cup and the 2016 Olympic Games on Brazilian International Travel Receipts." Tourism Economics, 24(4): 486–91.

10. New 7 Wonders of the World. (2020). "Christ the Redeemer." https://world.new7wonders.com/wonders/christ-redeemer-1931-rio-de-janeiro-brazil/; New 7 Wonders of Nature. (2020). Iguazú Falls. https://nature.new7wonders.com/wonders/Iguazú-falls-argentina-and-brazil/.

11. RPC Foz do Iguaçu. (December 30, 2019). "Parque Nacional Do Iguaçu Bate Recorde Anual Com Mais de 2 Milhões de Visitantes." *Oeste E Sudoeste.* https://g1.globo.com/pr/oeste-sudoeste/noticia/2019/12/30/parque-nacional-do-Iguaçu-bate-recorde-anual-com-mais-de-2-milhoes-de-visitantes.ghtml.

12. Cibele Munhoz Amato. (April 22, 2020). Iguaçu National Park. Personal communication with author.

13. Duda Menagassi. (April 18, 2017). "É hora de repensar o Plano de Manejo." *(O) eco.* https://www.oeco.org.br/reportagens/e-hora-de-repensar-o-plano-de-manejo/.

14. Paulo Roberto Russo. (December 3, 2019). ICMBio. Personal communication with author.

15. Thiago do Val Simardi Beraldo Souza, Brijesh Thapa, Camila Gonçalves de Oliveira Rodrigues, and Denise Imori. (June 3, 2019). "Economic Impacts of Tourism in Protected Areas of Brazil." *Journal of Sustainable Tourism* 27(6): 735–49; Carlos Eduardo Frickmann Young and Rodrigo Medeiros. (2018). *Quanto vale o verde: a importância econômica das unidades de conservação brasileira.* Rio de Janeiro: Conservação Internacional: Rio de Janeiro. https://www.funbio.org.br/wp-content/uploads/2018/08/Quanto-vale-o-verde.pdf.

16. Instituto Chico Mendes de Conservação da Biodiversidade. (February 14, 2019). "Visitação em Parques Nacionais bate novo recorde em 2018." https://www.icmbio.gov.br/portal/ultimas-noticias/20-geral/10216-visitacao-em-parques-nacionais-bate-novo-recorde-em-2018.

17. André de Almeida Cunha, Teresa Cristina Magro-Lindenkamp, and Stephen Ford McCool, eds. (2018). *Tourism and Protected Areas in Brazil: Challenges and Perspectives.* New York: Nova Science Publishers.

18. Carlos Eduardo Frickmann Young and Rodrigo Medeiros. (2018). *Quanto vale o verde: a importância econômica das unidades de conservação brasileira.* Rio de Janeiro: Conservação Internacional. https://www.funbio.org.br/wp-content/uploads/2018/08/Quanto-vale-o-verde.pdf.

19. Carol Da Riva. (2014). *Unidades de Conservação no Brasil: A contribuição do uso público para o desenvolvimento socioeconômico.* São Paulo: Instituto Semeia. http://www.semeia.org.br/admuploads/uploads/download.php?doc=UC_no_BR_A_contribuicao_do_uso_publico.pdf&tp=10&id=58; IPÊ. (2016). *Boas Práticas Na Gestão De Unidades De Conservação.* https://issuu.com/institutoipe/docs/revista_boas_pr__ticas_2016.

20. ICMBio. (2018). *Interpretação ambiental nas unidades de conservação federais.* https://www.icmbio.gov.br/portal/images/stories/comunicacao/publicacoes/publicacoes-diversas/interpretacao_ambiental_nas_unidades_de_conservacao_federais.pdf.

21. Rede Brasileira de Trilhas. (2020). https://www.facebook.com/sistemabrasileirodetrilhas/.

22. A Critica. (February 6, 2020). "Parque Nacional da Tijuca recebe quase 3 milhões de turistas em 2019." https://www.acritica.net/editorias/geral/parque-nacional-da-tijuca-recebe-quase-3-milhoes-d-20004d9aa06431b47e7/433090/.

23. Ian Cheibub. (August 19, 2019). "Brazil to open long-distance hiking trail in Atlantic forest—in pictures." *The Guardian.* https://www.theguardian.com/world/gallery/2019/aug/19/brazil-to-open-long-distance-hiking-trail-in-atlantic-forest-in-pictures.

24. Cibele Munhoz Amato. (June 2, 2020). Iguaçu National Park. Personal communication with author.

Chapter 3.3

Banff National Park, Canada

By Kaitie Worobec

With its jagged peaks, turquoise lakes, waterfalls, and abundant wildlife, Banff National Park is recognized as a jewel of the Canadian Rockies. Located 75 miles (120 kilometers) west of the city of Calgary, Alberta, the park encompasses an area of 2,564 square miles (6,641 square kilometers), and despite steep increases in visitor numbers, 96 percent of the park remains as wilderness.[1]

In 1885, after three Canadian Pacific Railway workers discovered a natural hot spring in the area, Banff was declared Canada's first national park; it was the third national park in the world.[2] The railway company later built the iconic Banff Springs Hotel and Chateau Lake Louise and set in motion the flow of tourists by actively advertising the park's majestic mountain scenery.

Banff National Park is one of four adjacent national parks that make up the Canadian Rocky Mountain Parks UNESCO World Heritage Site, declared in 1984. It holds the title of Canada's most visited national park, attracting 4.18 million visitors at its visitation peak in fiscal year (FY) 2017–2018.[3] It is the traditional territory of multiple Indigenous peoples, two communities (Banff and Lake Louise), a transcontinental railway, a four-lane highway, three major ski resorts, a sightseeing gondola, and dozens of hotels, meaning that this protected area has no doubt felt the effects of human use and development.[4]

The town of Banff is home to approximately 9,000 residents, and Lake Louise has a population of approximately 1,100, many of whom rely on tourism for their livelihood. The overall province-wide economic impact from tourism in Banff in 2017 was an estimated Can$2.7 billion (US$2.03 billion).[5] As Canada's most popular tourism destination, Banff has a thriving economy and very high employment rate.

High season in the area is June through September, with July and August being the peak months. Although visitation, along with hotel occupancy and room revenue, is significantly lower in the winter, the destination marketing organization (DMO) Banff & Lake Louise Tourism (B&LLT) and local operators are continually working to promote the winter experience.

Causes and Impacts of Overtourism

Despite efforts to manage the multiple facets of tourism growth, Banff National Park is being pointed to as a victim of overtourism. In recent years, a series of studies, conferences, and media reports have warned: "Banff 'bursting at the seams' as tourism soars," "Overtourism in Canada's Mountain National Parks," and "Banff National Park Needs Holiday from Humans."[6] Park visitation has risen sharply, up from 3.11 million in FY 2008–2009 to 4.09 million in FY 2018–2019.[7] This increase in visitors has largely been concentrated at a few locations, with some park infrastructure and facilities at key sites at or near capacity.

Perspectives on current visitation levels vary significantly; however, during Parks Canada's 2019 stakeholder engagement consultation, the prevailing view was that Banff National Park is "overcrowded."[8] Many participants cited experiences in and around the Banff and Lake Louise townsites, on park roads, and at popular day-use areas as examples of overcrowding. Additionally, the Canadian Parks and Wilderness Society (CPAWS) has observed an upsurge in human-wildlife conflicts in recent years.[9] Peter Zimmerman, parks program supervisor for CPAWS, noted, "Annual visitation has been edging up continuously for years, but the last five years [2013–2018] have seen it explode."[10]

Traffic congestion in and around Banff National Park is also increasing, leading to longer travel times, more greenhouse gas emissions, increased traffic in wildlife corridors, a diminished visitor experience, and reduced quality of life for residents. The town of Banff's threshold for the onset of congestion is considered to be approximately 24,000 vehicles per day (VPD), and this measure is often reached on a daily basis in July and August.[11] Data collected by the town indicates that 90 percent of days in July and August 2018 recorded more than 24,000 VPD.[12]

The Town of Banff conducted Community Social Assessments in 2006, 2014, and 2018 to understand the social well-being of residents. These assessments revealed that high visitation is taking a toll on local residents, with many expressing frustration, weariness, and resentment. "Living Where the World Visits" emerged as a key challenge.[13] According to Alison Gerrits, the town's community services director, one key finding from the 2018 assessment of more than eight hundred residents was that "the community is really wanting to have what they classify as difficult conversations, and part of that is about the sheer volume, the notion of capacity, and when is enough, enough."[14]

Solutions

Within Banff National Park, three organizations play major roles in managing tourism. Parks Canada is the steward of the protected environment and manages the national park, the Town of Banff manages the municipality, and B&LLT acts as the DMO. These three organizations work collaboratively to ensure the sustainability of Banff economically, socially, and environmentally. As Leslie Bruce, president and CEO of B&LLT, explained in an interview, "We believe being in Canada's first national park is our greatest asset and our greatest responsibility to act accordingly. We need to ensure we are not only promoting the destination, but ensuring its protection and sustainability for now and for future generations."[15]

Parks Canada manages Banff National Park under the federal government's National Parks Act, which was first enacted in 1930. This act was later amended to clarify that "conserving, restoring, and maintaining ecological integrity" is the government's top priority.[16] It is the only national park agency in the world that has fully implemented a continuous system-wide ecological monitoring and reporting program, consisting of more than seven hundred independent scientific measures, such as the health of park species and ecosystems.[17]

Like all Canadian national parks, Banff National Park is required to have a park management plan. This public document, prepared in consultation with Canadians and Indigenous peoples, sets out a long-term vision for the park and provides strategic direction to guide decision-making about the park for a five- to ten-year period.[18] The latest version of the Banff National Park management plan was completed in 2010, and the park is currently undergoing the stakeholder engagement phase of a new plan.

The Canadian federal government has also enacted legislation to restrict the town of Banff's growth so as to better manage human use and development and maintain ecological integrity.[19] As a result of the landmark 1996 Banff-Bow Valley Study that assessed the cumulative environmental effects of development and land use activities on the Bow River watershed within Banff National Park, the town's boundaries were strictly defined and commercial development capped at roughly 10 percent of the town's total area.[20] The town's population was also capped at eight thousand permanent residents, and under a need-to-reside clause, residents must be employed by or own a business within the national park to have the right to live in Banff.[21]

Banff's unique situation creates opportunities for the park and town to show leadership in addressing a range of problems linked to overtourism. The Town of Banff's 2019–2022 strategic plan states: "A limited land base creates opportunities for leadership in areas such as environmental sustainability, active transportation, and a focus on infrastructure renewal and innovation, rather than expansion. Banff should not build our way out of traffic and parking issues. Our community must embrace more creative solutions to mobility in the townsite."[22] As such, Banff strives to be a global environmental leader and model of sustainable tourism.

In collaboration with local and regional stakeholders, Parks Canada, the Town of Banff, and B&LLT implemented a number of short- and long-term initiatives in 2018 to manage vehicle volumes and visitation during critical periods while still ensuring a high-quality visitor experience and sustainable tourism within the ecological limits of the park.

These initiatives included infrastructure improvements, expanded public transit, increased staffing levels, enhanced wildlife messaging, and a proactive communication plan to inform and educate visitors, influence behavior, and disperse crowds.[23]

Marketing

One example of the collaborative approach between Banff's multiple management bodies is a joint summer communication plan, which was first developed in 2017 in anticipation of Canada's 150th anniversary and continued into 2018, 2019, and 2020. The goal was to encourage visitors to leave their vehicles behind and use instead the increased transit and walking infrastructure made available to residents and visitors. Marketing tactics included paid and organic social media, digital media, printed collateral, and a trip-planning website, explorethepark.ca. The website provides detailed transit and shuttle routes and fares, as well as details on bike rentals, easy-to-reach walking trails, and booking tours. The tool also helps reduce congestion at popular locations by outlining different times to visit key attractions and showcasing how to get around the park using public transit.[24]

The campaign was aimed at day visitors from Calgary, campground and hotel guests, RV users, the local business community, and residents. Not only were the campaign's advertising and its content measured for performance, but the Town of Banff monitored changes in parking, traffic flow, pedestrian traffic, and transit usage as well. Commenting on the success of the campaign, Mayor Karen Sorensen said, "We are strategic in using multiple ways to persuade visitors, even before they arrive in Banff, that public transit and local shuttles provide a more enjoyable way to experience the Park. And we are seeing results—we have maintained visitation levels, and are seeing large increases in transit use and drops in vehicle congestion at peak periods and at key locations."[25]

Managing Traffic Flows

The destination's strategy to manage traffic volume is very data-driven. For example, traffic monitoring sensors detect signals from discoverable Bluetooth and Wi-Fi technology embedded in devices and vehicles traveling between specific points within town. The information is used to calculate travel times and determine origin destination and length of stay, and notifications are sent to the traffic management team and streets on-call team when travel times exceed delay thresholds. The data also feed a "traffic dashboard," a one-stop shop for visitors and residents to obtain travel times, identify available parking spaces, and view webcams of twelve main road or intersection locations.[26] In the summer of 2019, an additional five hundred parking spaces were added at the Banff train station, a ten-minute walk from downtown, allowing visitors to leave their vehicle while exploring the town.

To measure foot traffic, ten overhead pedestrian counters are maintained throughout downtown Banff. The data from these sensors enable town managers to correlate daily and monthly pedestrian volumes to vehicle and transit ridership volumes and to other metrics such as hotel occupancy.

Promoting Sustainable Year-Round Tourism

Finally, as a DMO, B&LLT focuses its marketing campaigns and events during the winter (December through April) and shoulder seasons (May and late September through mid-November) to create a more sustainable year-round tourism economy. In 2018, the destination generated an additional 100,000 room nights in the winter and shoulder seasons and exceeded 50 percent average occupancy in November for the first time ever.[27] One tactic that has proved successful is a digital marketing campaign highlighting the full winter experience, inspiring visitors to weave other activities into their ski holiday—ice-skating, dogsledding, cross-country skiing, snowshoeing, sleigh rides, dining, and more—resulting in increased expenditures on accommodation, transportation, and activities over the winter and shoulder seasons.

Lessons Learned from Banff

In Banff, institutional collaboration has been crucial to success. The Town of Banff, B&LLT, and Parks Canada continually work together to strike a balance for sustainable tourism, positive visitor experience, and the protection of ecological integrity. The three organizations collaborate on both marketing and management strategies to ensure a coordinated and consistent approach. Collaboration with industry partners is also essential. The Town of Banff has a very engaged business community, and together, the town government and local businesses work to develop responsible practices around recycling, water usage, and waste management to ensure that their environmental impact is minimal.

To minimize the negative impacts from tourism, it is also important to maintain critical infrastructure investments and ensure appropriate quality standards exist to maintain and preserve key assets. In Banff's case, the town exists on a fixed boundary of less than 1.5 square miles (4 square kilometers), and this infrastructure must accommodate the impacts of four million visitors annually, in addition to its nine thousand residents. The Town of Banff therefore has a one-hundred-year capital asset plan to ensure that it is allocating appropriate resources every year to support infrastructure maintenance and renewal.

Last, data-driven decision-making is a key strategy to drive consensus. Parks Canada, the Town of Banff, and B&LLT all collect detailed data on everything from the number of users on trails to the number of minutes of every traffic delay. Once a destination has access to such rich data, it can be used to gain meaningful insights about the visitor and resident experience, which are essential to driving positive change and effectively managing overtourism.

Notes

1. Town of Banff. (2020). https://banff.ca/252/Learn-About-Banff.

2. Canadian Rockies Hot Springs. (2019). https://www.hotsprings.ca/banff-history.

3. Parks Canada Agency. (Accessed March 2020). "Attendance 2017–2018." https://pcacdn.azureedge.net/-/media/docs/pc/attend/Parks-Canada-Attendance-2017-18-FINAL.pdf?la=en&modified=20180604161223&hash=4363C45A912B2BDBBC28D6595A02A7A0767E4724.

4. Banff-Bow Valley Task Force. (October 1996). *Banff-Bow Valley: At the Crossroads—Summary Report*. https://web.archive.org/web/20031129185852/http://www.whyte.org/time/riveroflife/bveng.pdf.

5. Cathy Ellis. (April 9, 2018). "Banff Generates $7.5 Million a Day for Alberta." *Empire Advance*. https://www.empireadvance.ca/regional-news/banff-generates-7-5-million-a-day-for-alberta-1.23260285.

6. Cathy Ellis. (October 11, 2018). "Banff 'Bursting at the Seams' as Tourism Soars." *RMO Today*. https://www.rmotoday.com/banff/banff-bursting-at-the-seams

-as-tourism-soars-1573160; *Ward Cameron*. (June 13, 2019). "081 Overtourism in Canada's Mountain National Parks." The Mountain Nature and Culture Podcast. https://www.mountainnaturepodcast.com/081-overtourism-in-canadas-mountain-national-parks/; Bill Kaufmann. (August 27, 2018). "As Review of Public Access Nears, Banff National Park Needs Holiday from Humans, Say Conservationists." *Calgary Herald*. https://calgaryherald.com/news/local-news/as-review-of-public-access-nears-banff-national-park-needs-holiday-from-humans-says-conservationists.

7. Parks Canada Agency. (2019). https://www.pc.gc.ca/en/index.

8. Parks Canada Agency. (2019). "What We Heard—Banff National Park." https://www.pc.gc.ca/en/pn-np/ab/banff/info/gestion-management/involved/plan/entendu-heard.

9. CPAWS Representative. (May 2019). Personal communication with author.

10. Cathy Ellis. (October 11, 2018). "Banff 'Bursting at the Seams' as Tourism Soars." *RMO Today*. https://www.rmotoday.com/banff/banff-bursting-at-the-seams-as-tourism-soars-1573160.

11. Claudia Rustenberg. (October 9, 2018). *Briefing: Summer 2018 Transportation Overview*. https://banff.ca/DocumentCenter/View/6105/Summer-2018-Transportation-Overview?bidId=

12. Ibid.

13. Town of Banff. (2020). "Community Social Assessments." https://banff.ca/243/Community-Social-Assessments.

14. Cathy Ellis. (June 13, 2019). "Banff Residents Struggling to Maintain Quality of Life with Growing Tourism Economy." *RMO Today*. https://www.rmotoday.com/banff/banff-residents-struggling-to-maintain-quality-of-life-with-growing-tourism-economy-1574334.

15. Leslie Bruce. (May 2019). President and CEO, Banff & Lake Louise Tourism. Personal communication with author.

16. Parks Canada Agency. (2000). "*Unimpaired for Future Generations"? Protecting Ecological Integrity with Canada's National Parks. Vol. I: A Call to Action*. http://publications.gc.ca/collections/Collection/R62-323-2000-2-1E.pdf.

17. Dominique Tessier. (July 2019). Chief of Media Relations, Parks Canada Agency. Personal communication with author.

18. Parks Canada Agency. (April 10, 2019). "A Backgrounder on the Park Management Planning Process." https://www.pc.gc.ca/en/pn-np/ab/banff/info/gestion-management/involved/plan/primer.

19. Banff-Bow Valley Task Force. (April 12, 2005). *Banff-Bow Valley Study*. http://epe.lac-bac.gc.ca/100/200/301/parkscanada/banff_bow_valley-e/chapter_1-en.pdf.

20. Town of Banff. (2020). https://banff.ca/252/Learn-About-Banff.

21. Ibid.

22. Town of Banff. (April 2019). *2019–2022 Strategic Plan*. https://banff.ca/DocumentCenter/View/6102/Banff-Strategic-Plan-2019-2022?bidId=.

23. Dominique Tessier. (July 2019). Chief of Media Relations, Parks Canada Agency. Personal communication with author.

24. Banff & Lake Louise Tourism. (2020).https://www.banfflakelouise.com/explorethepark.

25. Banff & Lake Louise Tourism. (Accessed March 2020). "Alive with Stories:

2018 Annual Report." https://www.banfflakelouise.com/sites/default/files/bllt_ar2018_jun28_web.pdf.

26. Town of Banff. (2020). *Traffic Dashboard: Travel Times.* http://dashboard.banff.ca/.

27. Banff & Lake Louise Tourism. (2019). *Alive with Stories: Banff & Lake Louise Tourism - 2018 Annual Report.* https://www.banfflakelouise.com/sites/default/files/bllt_ar2018_jun28_web.pdf.

Chapter 3.4

Serengeti National Park, Tanzania

By David Blanton

The Serengeti ecosystem, a UNESCO World Heritage Site since 1981, is rightly regarded as one of the planet's greatest natural treasures. It is home to the world's largest terrestrial animal migration, the twice-yearly trek by more than 1.5 million wildebeest, zebras, and gazelles who move across the Serengeti's savannah, stalked by lions, cheetah, leopards, hyenas, and (while crossing the Mara River) crocodiles.[1]

This impressive cross-border ecosystem encompasses Kenya's Maasai Mara National Reserve, which includes 583 square miles, and Tanzania's Serengeti National Park, which with 5,695 square miles is ten times larger.[2] Both the Mara Reserve and Serengeti Park are surrounded by several other protected areas, as well as by human settlements and farmlands.

Despite its vast size, the Serengeti ecosystem has been experiencing overtourism, most acutely in the Maasai Mara and during the wildlife migrations. A 2018 report on the Maasai Mara portrays a sobering reality:

> Is unregulated traffic at key crossing points actually destroying the very spectacle that attracts tens of thousands of tourists to the

> Mara. . . . The answer is an unequivocal yes. There is absolutely no doubt that the number of vehicles witnessed at crossing points (up to 300 were reported in one case) totally disrupted the crossings—driving the animals away to quieter spots. . . . One cannot but speculate that the chaos at crossings is actually destroying the very spectacle that so many people come to see.[3]

Rise and Importance of Tourism

Kenya and Tanzania share a common colonial history as part (along with Uganda) of British East Africa. After independence in the early 1960s, however, the two countries followed different development models, which in turn affected the type of tourism that evolved in the Maasai Mara Reserve and Serengeti National Park.

From independence onward, Kenya's wide-open, laissez-faire capitalism and open door for private investment helped turn the country into Africa's most popular wildlife destination centered on wildlife viewing safaris. Kenya's global reputation was enhanced by wildlife celebrities, documentaries, books, and, most importantly, the 1985 Oscar-winning film *Out of Africa.* Kenya leveraged this image and encouraged ever-expanding tourism investment. By 1987, tourism was Kenya's top foreign exchange earner; today it ranks among the top three earners.[4]

Since its creation in 1961, the Maasai Mara National Reserve has been managed not by the central government but by local Maasai pastoralists through the Narok County Council. The aim was to return more benefits from park fees and tourism to communities living around the park. Initially, this model was highly successful: the Maasai Mara quickly became the most visited park in East Africa; private developers built dozens of permanent tented camps and safari lodges; and the local Maasai received benefits in terms of jobs, income, and expanded roads, schools, water holes, and other social services.

By the 1980s, however, the Mara Reserve was suffering from a host of intertwined problems: corruption, cronyism, and mismanagement of funds; decline in wildlife numbers; poor maintenance of roads and other infrastructure; agricultural encroachment; and lack of sound and sustainable management plans. Rather than providing broad community benefits, tourism earnings were being appropriated by a small elite of powerful businessmen and politically well-connected Kenyans, most of whom were not Maasai. As a result, the Maasai Mara, once hailed as a world leader in innovative community-based ecotourism, has become in recent decades "Kenya's poster child for tourism overdevelopment."[5]

With independence, Tanzania took a different path, pursuing a non-aligned socialist development model known as "ujamaa" that nationalized major parts of the economy, shunned private investment, built a robust system of national parks, and viewed international tourism as elitist and a poor tool for equitable economic development.[6] Well into the 1980s, Tanzania's tourism industry was small, largely government owned, and relatively underdeveloped. Across the vast Serengeti, only a handful of government lodges were built, greatly limiting tourism.[7]

By the early 1990s, however, Tanzania had abandoned its socialist experiment and began soliciting foreign investment for tourism and other key sectors. Over the coming decades, several larger private game lodges and a scattering of boutique luxury safari camps and lodges were built in the Serengeti, many by South African tour operators. Tourism numbers and earnings grew, and wildlife in the Serengeti generally thrived. Today, tourism is Tanzania's largest foreign exchange earner.[8]

The Serengeti has, however, faced challenges and conflicts with communities on its periphery over land rights, access to park resources, and equitable division of tourism earnings. For instance, a 2015 World Bank study found that few tourism benefits trickle down to Tanzanians not involved in tourism and that most of these benefits do not reach communities in and around Tanzania's main tourist attractions.[9]

Today, Tanzania's Northern Circuit,[10] which includes Serengeti National Park, is renowned for its high value wildlife viewing tourism. In 2018, the total contribution of tourism to Tanzania's gross domestic product was US$6.17 billion, whereas in Kenya that number was US$7.21 billion.[11] That same year, Tanzania received 1.49 million tourists, whereas Kenya received 2.02 million.[12] That difference equates to US$4,141 per tourist in Tanzania and US$3,569 in Kenya, an indication that Tanzania is attracting somewhat higher-value visitors.

In recent years, the number of annual visitors to Tanzania's Serengeti have often surpassed Kenya's Maasai Mara. In 2016, for instance, Serengeti National Park[13] received 201,728 visitors, whereas the Maasai Mara received 146,000.[14] Figures for 2019 show visitors to be about the same: 291,200 in the Maasai Mara and 270,946 in Serengeti National Park. Given the difference in the size of each park, that equates to a tourism density of about ten times greater in the Mara.

By 2020, according to Booking.com, there were 125 hotels in the Maasai Mara, and according to Hotels.com, Serengeti National Park and surrounding areas, although ten times larger than the Maasai Mara, had 62 accommodations.[15] These numbers again demonstrate that the density of accommodations in the Maasai Mara is much greater than in Tanzania's Serengeti.

These distinct tourism profiles within the Serengeti ecosystem—of higher density of lodges and visitors and lower tourism earnings in the Mara—result from historical and geographical difference. They help highlight why overtourism has become a more serious problem in Kenya's Mara Reserve than in Tanzania's Serengeti.

Causes and Impacts of Overtourism

Overtourism is one of a range of challenges facing the Serengeti ecosystem. The list of challenges also includes climate change, population growth, land degradation, human-wildlife conflict, poor guiding practices, overdevelopment of tourism infrastructure, human and agricultural encroachment, invasive species, poaching, financial mismanagement, and water shortages. Many but not all of these challenges are contributing directly to overtourism, others are contributing indirectly, and all, combined, are working to create a perfect storm of threats to the Serengeti ecosystem.

Overdevelopment of Infrastructure: Hotels and Roads

By the mid-1990s, two dozen lodges and permanent tented camps were clustered in two main areas of the Maasai Mara Reserve, causing problems with waste disposal, shortages of wood for cooking, degradation of roads, and overuse of viewing sites in the area.[16] Despite the problems, however, the pace of development has accelerated: by 2020, there were 122 hotels listed in the Maasai Mara.[17]

A 2018 study of tourism in the Maasai Mara found that "visitors and high concentration of lodges have partly contributed to the reduction in animals' numbers."[18] A World Bank study found that lodges built near watering holes compete with wildlife for prime habitat and fresh water, negatively impacting both "predator abundance and tourist satisfaction."[19] In addition, wildlife lodges often bring in nonnative ornamental plants for their gardens, displacing native species and degrading the nutrition of wildlife.[20]

Although roads are key to economic development, roads are being built that can fragment the Serengeti ecosystem.[21] The World Bank concludes that road expansion "ranks high in the list of factors contributing to wildlife loss," resulting in a net loss to the economy.[22] In Tanzania, a vigorous debate has been waged since 2010 over plans to construct an all-weather paved road across the northern part of Serengeti National Park. Proponents argue that the road will facilitate economic development and help reduce poverty, whereas opponents warn that the road will disrupt the biannual wildlife migration and increase already high levels of poaching.[23] Although plans for the paved road through Serengeti National Park have been put on hold, road construction is occurring on either side of the park—and it is feared that these roads will be connected across the park in the future. An alternate route south of the Serengeti has been proposed and shown to have greater economic and social benefits to communities; it would also not disrupt the wildlife migrations.[24]

Poorly Controlled Tourism Behavior and Numbers

Poorly regulated tourism—and at times overtourism—is directly impacting wildlife and habitat in the Serengeti ecosystem. Two of the major issues are land degradation by vehicles that drive off-road and interference with wildlife behavior and movement.

During the high season, about twenty-seven hundred people visit the Maasai Mara daily, and often they are not adequately managed. As cheetah researcher Femke Broekhuis wrote: "The Mara Reserve—with the exception of a conservancy called the Mara Triangle—doesn't limit the number of tourists that enter the park per day, and there are no restrictions on the number of tourist vehicles at a predator sighting. It's therefore not uncommon to see more than 30 tourist vehicles at a sighting."[25]

Vehicles crowd areas where wildlife are spotted, Serengeti National Park. Source: David Blanton.

Indeed, scientific research is revealing that wildlife is affected by too much tourism traffic, as well as by human encroachment around the park. A study of impalas in the Serengeti found that "human-induced disturbances have a pronounced effect on the behaviour and demography of impala populations in the Serengeti ecosystem," resulting in "a more female skewed sex ratio, lower observed reproductive and recruitment rate, and reduced time spent on restorative behaviour (i.e. resting)."[26]

Similarly, Broekhuis's research in the Mara Reserve found that tourism affects cheetahs' reproductive success, number of cubs reared, and ability to hunt. "This suggests that cheetahs aren't getting the protection they need, particularly from the impact of growing numbers of tourists," Broekhuis concluded.[27]

Declining Visitor Experience

Visitors have begun to note the problems with overcrowding. As a traveler in Serengeti National Park blogged, "A stressed out Cape buffalo charged our car because he was separated from his herd. A scared leopard was forced into hiding after being boxed in by a mass of tourist vehicles. [These] experiences were phenomenal for us; I can't help wondering: is it fair to the animals?"[28]

In a 2019 Tripadvisor post, a traveler to Serengeti National Park reported that "most crossings had 40+ vehicles watching. . . . It was a massive increase compared to when we were there in 2013, and really took away from the experience for me."[29] Another tourist in the same forum wrote, "I couldn't wait to get out of this area [Seronera, in central Serengeti]—entirely too crowded. Game viewing at its worst."[30] The Kenya Association of Tour Operators reports that because of the overcrowding, higher-end tourists and tour operators are avoiding the Mara during peak tourism seasons.[31]

Decline in Wildlife Populations

"The wildlife that has lured travelers to Kenya by the planeload is in dramatic decline. In the past three decades, the country has lost more than half of its wildlife," stated a 2019 World Bank study.[32] In the Maasai Mara Reserve, between 1989 and 2003, the number of giraffes declined by 75 percent, warthogs by 80 percent, hartebeest by 76 percent, and impala by 67 percent.[33] In addition, the number of wildebeest, the keystone species on which the health of the ecosystem depends, that migrate from Tanzania to Kenya fell from 588,000 in 1979 to 157,000 in 2016, a drop of 73 percent.[34]

There are several contributing factors. Human population around the Mara increased from 4.6 million in 1977 to 5.8 million in 2016.[35] Fencing is rapidly increasing, "threatening to lead to the collapse of the entire ecosystem in the near future."[36] But in recent years, a combination of tourism factors has coalesced to make overtourism glaringly evident in the Serengeti ecosystem, particularly in the Maasai Mara. Overcrowding, overbuilding, poor tourism practices, and lack of enforcement of regulations have all contributed to the decline in wildlife.

Solutions: The Way Forward

Reducing and sustainably managing tourism's impacts on the Serengeti ecosystem require an integrated, multinational approach, from wise government policies to best tourism practices on the ground. Also required is international monitoring, especially by UNESCO as the administrator of World Heritage Sites (see chapter 4.1). Models and on-the-ground measures to help control overtourism range from good governance to encouraging tourists to give back.

Good Governance

Within the Serengeti ecosystem, the Maasai Mara is most endangered. Kenya needs to rethink tourism, reform its policies, and build wildlife protection before it's too late. As a 2016 study concludes, "Our analysis shows that the future of Kenyan wildlife is in serious jeopardy without urgent, far-reaching and far-sighted changes to their current conservation and management."[37]

Especially promising within the Mara Reserve is the growth of Wildlife Conservancies.[38] Today, there are fifteen conservancies on private Maasai landholdings surrounding the reserve.[39] Tourism companies lease land for wildlife viewing tourism from the Maasai landowners on the condition that they do not live on the land, cultivate, or develop it. The Mara Triangle and other conservancies have demonstrated that they can offer high-quality, lower-density, responsible tourism. As cheetah researcher Broekhuis wrote, the Maasai Mara conservancies "are largely getting this right. Tourist numbers are limited to the number of beds per conservancy and only five vehicles are allowed at a sighting at any given time."[40]

Controlling Infrastructure Development

"Getting infrastructure 'right' is critical because infrastructure choices have long-lived and difficult-to-reverse impacts on land, wildlife, water, and future patterns of development," concluded a 2019 World Bank study of wildlife in Kenya.[41] Within the Serengeti ecosystem, tourist accommodations and roads are the most important types of infrastructure and have the most impact. Even with fewer tourism accommodations in Serengeti National Park than in the Maasai Mara, Tanzania has for the time being taken the position that there are "enough hotels inside the [Serengeti National] Park" and that any future hotel construction should be done outside the park.[42] Kenya would be wise to enforce a similar ban on new hotels within the Mara Reserve.

In terms of roads, the 2019 World Bank study on Kenya noted that the tools exist to build "smarter, greener" road networks. "Through careful and strategic planning," the study concluded, "spending on infrastructure can be rendered more effective and more conducive to growth and poverty reduction, and less impactful on wildlife and the economic opportunities that they bring."[43]

Educating Drivers, Guides, Tour Operators, and Tourists

It is also important to implement and enforce strict wildlife viewing guidelines, especially in areas where tourism numbers are high.[44] Drivers and guides must follow rules for wildlife viewing and numbers of vehicles permitted around wildlife. Travelers need to understand the impacts from harassing wildlife and off-road driving in restricted areas. Tour operators, guides, and drivers need to understand and follow best practices, and ideally, any gratuities should be tied to these practices. To date, voluntary codes have not been sufficient to curb bad practices. Authorities must educate visitors and tourism workers, monitor behavior, enforce regulations, and punish offenders.

Limiting and Dispersing Visitors

Both Kenya and Tanzania would be wise to selectively limit visitors to the Serengeti ecosystem where and when overcrowding occurs, especially during the biannual migrations. As Broekhuis wrote, "Ideally the Mara Reserve should restrict the number of tourists, especially during the peak tourist seasons."[45] Similarly, a former safari camp manager in Tanzania warned that "if Tanzania increases the number of tourists without control, the Serengeti will undoubtedly suffer from the 'invasion.'"[46]

In addition, pressure can be alleviated if tourism is dispersed beyond the Serengeti ecosystem. In Kenya, conservancies now cover more land than the national parks and reserves and have a higher density of wildlife, but they account for only "a meager 1.3 percent" of tourism earnings. This figure suggests, a World Bank study concluded, that conservancies offer "considerable potential and scope for expansion in a specialized market that caters to the high-value and low-volume tourists," thereby taking some pressure off the Mara Reserve.[47]

In Tanzania, the Northern Circuit, which includes Serengeti National Park, generates 85 percent of the country's total national park revenues. A 2015 World Bank study advocated the ambitious target of an eight-fold increase in tourist arrivals by 2025. To do that without damaging the Northern Circuit, it proposed diversifying tourism to lesser-used parks, such as Ruaha and the Selous in the southern part of the country.[48]

Like all iconic tourist attractions, however, the Serengeti ecosystem will inevitably continue to attract travelers. The governments therefore need to promote other parks through aggressive marketing, building infrastructure, and helping tour operators with tax breaks, permits, and other incentives. In Tanzania, for instance, the government is offering tour operators and hotel developers incentive packages to promote the southern parks.[49]

Building the Right Kind of Tourism

The right model of tourism for the Serengeti ecosystem is high-value, low-density (HVLD) nature tourism: fewer tourists generating higher per capita revenue. In the Serengeti ecosystem, Tanzania's tourism is mainly the HVLD model, whereas Kenya's is far more mixed.

Tanzania's HVLD model is not without challenges, however, and its long-term future may not be secure. For instance, Tanzanian tourism workers make on average only one-third as much as their counterparts in Kenya, and many resorts rely heavily on imported materials, equipment, and food.[50] In addition, Tanzania is eager to target new markets in China and India.[51] If these markets take off, the numbers could be staggering, and the kind of traveler being attracted could undermine HVLD tourism. For instance, there have been discussions about building golf courses, luxury resorts, and even a theme park catering to the Chinese market.[52]

By contrast, HVLD tourism attracts people who want more authentic experiences and are willing to pay for it. In addition, economic downturns don't greatly affect the HVLD market. According to the

World Bank, "During the 2008–09 recession, tourist numbers plummeted across the globe, yet tourist numbers in Tanzania were largely unaffected."[53] Long-term planning and good marketing are key. Both Kenya and Tanzania should take measures to strengthen and grow the important and more sustainable HVLD market segment.

Tourism Giving Back to Conservation and Local Communities

The tourism industry and travelers need to be part of the solution. Those who benefit from and use the Serengeti ecosystem should give back to support community and conservation efforts. A few of the best tourism companies do so already. An example is Friends of Serengeti, a project of Serengeti Watch. Friends of Serengeti enlists tour companies to provide support for community conservation and education programs, including conservation radio programming, microenterprises for women's groups, and tree planting. School programs include financing students to travel to the Serengeti National Park for their first-ever visit. Member companies inform their travelers about threats to the Serengeti ecosystem and encourage them to help fund projects. Many travelers are eager to help, but more companies need to be involved.[54]

Conclusion

Overtourism has become an important issue for the Serengeti ecosystem and must be considered as a long-term threat to protected areas and safari tourism. The income and jobs that nature-based tourism bring have justified the protection of wildlife in a region plagued by poverty, but without those benefits, wildlife and habitat will be critically endangered. It is said that wildebeest in the Serengeti are its "keystone species," the critical species upon which the ecosystem depends. To that we might now add, so too are tourists themselves. Wisely managed high-value, low-impact tourism can help save and sustain one of our greatest natural wonders, the Serengeti ecosystem.

Notes

1. UNESCO. (2018). State of Conservation. Serengeti National Park (United Republic of Tanzania). https://whc.unesco.org/en/soc/3680.

2. Sometimes called the Serengeti-Mara ecosystem or Greater Serengeti, here the term *Serengeti ecosystem* will be used to denote the entire protected area in both Kenya and Tanzania.

3. Macharia Kamau. (October 7, 2018). "How the Masai Mara Is Sinking under Its Own Global Success." *Standard Digital.* https://www.standardmedia.co.ke/article/2001298146/masai-mara-suffers-from-its-own-success.

4. Martha Honey. (2008). "Kenya: The Ups and Downs of Africa's Ecotourism 'Mzee.'" *Ecotourism and Sustainable Development: Who Owns Paradise?* Island Press; George Obulutsa. (January 10, 2020). "Kenya's Tourism Earnings Up 3.9%." *Reuters.* https://www.cnbcafrica.com/east-africa/2020/01/10/kenyas-tourism-earnings-up-3-9/.

5. Martha Honey. (2008). "Kenya: The Ups and Downs of Africa's Ecotourism 'Mzee.'" *Ecotourism and Sustainable Development: Who Owns Paradise?* Island Press.

6. Nyerere Centre for Peace Research. (2020). "Mwalimu Nyerere on Wildlife Conservation." https://www.juliusnyerere.org/about/mwalimu_nyerere_on_wildlife_conservation.

7. Martha Honey. (2008). "Tanzania: Whose Eden Is It?" *Ecotourism and Sustainable Development: Who Owns Paradise?* Island Press.

8. East African Business News Editor. (February 19, 2019)."Tanzania Earns US$2.44 Billion from Tourism in 2018." *EABW Digital.* https://www.busiweek.com/tanzania-earns-2-44-billion-from-tourism-in-2018/.

9. Jacques Morisset. (January 2015). *The Elephant in the Room: Unlocking the Potential of the Tourism Industry for Tanzanians.* The World Bank. https://www.worldbank.org/en/country/tanzania/publication/tanzania-economic-update-increasing-tourism-for-economic-growth.

10. Tanzania's Northern Circuit includes the Serengeti, Ngorongoro Crater, Kilimanjaro National Park, Lake Manyara, Tarangire National Park, Arusha National Park, Olduvai Gorge, and Mkomazi National Park.

11. World Travel and Tourism Council and Oxford Economics. (March 2020). *Travel & Tourism Economic Impact 2020 Tanzania.* London: WTTC. p. 18.

12. Tanzania Invest. (April 8, 2019). "Tanzania Tourism Revenues and Arrivals Up in 2018." https://www.tanzaniainvest.com/tourism/tanzania-tourism-revenues-and-arrivals-up-in-2018; Winnie Atieno and Bonface Otieno. (January 7, 2019). "Kenya Tourism Earnings Rise to Sh157 Billion as 2018 Arrivals Cross 2m Mark." *Business Daily.* https://www.businessdailyafrica.com/economy/Kenya-tourism-earnings-rise-to-Sh157bn-as-2018-arrivals-cross-2m/3946234-4924636-15p97f6z/index.html.

13. Serengeti National Park visitor density applies only to the park itself, which is surrounded by other protected areas with a total of 12,000 square miles. Tourism spills over into these areas as well.

14. Ministry of Natural Resources and Tourism. (2018). *The 2018 Tourism Statistical Bulletin.* United Republic of Tanzania. p. 36; Kenya National Bureau of Statistics. (2019). *Economic Survey 2019.* Nairobi: Kenya National Bureau of Statistics. https://www.knbs.or.ke/?wpdmpro=economic-survey-2019&wpdmdl=5243&ind=4kBTC7jtimzcAxu6PYiiqMRpHLnMJ1jSWs5EvyrKmAvAbR_FqgXXSIdusoAEI8k2.

15. Booking.com. (2020). "Hotels in Masai Mara National Reserve, Kenya." https://www.booking.com/region/ke/maasai-mara-national-reserve.html; Hotels.com. "Hotels in Serengeti National Park, Tanzania." https://www.hotels.com/de1688464/hotels-serengeti-national-park-tanzania/; In 2020, Tanzania's Ministry of Natural Resources

and Tourism put the number of licenses given for accommodations in the Serengeti at 199, but that includes 63 "mobile camps." Neither Booking.com nor Hotels.com include any mobile camps in their listings. Tanzania Ministry of Natural Resources and Tourism. (April 2020). "List of Licensed Tour Operators in Tanzania 2020." https://www.maliasili.go.tz/highlights/view/licensed-tour-operators-in-tanzania.

16. Martha Honey. (2008). "Kenya: The Ups and Downs of Africa's Ecotourism 'Mzee.'" *Ecotourism and Sustainable Development: Who Owns Paradise?* Island Press.

17. Booking.com. (2020). "Hotels in Masai Mara National Reserve, Kenya." https://www.booking.com/region/ke/maasai-mara-national-reserve.html.

18. Macharia Kamau. (October 7, 2018). "How the Masai Mara is sinking under Its Own Global Success." *Standard Digital.* https://www.standardmedia.co.ke/article/2001298146/masai-mara-suffers-from-its-own-success.

19. World Bank Group. (September 2015). *Tanzania's Tourism Futures Harnessing Natural Assets.* Environment and Natural Resources Global Practice Policy Note 95150. http://documents.worldbank.org/curated/en/204341467992501917/pdf/96150-REVISED-PN-P150523-PUBLIC-Box393206B.pdf.

20. Arne B. R. Witt, Sospeter Kiambi, Tim Beale, and Brian W. Van Wilgen. (2017). "A Preliminary Assessment of the Extent and Potential Impacts of Alien Plant Invasions in the Serengeti-Mara Ecosystem, East Africa." *Koedoe*, 59(1): a1426. https://doi.org/10.4102/koedoe.v59i1.1426.

21. Serengeti Watch. (2020.) "Threats Facing the Serengeti: Large Scale Development." https://serengetiwatch.org/large-scale-development/.

22. Richard Damania, Sébastien Desbureaux, Pasquale Lucio Scandizzo, Mehdi Mikou, Deepali Gohil, and Mohammed Said. (October 2019). *When Good Conservation Becomes Good Economics: Kenya's Vanishing Herds.* World Bank Group. Global Environmental Fund. http://documents.worldbank.org/curated/en/465881576053357383/When-Good-Conservation-becomes-Good-Economics-Kenya-s-Vanishing-Herds.

23. University of Copenhagen. (March 19, 2019). "When Development and Conservation Clash in the Serengeti." *ScienceDaily.* www.sciencedaily.com/releases/2019/03/190319112208.htm.

24. Serengeti Watch. (2020). "Threats Facing the Serengeti: The Serengeti Highway." https://serengetiwatch.org/serengeti-highway/#toggle-id-2.

25. Femke Broekhuis (August 13, 2018). "We Need to Limit Tourist Numbers to Save Cheetahs from Becoming an Endangered Species." *Quartz Africa.* https://qz.com/africa/1354703/tourist-maasai-mara-visits-threaten-cheetahs/.

26. T. Setsaas, L. Hunninck, C.R. Jackson, R. May, E. Røskaft. (October 2018). "The Impacts of Human Disturbances on the Behaviour and Population Structure of Impala (*Aepyceros melampus*) in the Serengeti ecosystem, Tanzania." ~~~I~~~Global Ecology and Conservation. %%%I%%%p. 16. https://www.sciencedirect.com/science/article/pii/S2351989418301306.

27. Femke Broekhuis. (August 13, 2018). "We Need to Limit Tourist Numbers to Save Cheetahs from Becoming an Endangered Species." *Quartz Africa.* https://qz.com/africa/1354703/tourist-maasai-mara-visits-threaten-cheetahs/.

28. Cathleen Steinbeiser. (November 28, 2016). "Impacts of Tourism in Serengeti National Park." SFS Life in the Field Blog. https://fieldstudies.org/2016/11/impacts-of-tourism-in-serengeti-national-park/

29. Noexpert. (September 24, 2019). "It's Not Really That Crowded Is It?" Tanza-

nia Forum. Tripadvisor. https://www.tripadvisor.com/ShowTopic-g293747-i9226-k12936570-Its_not_really_that_crowded_is_it-Tanzania.html.

30. Seronera. "Way Too Crowded in this Area." (March 18, 2013). Tripadvisor. https://www.tripadvisor.com/ShowUserReviews-g293751-d1593173-r154967997-Seronera-Serengeti_National_Park.html.

31. Macharia Kamau. (October 7, 2018). "How the Masai Mara Is Sinking under Its Own Global Success." *Standard Digital.* https://www.standardmedia.co.ke/article/2001298146/masai-mara-suffers-from-its-own-success.

32. Richard Damania, Sébastien Desbureaux, Pasquale Lucio Scandizzo, Mehdi Mikou, Deepali Gohil, and Mohammed Said. (October 2019). *When Good Conservation Becomes Good Economics: Kenya's Vanishing Herds.* World Bank Group. Global Environmental Fund. http://documents.worldbank.org/curated/en/465881576053357383/When-Good-Conservation-becomes-Good-Economics-Kenya-s-Vanishing-Herds.

33. J. O. Ogutu, H. P. Piepho, H. T. Dublin, N. Bhola, and R .S. Reid. (May 2009). "Dynamics of Mara-Serengeti Ungulates in Relation to Land Use Changes." *Journal of Zoology*, 278(1): 1–14. https://zslpublications.onlinelibrary.wiley.com/doi/10.1111/j.1469-7998.2008.00536.x.

34. Joseph Ogutu. (May 12, 2019). "Mara Wildebeest Migration Is Dying: Here's the Evidence." *The Star.* https://www.the-star.co.ke/news/2019-05-12-mara-wildebeest-migration-is-dying-heres-the-evidence/

35. Joseph Ogutu. (April 1, 2019). "People Are Taking a Huge Toll on the Plains of the Serengeti-Mara." *The Conversation.* http://theconversation.com/people-Mare-taking-a-huge-toll-on-the-plains-of-the-serengeti-mara-114389; J. O. Ogutu, H. P. Piepho, H. T. Dublin, N. Bhola, and R. S. Reid. (May 2009). "Dynamics of Mara-Serengeti Ungulates in Relation to Land Use Changes." *Journal of Zoology*, 278(1): 1–14. https://zslpublications.onlinelibrary.wiley.com/doi/10.1111/j.1469-7998.2008.00536.x.

36. Mette Løvschal, Peder Klith Bocher, Jeppe Pilgaard, Irene Amoke, Alice Odingo, Aggrey Thuo, and Jens-Christian Svenning. (January 25, 2017). "Fencing Bodes a Rapid Collapse of the Unique Greater Mara Ecosystem." *Scientific Reports*, 7(41450). https://www.nature.com/articles/srep41450.

37. Joseph O. Ogutu, Hans-Peter Piepho, Mohamed Y. Said, Gordon O. Ojwang, Lucy W. Njino, Shem C. Kifugo, and Patrick W. Wargute. (September 27, 2016). "Extreme Wildlife Declines and Concurrent Increase in Livestock Numbers in Kenya: What Are the Causes?" PLOS ONE. https://doi.org/10.1371/journal.pone.0163249.

38. Massai Mara Wildlife Conservancies Association. (2020). "Why Mara Conservancies." https://www.maraconservancies.org/why-mara-conservancies/.

39. In Kenya, there are around 160 conservancies spread across 28 counties, all under the umbrella of Kenya Wildlife Conservancies Association.

40. Femke Broekhuis. (August 13, 2018). "We Need to Limit Tourist Numbers to Save Cheetahs from Becoming an Endangered Species." *Quartz Africa.* https://qz.com/africa/1354703/tourist-maasai-mara-visits-threaten-cheetahs/.

41. Richard Damania, Sébastien Desbureaux, Pasquale Lucio Scandizzo, Mehdi Mikou, Deepali Gohil, and Mohammed Said. (October 2019). *When Good Conservation Becomes Good Economics: Kenya's Vanishing Herds.* World Bank Group. Global Environmental Fund. http://documents.worldbank.org/curated/en/465881576053357383/When-Good-Conservation-becomes-Good-Economics-Kenya-s-Vanishing-Herds.

42. Katare Mbashiru. (February 6, 2020). "Tanzania: Hunting Guidelines in Offing-Kanyasu." *Tanzania Daily News*. https://allafrica.com/stories/202002070476.html.

43. Richard Damania, Sébastien Desbureaux, Pasquale Lucio Scandizzo, Mehdi Mikou, Deepali Gohil, and Mohammed Said. (October 2019). *When Good Conservation Becomes Good Economics: Kenya's Vanishing Herds*. World Bank Group. Global Environmental Fund. http://documents.worldbank.org/curated/en/465881576053357383/When-Good-Conservation-becomes-Good-Economics-Kenya-s-Vanishing-Herds.

44. Femke Broekhuis. (August 13, 2018). "We Need to Limit Tourist Numbers to Save Cheetahs from Becoming an Endangered Species." *Quartz Africa*. https://qz.com/africa/1354703/tourist-maasai-mara-visits-threaten-cheetahs/.

45. Ibid.

46. Serengeti Watch. (2016). Personal communication with author.

47. Richard Damania, Sébastien Desbureaux, Pasquale Lucio Scandizzo, Mehdi Mikou, Deepali Gohil, and Mohammed Said. (October 2019). *When Good Conservation Becomes Good Economics: Kenya's Vanishing Herds*. World Bank Group. Global Environmental Fund. http://documents.worldbank.org/curated/en/465881576053357383/When-Good-Conservation-becomes-Good-Economics-Kenya-s-Vanishing-Herds.

48. Jacques Morisset. (January 2015). *The Elephant in the Room: Unlocking the Potential of the Tourism Industry for Tanzanians*. The World Bank. https://www.worldbank.org/en/country/tanzania/publication/tanzania-economic-update-increasing-tourism-for-economic-growth.

49. Adam Ihucha. (January 26, 2020). "Tanzania Plans Public Private Tourism Partnership." *eTN Tanzania*. https://www.eturbonews.com/540972/tanzania-plans-public-private-tourism-partnership/.

50. Jacques Morisset. (January 2015). *The Elephant in the Room: Unlocking the Potential of the Tourism Industry for Tanzanians*. The World Bank. https://www.worldbank.org/en/country/tanzania/publication/tanzania-economic-update-increasing-tourism-for-economic-growth.

51. Ministry of Natural Resources and Tourism. (2020). *The 2017 Tourism Statistical Bulletin*. http://hat-tz.org/hattzorg/wp-content/uploads/2018/09/Statistical-Bulletin-Booklet.pdf.

52. XinhuaNet. "Tanzania's Conservation Authority in New Drive to Attract More Tourists." (November 22, 2019). http://www.xinhuanet.com/english/2019-11/22/c_138573601.htm; Zephania Ubwan. (August 5, 2019). "Chinese Investor Plans 300-Room Hotel." *The Citizen*. https://www.thecitizen.co.tz/news/-Chinese-investor-plans-300-room-hotel/1840340-5224484-dc7e2az/index.html.

53. World Bank Group. (September 2015). *Tanzania's Tourism Futures Harnessing Natural Assets*. Environment and Natural Resources Global Practice Policy Note 95150. http://documents.worldbank.org/curated/en/204341467992501917/pdf/96150-REVISED-PN-P150523-PUBLIC-Box393206B.pdf.

54. Friends of Serengeti. (2020). www.friendsofserengeti.org.

Chapter 4.1

Overtourism at World Heritage Sites

By Martha Honey

By mid-2019, the United Nations Educational, Scientific and Cultural Organization (UNESCO) had officially inscribed on the World Heritage Site (WHS) list 1,121 natural and cultural properties in 167 countries around the world.[1] The WHS designation means that these are "places on Earth that are of Outstanding Universal Value to humanity and as such . . . [should] be protected for future generations to appreciate and enjoy."[2] World Heritage designation can bring significant international attention and assistance, as well as a growth in tourism, thereby contributing positively to economic development. Tourism can, however, if not well controlled and managed, negatively impact the site's natural and built environment, as well as the surrounding area, including the host communities.

Take, for instance, the Old City (within the city walls) of Dubrovnik, Croatia, which was inscribed as a World Heritage Site in 1979, becoming Croatia's first UNESCO cultural designated site. That same year, Plitvice Lakes National Park was designated as Croatia's first natural heritage site. For several decades afterward, Dubrovnik's international designation failed to boost tourism. In fact, the regional conflicts in the 1980s and 1990s led to a decline in tourism in Dubrovnik, as well as all of Croatia. As the *Washington Post* reported in a 1994 travel piece,

"Dubrovnik's streets were filled almost entirely with local people—the rare exception being a uniformed United Nations peace-keeper from Denmark or elsewhere."[3]

After 1995, as peace settled in, Dubrovnik's tourism tide turned, doubling from 606,000 overnight visitors in 2011 to 1.2 million in 2018—twenty-nine times more than the city's resident population of 42,000. [4] This figure did not include cruise passengers, who were arriving in increasing numbers, peaking at 1.1 million in 2013. In addition, Dubrovnik's tourism numbers have been bolstered in recent years by another phenomenon: the fantasy TV blockbuster series *Game of Thrones* that hit the small screen in 2011. Dubrovnik was one of the main filming locations for the TV series, standing in as the capital of the Seven Kingdoms. It's estimated that *Game of Thrones* attracted more than 244,000 tourists to the city between 2012 and 2015.[5] Tourist guide Ivan Vukovic calls Dubrovnik's fame both a "blessing and a curse," adding that "it's an amazing thing because [tourism] brings us so many people, but now we do not know what to do with those people."[6]

Dubrovnik is hardly alone. As the chapters in this book attest, a number of UNESCO World Heritage Sites are experiencing overtourism. UNESCO's 2018 State of Conservation (SOC) reports, which contain conservation assessments of 157 World Heritage Sites, found that almost one-third were facing challenges associated with "impacts of tourism/visitor/recreation," or "major visitor accommodation and associated infrastructure."[7] Although the UNESCO designation often contributes to destinations' tourism popularity, it is typically a combination of factors, as Dubrovnik illustrates, that bring these unique places to the overtourism tipping point. And WHS status also brings a special urgency to deal with overtourism, as it designates places being of "Outstanding Universal Value to humanity." UNESCO also provides some tools for helping these sites tackle overtourism.

UNESCO's World Heritage Convention, Tourism, and Overtourism

In 1972, the UNESCO General Conference adopted the World Heritage Convention, and in 1978, the first twelve sites were inscribed on the World Heritage List; by 2020, the number had grown more than ninety-fold, to 1,121. Of these, 869 are cultural, 213 are natural, and 39 are mixed properties. This total also includes 39 properties that are transboundary and 53 that are on the World Heritage in Danger list (see details below). This World Heritage list has always been weighted toward Europe and North America; at present, 47 percent of the sites are located in these two regions.[8]

Today, 193 countries, known as state parties, adhere to the World Heritage Convention and pledge to protect their natural and cultural heritage. Of these, 167 have one or more World Heritage Sites; the rest do not have any. To be included on the World Heritage List, sites must be of Outstanding Universal Value (OUV), which, according to UNESCO, means that they have "cultural and/or natural significance which is so exceptional as to transcend national boundaries and to be of common importance for present and future generations of all humanity."[9] To be considered of OUV, a property must meet one or more of the ten World Heritage Criteria. This list of cultural and natural criteria includes, for example, that sites "represent a masterpiece of human creative genius," "bear a unique or at least exceptional testimony to a cultural tradition or to a civilization which is living or which has disappeared," and "contain superlative natural phenomena or areas of exceptional natural beauty and aesthetic importance."[10] These criteria, defined by the World Heritage Committee, the main decision-making body, are detailed in the *Operational Guidelines for the Implementation of the World Heritage Convention*, which, together with the convention itself, are the main working documents on World Heritage.[11] The criteria are regularly revised by the committee "to reflect new concepts, knowledge or experiences."[12]

Inscription Process

Only countries that have ratified the World Heritage Convention can submit nomination proposals for properties on their territories to be considered for the World Heritage List.[13] The lengthy nomination process can take multiple years to complete, depending on the complexity of the site and the quality of the local team developing the inscription application. Ideally, a strong local coalition is involved in the application progress.

One successful case is the English Lake District and Cumbria, which was inscribed as a World Heritage Site in 2017 following a local initiative with a strategic focus on sustainable tourism. Its campaign for inscription included a coalition of local businesses and organizations, district councils, and park authority. This partnership developed a new English Lake District World Heritage Site brand that enables tourism businesses and other sectors from across the region to be associated with "an internationally recognised badge of quality and specialness."[14]

WHS inscription is not immune to political and economic pressures, however. A study of Nigeria's "Sacred Grove" in Osun-Osogbo demonstrates that "lobbying, combined with political and economic interests, played a central role in its 2005 listing." According to the study, "the inclusion of the Sacred Grove of Osun-Osogbo on the World Heritage list is the result of nearly 15 years of efforts on the part of Osun state to give itself cultural and historical legitimacy."[15]

World Heritage Funding and Role of Tourism

The World Heritage Committee grants international assistance under the World Heritage Fund, which contains compulsory and voluntary contributions from the state parties.[16] Each state party is required to contribute a fixed amount to the fund annually, based on the party's gross domestic product. The fund is modest, having about US$4 million a year, which is used to support activities for specific sites based on requests from state parties. The fund is also used for monitoring and evaluating sites and can be used, if requested, to help underwrite the costs of the nomination process for new sites. In addition, countries can make voluntary contributions into a trust fund earmarked for specific purposes. Japan, for instance, has contributed donations to UNESCO's Silk Road project in Asia.

Because funds available from UNESCO are limited, tourism and international development assistance are the key tools for providing resources to protect World Heritage Sites. As the 2016 book, *The History of UNESCO*, explained, WHS designation offers not only "increased protection, enhanced political prestige and public awareness," but also "economic development through international aid and tourism expenditure."[17] Similarly, in a 2018 issue brief published by the International

Studies Program at Old Dominion University, Talor Stone wrote, "A sizable lobbying industry has grown around the awards because World Heritage listing has the potential to significantly increase tourism revenue in connection to the sites selected. Inscription onto the list has the potential to generate millions of visitors to the site every year resulting in massive financial flows caused by tourism."[18] Stone stated further, "In many cases, [overtourism] has led to the exploitation of sensitive sites for their income generating potential at the expense of their protection and preservation."[19]

Causes and Impacts of Overtourism

Although designation as a World Heritage Site can bring increased visitation, more jobs, and greater spending and investment to the site and its host country, the increase in tourism can also cause stress on built infrastructure, create disruptions to cultural heritage, change urban landscapes, and damage the natural environment. "The very reasons why a property is chosen for inscription on the World Heritage List are also the reasons why millions of tourists flock to those sites year after year," wrote Francesco Bandarin, former director of the UNESCO World Heritage Centre.[20] The tipping point between sustainable tourism and overtourism varies from one WHS to another, depending on the site's geographical location, economic capacity, physical condition, and management.

World Heritage Site Designation

In some instances, the simple designation as a WHS appears to trigger, as happened in George Town, Malaysia, "a tide of invasive tourism."[21] The recognition of UNESCO World Heritage status in 2008 was intended to preserve from decay, destruction, and sale to developers the remaining historic Chinese "clan jetties" on the outskirts of George Town on Penang Island. The site's OUV is as the "last intact bastions of Malaysia's old Chinese settlements" in this once bustling seafront trading community. After the WHS designation, George Town's clan jetties—long dismissed as a gambling-ridden, squatters' slum—were suddenly a top tourism attraction.

According to an article in *The Guardian*, with WHS designation George Town "experienced a rebirth as a tourist haven" and by 2017 was receiving six to seven million stayover tourists each year. "Historic homes are now commercial stalls branded with neon signs; one-time fishermen peddle T-shirts, magnets and postcards; tour buses deposit vacationers from early in the morning until well after sunset." The article continued, "The daily intrusion has clearly taken a toll: windows are boarded, 'no photo' signs are pervasive, and tenants quickly vanish at the sight of a foreign face."[22]

Local jetty resident Chew Siew sees WHS status as a mixed blessing. "We would be gone today if not for the UNESCO listing," she said. But although it may have spared the jetties from total disrepair and eventual sale to developers, it has also "affected our privacy. Our jetty has become commercialised. People are moving." She adds that during holidays "it's not even a place to live."[23]

In another case, WHS designation in 1997 triggered rapid tourism development and gentrification in Panama City's historic Casco Viejo district, once a run-down neighborhood on the edge of the capital. In the wake of WHS designation, poor residents were forcibly evicted, and the center of the historic district was transformed into a tourism zone, owned largely by wealthy foreigners. As Chloé Maurel wrote, "Tourism in Panama City has increased exponentially since the heritage listing, homogenising the urban landscape and exacerbating inequalities."[24] In her assessment, the "issues of heritage are closely linked to economic, social and political issues, and result in power disparities."[25]

Cluster of Causes

Although World Heritage designation has led to a tourism explosion in some instances, in other cases, including in a number of European cities with World Heritage Sites, overtourism is triggered by a cluster of factors (as detailed throughout this book). They can include, as the World Heritage Committee wrote in 2018:

> The development of new technologies, increasing investment in tourism infrastructure, the growth of peer-to-peer and shared usage platforms, low cost carriers, and larger airplanes and cruise ships is resulting in the transformation of the tourism marketplace and

> bringing an increasing number of visitors, and therefore increased pressures and threats, to World Heritage properties. New international source markets are emerging with increasing numbers of tourists travelling internationally. Domestic tourism is also increasing in many destinations.[26]

The roots of overtourism in Europe can be traced to a certain extent to the economic recession that began in 2008. The recession led to two key outcomes. First, governments turned to tourism to revitalize their economies, and their destination marketing organizations pumped money into heavy marketing campaigns to attract international visitors. The emphasis was on increased numbers rather than increased spending or longer stays, with little attention paid to managing the incoming tourism tidal wave. Second, the recession coincided with the introduction of low-cost airlines across Europe. Within a few years, a number of World Heritage Site cities in Europe— Prague, Krakow, Amsterdam, Barcelona, Venice, Santorini, Florence, Paris, Riga (Latvia)— were suddenly seeing big tourism numbers. As travel became more affordable and accessible, even farther away places with World Heritage Sites, like Iceland and Thailand, saw their tourism skyrocket as they undertook international marketing campaigns and were connected by budget airlines. In turn, the cheap flights helped set off other trends, most importantly the rapid rise of short-term, household-based vacation rentals and most notably Airbnb.

Bucket Lists

In 1997, a Dutch woman set up Worldheritagesite.org, a portal for "WHS-baggers," and began traveling to World Heritage Sites around the world. "We do it because it's a list of things you can tick off, like bird-watching or plane-spotting," she emailed while bagging her 310th site.[27] Today, the market is crowded with a growing number of travel companies capitalizing on the bucket list craze. A quick Internet search turned up the following headlines:

- "10 UNESCO World Heritage Sites to Add to Your Travel Bucket List"[28]
- "25 UNESCO Sites to Put on Your Bucket List"[29]

- "7 UNESCO World Heritage Sites for Every Travel Bucket List"[30]
- "Jetsetter's 9 UNESCO World Heritage cities for your bucket list"[31]
- "New UNESCO World Heritage Sites You Can Visit."[32]
- "Top Historical Places: 10 UNESCO World Heritage Sites Around the World"[33]

The lists go on, compiled and promoted by both tour operators and the travel media. It appears likely that more and more companies will make itineraries of World Heritage Sites, which will, in turn, increase impacts, especially on sites where there are not effective management plans.

Lack of Good Management

Although overtourism in World Heritage Sites has arisen due to a variety of factors, the most common challenge is a lack of or insufficient management planning, which can both cause and accelerate negative impacts. Although sites must have conservation management plans, they are not required to have tourism management plans, either as part of their conservation plan or as a separate document. Tourism management plans outline sustainable tourism development goals, such as monitoring the number of visitor arrivals and identifying specific areas zoned for tourists, that are the foundation of the UNESCO organization.[34]

A 2017 study of 229 natural World Heritage Sites, commissioned by UNESCO and conducted by Griffith University in Australia, surveyed the nature of management planning at these sites. The study found that for 105 sites (46 percent), the "research team could not locate a clearly accessible and in-date [current] tourism plan, either as part of a general management or a stand-alone tourism plan."[35] Of these, 18 sites (8 percent) had management plans that had expired. Only 11 sites (5 percent) had public, up-to-date, stand-alone tourism plans that were not part of larger general management plans. The study noted that "newer/more recently inscribed World Heritage sites were more likely to have extensive planning or at least moderate tourism planning, compared with sites that had been listed a very long time ago" and concluded that "possibly this reflects the more recent nature of the tourism growth phenomenon, or the greater expectations associated with tourism planning in more recent years."[36]

Dan Peltier, in a *Skift* analysis of this study, noted, "Cultural sites, such as Vatican City or the Pyramids of Giza, weren't part of the Griffith University study and the number of UNESCO sites with tourism management plans would likely be higher if those sites were included in the study." In either case, he concluded, "UNESCO sites—of all places—should be the destinations taking leadership roles in tourism management planning as they're often the first place many tourists think about when deciding what to do on a trip, But data show that's far from reality for a variety of political and economic reasons."[37]

Solutions: Strengthening Current Policies and Programs

UNESCO World Heritage Sites have developed, particularly in this era of overtourism, a dual personality. On the one hand, these iconic sites "are deemed deserving of the ultimate layer of protection—to be placed beyond the reach of polluters, developers, looters, bombers, and the ravages of time. The World Heritage seal is a guarantee of preservation," wrote Simon Usborne in *The Independent*.[38]

As demand for protection has increased and the World Heritage Center's budget and staff have both been cut, however, preservation has become more difficult and complex. As a result, Usborne contends that "many within the conservation community" believe that UNESCO has become a "moribund organisation . . . teetering on its once sound foundations as its principles and priorities crumble under the weight of bureaucracy and outside influence."[39] The takeaway from these two opposite-side-of-the-coin images is that UNESCO Heritage Sites policies and practices need to be reformed and revitalized, and now more urgently than ever. The following are some of the most essential steps to strengthen and expand upon current UNESCO sustainable tourism initiatives and effectively address the impacts of overtourism.

Reporting Requirements

Governments (state parties) are required to report to the World Heritage Committee on the state of conservation and the various management and protection measures in place at inscribed sites. These reports allow the committee to assess conditions at the sites and, if necessary, decide on the specific measures needed to resolve any issues that are impacting OUVs.

One category of reporting is reactive monitoring via SOC reports for sites where an event or action threatens or could threaten the site's OUV. Based on a common template, SOC reports must include evidence that the site is working to correct the problem. The World Heritage Committee then reviews each SOC report and takes actions that range from requesting more information, to sending a mission to evaluate the site, to determining that the site is not seriously deteriorated and no further action is needed, to deciding the site should remain on the World Heritage List but take certain mitigation measures, to putting the site on the World Heritage in Danger list (see next section), or, if mitigation is deemed not possible, to removing the site from the World Heritage List. (Removal has happened just twice in forty years.) These SOC reports and the World Heritage Committee decisions are part of the rich "trove of reliable data on the state of conservation of World Heritage properties [compiled] since 1979" and are "one of the most comprehensive monitoring systems of any international convention."[40]

The 2018 SOC reports, for instance, contain conservation assessments of 157 World Heritage Sites, including the Galapagos Islands, Stonehenge, Grand Canyon National Park, and Dubrovnik, as well as all sites on the World Heritage in Danger list. In assessing these SOC reports for 2018, the World Heritage Committee identified "64 different factors affecting these properties," including management systems and management plans, which topped the list (74 percent), housing (32 percent), and illegal activities (25 percent). It also found that 22 percent of the sites were experiencing negative "impacts of tourism" and that 14 percent faced "major visitor accommodation and associated infrastructure" problems.[41]

Although tourism-related problems are significant, Peltier noted in a *Skift* article that the 2018 SOC reports contain only a few of the WHS destinations in Europe that are most often listed in the media and academic studies as suffering from overtourism. Peltier wrote that "sites that are dealing with overtourism in Europe, such as Venice, Pompeii, and Barcelona's Park Güell, are mostly absent from UNESCO's 2018 list of [State of Conservation] reports."[42] (However, as described below, by 2020 the World Heritage Committee was actively considering putting Venice on the World Heritage in Danger list.) Peltier's observation reflects the more restrictive measurements of overtourism used by the World Heritage Committee than by other organizations or the media. It also raises the question of whether the committee needs a new category and legal instruments specifically designed to assess and address overtourism. (See discussion below.)

There are also calls for UNESCO to make the reporting requirements for World Heritage Sites more frequent and rigorous. As the Griffith University study stated, "The periodic reporting may not be specific enough and only occurs every six years. More frequent reporting could be beneficial."[43]

World Heritage Sites in Danger and UNESCO's Definition of Overtourism

The World Heritage Sites that are deemed most at risk are identified by UNESCO on the list of World Heritage Sites in Danger. According to the UNESCO World Heritage Center website, "Armed conflict and war, earthquakes and other natural disasters, pollution, poaching, uncontrolled urbanization and *unchecked tourist development* pose major problems to World Heritage sites" (italics added).[44] In addition, a common thread running across most of the endangered sites is a lack of proper management.

The World Heritage Sites in Danger list is designed to draw international attention to a site where one or more of the OUV of the property are threatened and to encourage good management. An endangered listing is not intended to be punitive, but rather to help attract assistance, expertise, support, and money to fix the problems. Typically, although not exclusively, sites on the World Heritage in Danger list are in the developing world. By 2020, there were fifty-three World Heritage Sites on the list of World Heritage Sites in Danger, up from fewer than ten sites between 1972 (when the list was created) and 1992.

Dubrovnik is one European city that has been on the World Heritage Sites in Danger list. In 1991, the walled city was declared endangered after the regional conflict that "threatened to destroy monuments that had withstood the passing of centuries as well as several earthquakes."[45]

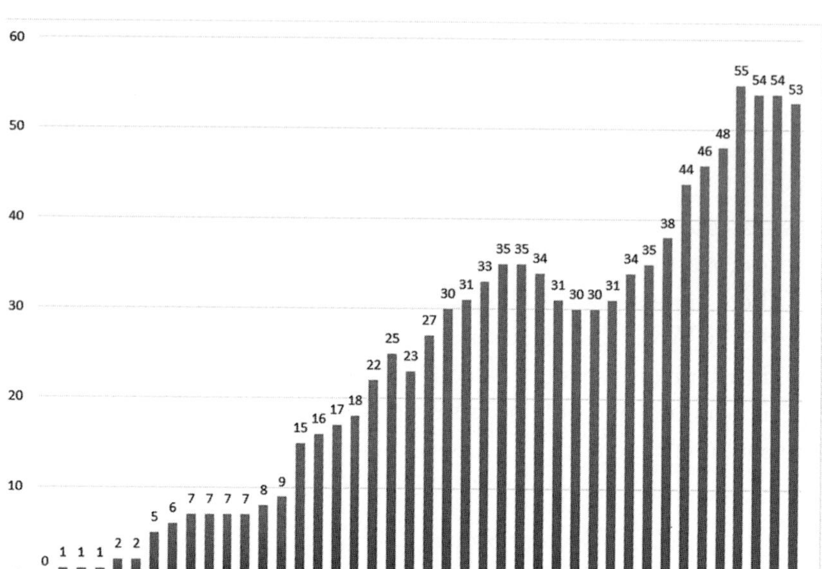

World Heritage Sites in Danger, 1979–2018. Source: Prepared by CREST with data from UNESCO.

Then, in 1998, Dubrovnik was officially removed from the list after the city government, with financial and expert support from the World Heritage Committee and UNESCO, had restored its Franciscan and Dominican cloisters and rebuilt palaces.[46] "That's how the system is meant to work," wrote Usborne in *The Independent*. "Since its inception, . . . UNESCO World Heritage has become a global brand whose seal is slapped on the planet's most precious places."[47]

In terms of negative impacts from tourism, the assessment process is shaped by the World Heritage Committee's evaluation of what puts a site's OUVs at risk. Because the focus is on the criteria for inscription, infrastructure—damage to built or natural structures— is emphasized rather than either excessive arrival numbers or softer societal issues such as deteriorating quality of life for residents or the visitor experience. As Peltier wrote, UNESCO's World Heritage Committee doesn't view "the risk from crowds as the most problematic" to OUVs, and therefore its "latest Endangered List leaves out many sites plagued by overtourism."[48]

Despite overtourism, Dubrovnik has dodged the endangered list in recent years, but UNESCO is currently studying whether to add Plitvice Lakes National Park, also in Croatia, to the list. The local community has been under threat from both "overwhelming numbers of visitors"[49]—a twelve-fold increase between 2010 and 2017—and the irresponsible construction of holiday homes and restaurants. According to the World Wildlife Fund, "Due to a series of wrong decisions, Croatia's most precious natural pearl is critically endangered! The uncontrolled increase of tourists visiting the park and irresponsible construction projects in the area are putting at risk Plitvice's unique waters and nature" and could "cause its deletion from the UNESCO World Heritage List."[50] In fact, that is not likely to happen: only two sites have ever been removed from the list.[51]

UNESCO's deliberative process is focused on the impacts of tourism-related construction at the Plitvice National Park rather on uncontrolled tourism numbers. As the World Heritage Center's 2018 SOC report on Plitvice stated, "The Committee considered that the significant and unsustainable expansion of tourist facilities inside the property with associated visual impacts as well as pressures on the property's sensitive hydrogeology presented a potential danger to its OUV. It further noted that in the absence of substantial progress in addressing these threats the possible inscription of the property on the List of World Heritage in Danger would be considered."[52] Plitvice's inscription on the list, if it happens, is years away, however. UNESCO has identified the problems and will give Croatia one to two years to correct the situation, followed by another monitoring and observation report. It is typically up to three years between when an issue is identified and when it is put on the endangered list.

In practice, countries work hard not to be put on the list because it often has a negative effect on tourism. In 2014 and 2015, for instance, the Australian government engaged in "frantic diplomatic efforts" to stop UNESCO from adding the Great Barrier Reef to the World Heritage in Danger list. The Great Barrier Reef, off Australia's northeast coast, is home of the world's largest collection of coral reefs and generates an estimated $56 billion in tourism revenue annually.[53]

UNESCO's immediate concern was the Australian government's plans to open the gigantic new Carmichael coal mine in Queensland

and expand the coal terminal at the Abbot Point port where the dredging activity falls within the Great Barrier Reef World Heritage Area.[54] The project, as approved by Australia's conservative government, would allow dredging within the Great Barrier Reef, ship millions of tons of coal through reef waterways, and permit dumping on the reef of the "spoil" or sludge from the port dredging project. Before this controversy arose, UNESCO had already expressed concerns of declining conditions on the reef—which has lost 50 percent of its coral cover since the mid-1980s—due to pollution, the strongest El Niño in decades, a plague of coral-eating starfish, and most importantly, the impacts of climate change, including acidification and warming waters.[55] The Australian government feared that the endangered listing would be both a political embarrassment and an economic blow to its highly lucrative tourism industry based on the Great Barrier Reef.[56]

Leading environmental organizations, including Greenpeace Australia Pacific, urged the World Heritage Committee to put the Great Barrier Reef on the endangered list so as to force the Australian government to halt the coal project and to act on climate change.[57] The government's lobbying efforts appeared to prevail, however. In 2017, the World Heritage Committee decided to accept Australia's revised management plan and not to put the Great Barrier Reef on its endangered list; at the same time, the committee expressed "serious concern" about the health of the reef.

Then, in September 2019, the state of Queensland approved the groundwater plan for the Carmichael mine, thereby removing the last legal hurtle to opening the coal mine. When opened, the Carmichael mine, one of the largest in the world, is expected to extract up to sixty million tons of coal a year ,which will be shipped mainly to India.[58]

Some leading cultural sites have also managed to evade a World Heritage in Danger listing. In Europe, Venice, a World Heritage Site since 1987, is widely viewed as the poster child for overtourism, yet the city has managed to stay off the endangered list. Today, tourism dominates all aspects of life in Venice, with an estimated 22 million visitors annually generating about $2 billion in revenue. In contrast, its resident population is only about 55,000, down from 175,000 in 1951.[59] By 2017, according to *The Guardian*, a thousand residents were moving out of

Crowds in the wake of a cruise ship arrival, St. Mark's Plaza, Venice, Italy. Source: fdecomite (Flickr).

Venice to the mainland each year, "unable to afford rapacious rents or find a niche beyond tourism."[60]

Although overtourism in Venice seems obvious to residents, tourists, civic and environmental organizations, and the media, the World Heritage Committee has not placed the iconic Italian canal city on the endangered list, in part because its process demands a more stringent definition of overtourism. It requires scientific evidence that the OUV of the city's architecture and infrastructure and lagoon—not just residents' quality of life—is being impacted by tourism.

Over the years, the Italian national and municipal governments have managed to keep Venice off the endangered list by producing a stream of studies and plans for dealing with cruise tourism, visitor congestion and flows, and other tourism issues. Then, in late 2019, following the city's worst flooding in recorded history, UNESCO declared that the city was "highly vulnerable" to the twin threats of overtourism and climate change (see chapter 2.1).[61] The World Heritage Committee identified what it sees as the main overtourism issues, including the lack of an effective management system and the damage caused by cruise ships that navigate in and out of the Venetian Lagoon. It further requested a report on the "short term outcomes achieved by" Venice's project for tourism governance, which is built on the World Heritage Centre's Sustainable Tourism Programme (see below).[62] The committee pledged to revisit the topic of the endangered list "with a view to considering the inscription" of Venice "if the implemented mitigation measures and the adapted management system does not result in significant and measurable progress in the state of conservation of the property."[63]

Other World Heritage Sites that are clearly not effectively addressing impacts from overtourism, including Angkor Wat in Cambodia, Mount Everest in Nepal, and the Galapagos Islands in Ecuador (see chapter 5.2), are not currently on the endangered list.

Although some critics fault UNESCO for not moving more quickly to inscribe iconic sites on the World Heritage in Danger list, often—although not always—the threat of inclusion on the list and the World Heritage Committee's consultative process can bring reforms. But given the multidimensional impacts of overtourism, it seems that the committee does need to broaden its definition beyond narrowly assessing the impacts on the built and natural assets. It would appear to be beneficial if the committee created an additional category and legal tools to evaluate and address the impacts of overtourism.

Ensuring Effective Tourism Management

There is a need to reform the World Heritage Convention's operational guidelines to require that sites produce and maintain up-to-date tourism management plans, either as stand-alone documents or as part of their overall management plan. "Tourism is growing at a fast pace and visitation to World Heritage Sites is increasing, leading to a wide recognition of the need to manage visitors," according to the 2017 Griffith University study of natural World Heritage Areas.[64] Proper management must be based on data such as visitor trends, arrival numbers, visitor nights, and tickets sold. The Griffith University study found a relatively low level of discussion of the need for assessing visitor carrying capacity or surveying visitor trends, behaviors, or perceptions, all critical for measuring impacts and improving visitor experience. In addition, the study concluded, "at present, even those

[World Heritage Sites] that undertake some form of monitoring use different approaches and methods" in collecting visitor information. It seems imperative that the World Heritage Centre create a common methodology and template for data collection that is applied across all sites.

For sites that are experiencing overtourism, there is a need to put in place measures to limit impacts of visitation by, for instance, encouraging visitation in nonpeak seasons, adjusting entrance fees, regulating the supply of accommodation and parking spaces, adding visitor centers, limiting souvenir shops, and requiring advance bookings. In addition, sites could charge visitors a higher price for shorter stays and could limit the number of visitors per day, based on sound visitor data and a solid management plan. In Dubrovnik, for instance, cruise passenger arrivals peaked at 1.1 million in 2013, but by 2018, passenger arrivals had dropped by nearly half, to 669,300, following measures spearheaded by the mayor, who worked with the cruise lines to set limits on the number of ships and passengers in port at any given time.[65]

The Mogoa Caves, located along the ancient Silk Road in Gansu Province, China, are a success story in terms of visitor management. Composing the world's largest treasure house of Buddhist art, the caves, also known as the Thousand Buddha Grottoes, became China's first World Heritage Site in 1987. Although they had "withstood hundreds of years of natural disasters, bandits, explorers, and armed conflict," in recent years they became "subject to the ravages of a far more insidious threat: modern tourism."[66] Tourism to the site soared to over a half million annually, stimulated by the WHS brand, improved transportation, and the Chinese government's "One Belt, One Road" initiative to heavily market Silk Road attractions for national and international tourists.

Beginning in 2014, the Dunhuang Academy, with help from international experts from the Getty Institute, developed and introduced new measures to control daily visitor volumes and prevent damage to the caves' Buddhist frescoes. Visitors are now required to make advance reservations, go first to a visitor center to see a film that includes information formerly left to guides to cover once in the caves, and are permitted to spend only five minutes in the caves. No flashlights or camera flashes are permitted, and there is only low-level lighting in the caves. It's working: carbon dioxide levels in caves are down, and bright light flashes are no longer damaging the frescos.

Finally, there is currently no World Heritage database for tourism or general management plans. As the Griffith University study recommends, one should be created.[67]

This type of database could showcase case studies of how individual World Heritage Sites are addressing overtourism. In turn, such a database would increase information exchange among World Heritage Sites, facilitate use of best practices, and provide information to academics, nongovernmental organizations, the media, and the public about policies, strategies, and technologies for addressing overtourism.

Enhancing Sustainable Tourism

Since the early 2010s, the World Heritage Centre has promoted sustainable tourism, including playing an active role in the United Nations 2017 International Year of Sustainable Tourism for Development. Earlier, in 2010, the World Heritage Committee requested UNESCO to redevelop its Sustainable Tourism Programme in recognition that preservation and protection of World Heritage Sites need to be managed based on the principles and best practices of responsible tourism. As a World Heritage Policy Compendium document stated, "If undertaken responsibly, tourism can be a driver for preservation and conservation of cultural and natural heritage and a vehicle for sustainable development. But if unplanned or not properly managed, tourism can be socially, culturally and economically disruptive, and have a devastating effect on fragile environments and local communities."[68] The Sustainable Tourism Programme seeks to "contribute a new paradigm" that takes "a holistic and strategic approach to World Heritage properties and destinations that will include bottom-up as well as top-down measures to ensure sustainability that reflects not only high-level goals but also local needs and the ability to attain these goals."[69] It has four focus areas: (1) policy and strategy, (2) tools and guidance, (3) capacity building, and (4) Heritage journeys.[70]

The Sustainable Tourism Programme is financed with extrabudgetary funds raised from governments and the private sector. These funds are used for site-specific projects, as well as systems-wide work that creates tools and strategies to be used for all sites. For instance, at present the World Heritage Centre is working with both Ha Long Bay in Vietnam and World Heritage Sites in Uzbekistan to develop management plans that integrate a sustainable tourism development perspective. At the same time, the Sustainable Tourism Programme is developing a range of tool kits that can be used by WHS destinations around the world. Under development is a "How to Guide" tool kit that includes "best practice approaches for sustainable economic development at WHSs through tourism."[71] A handful of World Heritage sites, including the Rice Terraces in Bali and Fortress of Suomenlinna on islands off the coast of Helsinki, Finland, have used the UNESCO tools to develop sustainable tourism strategies.

In addition, in 2018, the World Heritage Centre and National Geographic, with a grant from the European Union, launched World Heritage Journeys, a collection of vacation packages that are showcased on the VisitWorldHeritage.com website. The site features "unique and authentic World Heritage travel experiences recommended by local experts and curated by National Geographic."[72]

The website, which encourages visitors to understand the significance of each site and to travel deeper, travel differently, and stay longer, initially featured thirty-four World Heritage destinations in Europe. It has since added Buddhist cultural and historic sites and will expand to incorporate other sites in Southeast Asia. Eventually, it will feature WHS properties and destinations around the globe.

Some heritage experts say the goal of sustainability is still more aspirational than practical, however. "Currently UNESCO has no clear guidelines or effective methods to control the commercialisation of world heritage sites, and its talk on sustainability is more a verbal exercise than enforceable," said Clement Liang, a member of the Penang Heritage Trust in Malaysia who lobbied for the George Town clan jetties to be designated a WHS.[73]

The World Heritage Centre is, however, considering the possibility of creating a voluntary eco-certification program or at least a set of uniform sustainability guidelines for World Heritage Sites, which would build on existing management and assessment tools. It would be a welcome development, especially if it conforms to the standards of the Global Sustainable Tourism Council, which manages the global baseline criteria for sustainable tourism and acts as the international accreditation body for sustainable tourism certification programs.

Conclusion

Filmmaker Ken Burns has hailed the national parks in the United States as "America's best idea." Similarly, UNESCO's World Heritage Sites are widely recognized as a crowning achievement of the UN's global community. Many of these thousand-plus ironic, irreplaceable treasures would not have survived without the WHS designation and the support the UNESCO brand has generated, including from a rising tide of tourism.

These natural and cultural sites, many located in poor and unstable regions, have long faced an array of chronic challenges, and now, for a growing number, curbing overtourism has been added to the mix. Although UNESCO's World Heritage Centre has rich institutional experience and tools in place and in development that can help address overtourism, the institution itself is limited by insufficient funding, slow bureaucratic processes, and a narrow mandate for assessing and responding to overtourism. In addition, nearly half the World Heritage Sites do not have tourism management plans to guide their decision-making and operations. The World Heritage Centre's authority is also constrained by competing agencies and private entities that, many times, have authority for development of the inscribed site and its surroundings.

As demonstrated above, tourism can be a blessing and a curse for World Heritage Sites. Finding the right balance at each site—determining that tipping point between vigorous, healthy tourism that contributes sustainably to local development and preservation and overtourism that degrades and undermines the site's outstanding universal values—remains the central challenge.

Notes

1. Special thanks to Peter DeBrine, senior project officer, UNESCO's World Heritage and Sustainable Tourism Programme, for giving several long interviews and

offering his insights and analysis based on is professional expertise. The content and conclusions are, however, that of the author.

2. World Heritage Convention. (2020). "The Criteria for Selection." UNESCO. http://whc.unesco.org/en/criteria/.

3. James T. Yenckel. (May 22, 1994). "Croatia: Would Anyone Really Want to Go to War-Torn Croatia? And Just How Safe Is It?" *Washington Post*. https://www.washingtonpost.com/archive/lifestyle/travel/1994/05/22/croatia-would-anyone-really-want-to-go-to-war-torn-croatia-and-just-how-safe-is-it/5f1f1414-5c69-469b-82ef-e9b9bb400523/.

4. Jennifer Luty. (December 5, 2019)." Number of Overnight Tourist Arrivals in Dubrovnik, Croatia from 2011 to 2018." *Statistica*. https://www.statista.com/statistics/886613/dubrovnik-tourist-arrivals/.

5. Jonnalyn Cortez. (April 16, 2019). "Dubrovnik's Overtourism: A Price to Pay with 'Game of Thrones' Fame." *Business Times*. https://www.btimesonline.com/articles/110767/20190416/dubrovniks-overtourism-a-price-to-pay-with-game-of-thrones-fame.htm.

6. Ibid.

7. UNESCO. (2018). *State of Conservation: 2018 Data Table of Threats*. https://whc.unesco.org/en/soc/action=list&soc_start=2018&mode=table.

8. Aurélie Élisa Gfeller and Jaci Eisenberg, (2016). "UNESCO and the Shaping of Global Heritage." Palgrave Macmillan. *The History of UNESCO: Global Actions and Impacts*. London, UK: Basingstoke. p. 286.

9. World Heritage. (2020). "Outstanding Universal Value." https://worldheritage.gsu.edu/outstanding-universal-value/.

10. World Heritage Convention. (2020). "The Criteria for Selection." UNESCO. http://whc.unesco.org/en/criteria/.

11. World Heritage Convention. (2020). "The Operational Guidelines for the Implementation of the World Heritage Convention." UNESCO. https://whc.unesco.org/en/guidelines/.

12. Ibid.

13. World Heritage Convention. (2020). "How Is a Site Inscribed on the List?" UNESCO. https://whc.unesco.org/en/faq/24.

14. Cumbria Tourism. (2020). "Celebrating World Heritage Status in Cumbria & How to Maximise the Opportunity for Your Business." https://www.cumbriatourism.org/what-we-do/projects/world-heritage-site/.

15. Chloé Maurel. (January 11, 2017). "The Unintended Consequences of UNESCO World Heritage Listing," *The Conversation*. https://theconversation.com/the-unintended-consequences-of-unesco-world-heritage-listing-71047.

16. World Heritage Committee. (2020). "World Heritage Fund." UNESCO. https://whc.unesco.org/en/world-heritage-fund/.

17. Aurélie Élisa Gfeller and Jaci Eisenberg, (2016). "UNESCO and the Shaping of Global Heritage." Palgrave Macmillan. *The History of UNESCO: Global Actions and Impacts*. London, UK: Basingstoke. p. 286.

18. Talor Stone. (2018). "The Process of Selecting UNESCO World Heritage Sites." Old Dominion University 2018 Issue Brief, UNESCO.

19. Laignee Barron. (August 30, 2017). "'Unesco-cide': Does World Heritage Status Do Cities More Harm Than Good?" *The Guardian*. https://www.theguardian.com/cities/2017/aug/30/unescocide-world-heritage-status-hurt-help-tourism.

20. Arthur Pederson. (2002). *Managing Tourism at World Heritage Sites: A Practical Manual for World Heritage Site Managers.* UNESCO World Heritage Centre. p. 3. https://unesdoc.unesco.org/ark:/48223/pf0000128679.

21. Laignee Barron. (August 30, 2017). "'Unesco-cide': Does World Heritage Status Do Cities More Harm Than Good?" *The Guardian.* https://www.theguardian.com/cities/2017/aug/30/unescocide-world-heritage-status-hurt-help-tourism.

22. Ibid.

23. Ibid.

24. Chloé Maurel. (January 11, 2017). "The Unintended Consequences of UNESCO World Heritage Listing," *The Conversation.* https://theconversation.com/the-unintended-consequences-of-unesco-world-heritage-listing-71047.

25. Ibid.

26. World Heritage Committee. (June 15, 2018). "Convention Concerning the Protection of the World Cultural and Natural Heritage." *UNESCO.* https://unesdoc.unesco.org/in/documentViewer.xhtml?v=2.1.196&id=p::usmarcdef_0000265326&file=/in/rest/annotationSVC/DownloadWatermarkedAttachment/attach_import_244ce5d0-fad0-4fbc-b81d-2245fdb2ab4b%3F_%3D265326eng.pdf&locale=en&multi=true&ark=/ark:/48223/pf0000265326/PDF/265326eng.pdf#%5B%7B%22num%22%3A48%2C%22gen%22%3A0%7D%2C%7B%22name%22%3A%22XYZ%22%7D%2C68%2C418%2C0%5D.

27. Simon Usborne. (April 29, 2009). "Is Unesco Damaging the World's Treasures?" *The Independent.* https://www.independent.co.uk/travel/news-and-advice/is-unesco-damaging-the-worlds-treasures-1675637.html.

28. Brian Major. "10 UNESCO World Heritage Sites to Add to Your Travel Bucket List." *Vacation.com.* https://www.vacation.com/destinations-cultures/10-unesco-world-heritage-sites.

29. Rachael Funk. (March 2020). "25 UNESCO Sites to Put on Your Bucket List." *Great Value Vacations.* https://www.greatvaluevacations.com/travel-inspiration/bucket-list-unesco-sites.

30. Amanda DiSilvestro. (April 6, 2017). "7 UNESCO World Heritage Sites for Every Travel Bucket List." *Discovery Corps.* https://discovercorps.com/blog/unesco-travel-bucket-list/.

31. Jessica Yadegaran. (February 26, 2019)."Travel: Jetsetter's 9 UNESCO World Heritage Cities for Your Bucket List." *Mercury News.* https://www.mercurynews.com/2019/02/26/travel-9-unesco-world-heritage-cities-to-add-to-your-bucket-list/.

32. Meredith Rosenberg. "New UNESCO World Heritage Sites You Can Visit." *Travel Channel.* https://www.travelchannel.com/interests/history/photos/new-unesco-world-heritage-sites.

33. Annette White. (April 23, 2017). "Top Historical Places: 10 UNESCO World Heritage Sites Around the World." *Bucket List Journey.* https://bucketlistjourney.net/10-unbelievable-experiences-at-unesco-world-heritage-sites-around-the-world/.

34. Arthur Pederson. (2002). *Managing Tourism at World Heritage Sites: A Practical Manual for World Heritage Site Managers.* UNESCO World Heritage Centre. p. 3. https://unesdoc.unesco.org/ark:/48223/pf0000128679.

35. Susanne Becken and Cassandra Wardle. (January 2017). "Tourism Planning in Natural World Heritage Sites." Research Report, Griffith University. https://www

.griffith.edu.au/__data/assets/pdf_file/0034/18889/UNESCO-WHA-Report13Finalfinal-1.pdf.

36. Ibid.

37. Dan Peltier. (April 19, 2017). "Nearly Half of UNESCO Sites Don't Have Plans to Manage Overtourism Challenges." *Skift.* https://skift.com/2017/04/19/nearly-half-of-unesco-sites-dont-have-plans-to-manage-overtourism-challenges/.

38. Simon Usborne. (April 29, 2009). "Is Unesco Damaging the World's Treasures?" *The Independent.* https://www.independent.co.uk/travel/news-and-advice/is-unesco-damaging-the-worlds-treasures-1675637.html.

39. Ibid.

40. World Heritage Convention. (2020). "State of Conservation Information System." *UNESCO.* https://whc.unesco.org/en/soc/action=list&soc_start=2018&mode=table.

41. World Heritage Committee. (June 15, 2018). " Convention Concerning the Protection of the World Cultural and Natural Heritage: Forty-Second Session." *UNESCO.* https://unesdoc.unesco.org/in/documentViewer.xhtml?v=2.1.196&id=p::usmarcdef_0000265326&file=/in/rest/annotationSVC/DownloadWatermarkedAttachment/attach_import_244ce5d0-fad0-4fbc-b81d-2245fdb2ab4b%3F_%3D265326eng.pdf&locale=en&multi=true&ark=/ark:/48223/pf0000265326/PDF/265326eng.pdf#%5B%7B%22num%22%3A48%2C%22gen%22%3A0%7D%2C%7B%22name%22%3A%22XYZ%22%7D%2C68%2C418%2C0%5D.

42. Dan Peltier. (July 31, 2018). "UN Agency's Latest Endangered List Leaves Out Many Sites Plagued by Overtourism." *Skift.* https://skift.com/2018/07/31/un-agencys-latest-endangered-list-leaves-out-many-sites-plagued-by-overtourism/.

43. Susanne Becken and Cassandra Wardle. (January 2017). "Tourism Planning in Natural World Heritage Sites." Research Report, Griffith University. https://www.griffith.edu.au/__data/assets/pdf_file/0034/18889/UNESCO-WHA-Report13Finalfinal-1.pdf.

44. World Heritage Convention. (2020)."World Heritage in Danger." UNESCO. https://whc.unesco.org/en/158/.

45. World Heritage Convention. (December 1, 1998)."World Heritage Committee Removes Old City of Dubrovnik and Wieliczka Salt Mine from its List of Endangered Sites." UNESCO. https://whc.unesco.org/en/news/147.

46. Ibid.

47. Simon Usborne. (April 29, 2009). "Is Unesco Damaging the World's Treasures?" *The Independent.* https://www.independent.co.uk/travel/news-and-advice/is-unesco-damaging-the-worlds-treasures-1675637.html.

48. Dan Peltier. (July 31, 2018). "UN Agency's Latest Endangered List Leaves Out Many Sites Plagued by Overtourism." *Skift.* https://skift.com/2018/07/31/un-agencys-latest-endangered-list-leaves-out-many-sites-plagued-by-overtourism/.

49. Ibid.

50. Aurélie Élisa Gfeller and Jaci Eisenberg, (2016). "UNESCO and the Shaping of Global Heritage." Palgrave Macmillan. *The History of UNESCO: Global Actions and Impacts.* London, UK: Basingstoke. p. 286.

51. World Heritage Convention. (2020)." World Heritage List of Statistics: Total Number of Properties Inscribed per Year (Cumulative)." UNESCO. https://whc.unesco.org/en/list/stat/#s7.

52. World Heritage Convention. (2020)."State of Conversation: Plitvice Lakes National Park (Croatia)—Analysis and Conclusion by World Heritage Centre and the Advisory Bodies in 2018." UNESCO. http://whc.unesco.org/en/soc/3753/.

53. Colin Packham and Benjamin Cooper. (July 5, 2017). "UNESCO Leaves Great Barrier Reef Off 'In Danger' List." *Reuters.* https://www.reuters.com/article/us-australia-environment-reef/unesco-leaves-great-barrier-reef-off-in-danger-list-idUSKBN19R07S.

54. Australian Government: Great Barrier Reef Marine Park Authority. (December 12, 2013). "Abbot Point Capital Dredging Project." http://www.gbrmpa.gov.au/about-us/consultation/consultation-completed/abbot-point-capital-dredging-project.

55. Paul Farrell. (August 30, 2014). "Unesco Wants Great Barrier Reef on Danger List over Dredging Fears." *The Guardian.* https://www.theguardian.com/environment/2014/may/01/unesco-wants-great-barrier-reef-on-danger-list-over-dredging-fears.

56. Oliver Milman. (May 14, 2015). "Australia Lobbies Unesco to Stop It from Listing Great Barrier Reef as 'In Danger.'" *The Guardian.* https://www.theguardian.com/environment/2015/may/14/australia-lobbies-unesco-stop-listing-great-barrier-reef-as-in-danger.

57. Colin Packham and Benjamin Cooper. (July 5, 2017). "UNESCO Leaves Great Barrier Reef Off 'In Danger' List." *Reuters.* https://www.reuters.com/article/us-australia-environment-reef/unesco-leaves-great-barrier-reef-off-in-danger-list-idUSKBN19R07S.

58. Markku Björkman. (September 17, 2019). "Australia Approved to Open a Huge Coal Mine." MiningMetalnews.com. https://www.miningmetalnews.com/20190918/1298/australia-approved-open-huge-coal-mine.

59. Martha Honey. (2019). *Cruise Tourism in the Caribbean: Selling Sunshine.* London and New York: Routledge Press. pp. 53–57.

60. Lisa Gerard-Sharp. (May 26, 2017). "Venice World Heritage Status Under Threat." *The Guardian.* https://www.theguardian.com/travel/2017/may/26/venice-tourists-cruise-ships-pollution-italy-biennale.

61. Martha Honey, editor. *Cruise Tourism in the Caribbean: Selling Sunshine.* London and New York: Routledge Press. pp. 53–57.

62. UNESCO. (2019). *Decisions Adopted during the 43rd Session of the World Heritage Committee.* https://whc.unesco.org/en/decisions/7524.

63. Ibid.

64. Susanne Becken and Cassandra Wardle. (January 2017). "Tourism Planning in Natural World Heritage Sites." Research Report, Griffith University. https://www.griffith.edu.au/__data/assets/pdf_file/0034/18889/UNESCO-WHA-Report13Finalfinal-1.pdf

65. Jennifer Luty. (April 25, 2019). "Number of Cruise Passengers in Dubrovnik, Croatia from 2009 to 2018." *Statista.* https://www.google.com/search?q=Jennifer+Luty.+(April+25%2C+2019).+%22Number+of+cruise+passengers+in+Dubrovnik%2C+Croatia+from+2009+to+2018.%22+Statista.&rlz=1C1CHBF_enUS859US859&oq=Jennifer+Luty.+(April+25%2C+2019).+%22Number+of+cruise+passengers+in+Dubrovnik%2C+Croatia+from+2009+to+2018.%22+Statista.&aqs=chrome..69i57.2434j0j8&sourceid=chrome&ie=UTF-8.

66. Craig Lewis. (August 3, 2016). "Record Tourist Numbers Threaten Ancient Bud-

dhist Art in China's Mogao Caves." *Buddhistdoor*. https://www.buddhistdoor.net/news/record-tourist-numbers-threaten-ancient-buddhist-art-in-chinas-mogao-caves.

67. Susanne Becken and Cassandra Wardle. (January 2017). "Tourism Planning in Natural World Heritage Sites." Research Report, Griffith University. https://www.griffith.edu.au/__data/assets/pdf_file/0034/18889/UNESCO-WHA-Report13Finalfinal-1.pdf.

68. World Heritage Tourism Programme. (May 11, 2012). "Convention Concerning the Protection of the World Cultural and Natural Heritage: Thirty-Sixth Session." UNESCO. https://whc.unesco.org/archive/2012/whc12-36com-5E-en.pdf.

69. Ibid.

70. World Heritage Convention. (2020). "Sustainable Tourism. UNESCO World Heritage and Sustainable Tourism Programme." UNESCO. https://whc.unesco.org/en/tourism/.

71. Takamitsu Jimura. (January 15, 2019). *World Heritage Sites: Tourism, Local Communities, and Conservation Activities*. UK: Oxfordshire, CABI. p. 40.

72. World Heritage Journeys. (2020). https://visitworldheritage.com/en/eu/about-this-website/ee61b67f-c728-4851-b26e-9730ce3aa0f3.

73. Laignee Barron. (August 30, 2017). "'Unesco-cide': Does World Heritage Status Do Cities More Harm Than Good?" *The Guardian*. https://www.theguardian.com/cities/2017/aug/30/unescocide-world-heritage-status-hurt-help-tourism.

Chapter 4.2

Machu Picchu and the Inca Trail, Peru

By Louise Norton

Until 1911, the ancient Inca city of Machu Picchu, located nearly eight thousand feet high in the Andes Mountains, Peru, was unknown to the outside world, covered in cloud forest and unvisited except by Andean bears and a few local farmers and looters. Today, the Historic Sanctuary of Machu Picchu (HSMP), which encompasses the main archaeological sites of the citadel and the trails within the surrounding national park, is among South America's foremost visitor attractions. The number of international visitors to Machu Picchu has risen dramatically this century, from 294,437 visitors in 2000 to 1,174,435 visitors in 2019. Including domestic tourists, there were 1,543,036 visitors to the site in 2019.[1] As Gustavo Pinto from World Tourism Market wrote, "Machu Picchu may be the most iconic Latin American case in the current battle against overtourism worldwide."[2]

Growth of Tourism

Starting in the 1960s, the region of Cusco, and particularly Machu Picchu, has grown to become the center of Peru's tourism industry. In 1965, a US firm recommended that the Peruvian government focus its tourism development resources on infrastructure and marketing for a

Lima–Cusco–Machu Picchu circuit for international visitors.[3] Peru's tourism industry suffered in the 1980s and early 1990s due to armed internal conflict, but revived during the mid- to late-1990s when the government again put Machu Picchu at the forefront of its tourism promotion campaign.[4]

Today, public-sector responsibility for tourism at the national level lies with the Ministry of Exterior Commerce and Tourism. This government ministry, through its marketing agency PROMPERÚ (the Commission for the Promotion of Exports and Tourism of Peru), has been successful in promoting Machu Picchu as the country's primary destination, with 58 percent of foreign tourists mentioning it as their main reason for visiting Peru.[5] Now, Machu Picchu is considered an "iconic" site, an essential stop on any itinerary of Peru and fundamental to Peru's national identity.[6]

Causes and Impacts of Overtourism

Growth in tourism has radically altered the previously large rural barter economy of the Sacred Valley leading to Machu Picchu.[7] Today, the wealth extracted from Machu Picchu is not Inca figurines but tourist dollars. Due to its remote location and restricted access, popularity, economic importance, and the large number of stakeholders involved in the site, however, Machu Picchu is a challenging site to manage.

World Heritage Site Designation

In 1983, the Historic Sanctuary of Machu Picchu was registered as a UNESCO World Heritage Site (WHS).[8] Machu Picchu has the unusual distinction of being one of only thirty-eight sites designated as both a cultural and natural WHS, in recognition of both the diverse ecology of the area and the unique archaeology. The HSMP covers an area of 35,592 hectares and contains nine ecological zones.[9] This expansive area, however, presents many difficult ecological and social management issues, such as accidental forest fires and agricultural deforestation.[10]

As is commonplace for all sites of this designation, UNESCO status stipulates that responsibility for the preservation of sites is "shared by the international community."[11] Unfortunately, as has happened at other World Heritage Sites, the publicity generated by this UNESCO designation contributed to the steep rise in visitation numbers, from a mere 120,122 at the time of designation to 1,543,036 in 2019.[12] As Sandra Doig, deputy director of PROMPERÚ, said: "Machu Picchu is a great attraction, but we are worried about its sustainability . . . it is affected by too many people in the citadel at the same time."[13]

Challenges in Gateway Communities

Similarly, the mountain towns around Machu Picchu have grown to accommodate the expanding tourism. Most visitors travel to Machu Picchu by train from the city of Cusco. To arrive in Cusco, most visitors pass through the small, one-runway airport, which services flights only from Lima and a few other nearby cities. To accommodate the ever-growing arrivals, the government of Peru has now broken ground on an international airport near the town of Chinchero, located between Cusco and Machu Picchu. The new airport will be able to receive six million passengers per year, 60 percent more than the current airport, which is already functioning over its capacity limits.[14] Concerns are being raised not only for Chinchero, but for the nearby Inca town of Ollantaytambo, impacted by low-flying planes passing over and demand on its railway station, the departure point for most trains to Machu Picchu.

Machu Picchu Pueblo, also known as Aguas Calientes, located at the foot of Machu Picchu, has grown to service visitors around the train station. The residents of this small town now face a similar overtourism problem by living in the shadows of this world wonder. As one resident divulged to *Conde Nast Traveler*, "Tourism, for us, is the main economic activity in the region. The problem is we don't have too much infrastructure to ensure the wealth is properly distributed."[15] Although a vital economic livelihood, these locals are now confronted with constant crowding and rising food prices at restaurants and grocery stores.

Challenges along the Inca Trail

Since the 1970s, travelers seeking a pilgrimage or physical challenge have opted to hike to the citadel of Machu Picchu along the Inca Trail, a multiday hike along ancient Inca paths through the larger protected lands within the HSMP. The number of hikers rose gradually but remained below twenty thousand yearly until the mid-1990s, when Peru saw a steep boom in tourism. By 1998, sixty-six thousand hikers were trekking the Inca Trail.[16] As more hikers came, guided treks became more common, and villagers living along the trail began selling food and refreshments and offering themselves as porters. Welfare of these workers became an issue. One porter explained, "We would have to carry over 50 kilos (110 pounds) and there was never enough food. We had to sleep out in the open, with no tents. If it rained, we would try to find a cave so we could stay dry."[17] Due to the high foot traffic, other issues observed during the 1990s included erosion, disruption of wildlife, trampling of vegetation, crowded campsites and, most worrying to a government focused on the aesthetic appeal of the trail, solid and human waste.[18] The state began to view the trail as both a development project and an environmental problem.

Challenges to Management

To combat these issues, multiple governmental organizations and departments are represented on the unit that is charged with managing the site, the Unidad de Gestion de Machu Picchu. The interests of these separate departments and many stakeholders involved in the site, including local government ministries, residents, tour agencies, rail companies, and others, are diverse and conflicting. As such, reaching consensus is difficult, and new regulations, when introduced, are difficult to enforce. For example, a United Nations World Tourism Organization publication noted that "the [previous] restriction of access to only 2,500 persons per day, so as to avoid deterioration of the site, [was] being ignored" by the regional directorate of the Ministry of Culture, which does not want to turn any visitors away.[19] In ignoring its own limits and regulation, this ministry risks not only potential deterioration of the site but also congestion and resulting visitor dissatisfaction.

Solutions

Since 2000, a range of efforts to control visitor numbers and behavior, regulate porters and guides, improve infrastructure, and control sprawl have been implemented. These reforms have been aimed at addressing

Visitors to Machu Picchu's citadel. Source: Louise Norton.

the problems within the Inca Trail, Machu Picchu citadel, and surrounding gateway communities.

Inca Trail

Passed in 2000 and implemented in 2001, the "Regulation of Tourist Use of the Network of Inca Trails in the Machu Picchu Historic Sanctuary" was created as a multistakeholder legislation effort to better manage the harmful impacts of tourism on the Inca Trail. This legislation has been effective in reducing erosion, in controlling litter, and, to some extent, in improving working conditions for the porters employed by agencies to carry camping equipment on the Inca Trail.[20]

The 2001 rules for the Inca Trail mandated an increase in entrance fees from $17 to $50 per hike, to be split between the Ministry of Culture and the Natural Resources Institute, which supported cultural and natural heritage management within the HSMP. In 2008, these fees were raised again to $86 per hiker.[21]

Permits were restricted to five hundred each day, including porters, cooks, and guides, so that only around two hundred tourists hike each day. Independent hikes without a guide are no longer permitted, and most tourists therefore book their trek with a licensed agency.[22] The Inca Trail now attracts a controlled number of higher-value visitors while reducing environmental damage within the HSMP. These restrictions, however, have come at the cost of making the trail less accessible to domestic tourists and low-budget backpackers and reducing livelihood opportunities for local inhabitants. Despite the permit limits, overall numbers on the trail increased from 103,700 in 2001 to 132,500 in 2008, while the number of porters also increased to support high-end trekking.[23]

In 2001, legislation was also put into place to ensure better working conditions for porters. These regulations limited the amount of cargo each porter is allowed to carry to 25kg, and there is now a control post at the beginning of the trail to weigh the load each porter carries.[24] Tour agencies, however, frequently circumvent weight regulation by having the guide or porter, or even the tourists, carry part of the porters' load through the weighing point.[25] Although most porters acknowledge that loads are usually much lighter than they were before regulations were introduced and that pay and working conditions are better, some porters still do not make a living wage and often deal with discriminatory conditions on treks. As one porter said, "The informal [tour] agencies still mistreat the porters [by imposing] excessive weight, but there are companies that obey the Ministry of Work rules."[26] In response to these issues, a US nonprofit organization called the Porter Voice Collective is campaigning to raise funds for a documentary on the conditions of porters on the Inca Trail.[27]

Solid waste is also weighed to prevent porters dumping garbage within the sanctuary, and villagers along the Inca Trail are employed to collect garbage from the fields.[28] The system for disposal of solid waste has also improved in recent years due to the public-private investment in waste processing.[29]

With visitor numbers now limited, tour operators have begun to open up "alternative Inca Trails" throughout the region, such as the two-day trek to Choquequirao, one of the larger Inca sites outside of Machu Picchu.[30] These alternative routes spread economic, sociocultural, and environmental impacts, both good and bad, to other areas.

The Citadel

Despite the advances made on the Inca Trail during the early 2000s, UNESCO continued to raise concerns about the preservation of Machu Picchu's citadel. In response, in July 2017, legislation was introduced by the Ministry of Culture and approved by the Ministry of Tourism to "regulate sustainable use and tourist visitation to ensure the conservation of the Inca citadel of Machu Picchu,"[31] which marked the beginning of the government's recent efforts to regulate overtourism.

First tested in 2017 and permanently implemented in 2019, tickets to the citadel, costing $46, now permit entry for just four hours. Previously for this price tourists could spend all day, from sunrise until 5:30 p.m., at the site. The new rules also introduce specific time slots for entering the citadel, where visitors must arrive within a one-hour window selected. Officially the daily limit of tickets before July 2017 was 2,500, but with the changes, the limit has increased to 5,940, split between different time slots.[32] Guides have reported that the new rules have eased congestion in some areas, although others have remained overcrowded.[33]

Once inside, regulations now control the circuits along which tourists must travel, enforce one-way routes around the site, and restrict reentry.[34] Additionally, regulation defines areas for tour guides to stop their groups for interpretation and expands the restrictions on visitor behavior, such as making "loud or annoying noises such as clapping, screaming, whistling, singing, among other actions, because it disturbs the tranquility and the sacred character of the Machu Picchu Sanctuary."[35] Visitors are required to hire a guide, although this rule is not yet enforced.

In interviews, Machu Picchu guides expressed mixed opinions about the effectiveness of the new timed-entry regulations.[36] They were concerned about the level of visitor satisfaction, long bus lines, congestion at bottlenecks, and erosion, because the regulations still allow for double the number of daily visits as before.

Guides also said that, due to the reduced time slots, they were under a great deal of pressure to shuffle tourists through the site quickly. As one guide said, "You have to move fast because there are lots of people.

There isn't much freedom to explain the real concept of Machu Picchu, that it is a sacred temple, connected with Mother Earth."[37]

In addition, tourists without guides can get confused due to a lack of signage and often miss large parts of the site. Frequently, they are refused reentry to the site after leaving to use the bathroom or taking the wrong circuit. The guide explained, "Visitors are impressed. They are satisfied, but they leave with criticisms of the management of Machu Picchu. There is not good planning to receive the visitors."[38]

This guide and others believed that the new regulations were aimed at maximizing economic benefits rather than protecting the environment and people: "The temple has been used very commercially; they haven't really valued it. It annoys me that the guards inside blow their whistles every minute; it is really annoying. Other Inca routes should be unblocked and used more."[39]

Gateway Communities

In a ripple effect, reducing time slots but increasing visitor numbers has accentuated problems of uncontrolled construction in the access towns of Machu Picchu Pueblo and Ollantaytambo.

The municipality (local government) of Machu Picchu Pueblo receives 10 percent of ticket receipts from Machu Picchu "to improve the infrastructure, including sewage and waste treatment."[40] In other gateway communities with ticketed archaeological sites, such as Ollantaytambo, Pisac, Chinchero, and the city of Cusco, the municipalities have, since 2006, also received a portion of the ticket income to their sites.[41] This income is supposed to be spent on projects related to tourism, such as controlling traffic and processing solid waste.

In 2019, Ollantaytambo received 6,500,000 soles (US$1,815,769) from ticket sales to archaeological sites in the area.[42] Despite this funding, tourism management in and around Ollantaytambo continues to be haphazard due to a combination of factors, including local government corruption and mismanagement, which hampers the development of infrastructure and services.[43] The traffic in Ollantaytambo is exacerbated by vehicles ferrying tourists to the train station, with delays of an hour just to enter the town. Access routes to the station area are narrow, and efforts to expand have been problematic. The whole town is a terraced Inca archaeological site, bounded on one side by a river and the others by mountains, so there is a lack of unprotected land for development. There is also local resistance to relocation of the station.[44] After the district mayor pushed through a popular project to build a bypass to reduce the heavy traffic entering town, he was accused by the Ministry of Culture of illegally destroying Inca terraces to construct the road.[45]

And in Chinchero, more than one hundred thousand people have currently petitioned against the development of the airport, claiming that it will "endanger the conservation of one of the most important historical and archaeological sites in the world" and "affect the integrity of a complex Inca landscape and will cause irreparable damage due to noise, traffic and uncontrolled urbanization" in Chinchero and neighboring communities.[46] Given the looming construction, the World Monuments Fund in 2019 listed Machu Picchu as one of twenty-five cultural heritage sites at risk, and UNESCO has urged the government to reconsider the plan.[47]

Evaluation and Future Steps

The government must listen to stakeholders as it continues to implement legislation, reconsidering both the increased limit on visitor numbers and the proposed international airport at Chinchero. To manage excessive tourism, planning by the public sector with stakeholder participation is essential, and interventions must be chosen according to their suitability for the destination.[48] Strategies already being implemented in Peru include place-demarketing (such as increasing price to decrease demand) and diversification of the tourism product by developing and promoting alternative routes to Machu Picchu and other sites.[49] For example, access to the cultural site of Kuelap in North of Peru has been improved, and there are plans to build a cable car to Choquequirao, a sister site of Machu Picchu, although these plans are controversial.

At the same time, while Peru works to attract high-value (versus high-volume) international tourism, these policies should not adversely impact Peruvians' ability to access their own cultural and natural heritage. The current tiered pricing structure must be protected, as should programming to ensure that the Machu Picchu's history is utilized for the benefit of cultural preservation.

Machu Picchu continues to be a challenging site to manage due to its popularity, remote location, and the difficulty of coordinating solutions to the satisfaction of the many different stakeholders. Peru has worked to exploit this wonderful tourism resource and gradually introduced measures to protect its heritage. Stakeholders must not be blinded by economic benefits and lose sight of the ideal of sustainability in building on this success as the country works to extend the benefits brought by tourism to other regions.

Notes

1. Observatorio Turístico del Perú. (2019). http://www.observatorioturisticodel peru.com/mapas/impne.pdf.

2. Gustavo Pinto. (2019). "What Overtourism Are We Talking about in Latin America?" World Tourism Market. https://news.wtm.com/what-overtourism-are-we-talking-about-in-latin-america/.

3. L .C. Desforges. (1997). "Travelling to Peru: Representation, Identity and Place in British Long-Haul Tourism." PhD thesis. University College London. https://discovery.ucl.ac.uk/id/eprint/1317646/1/265333.pdf.

4. Ibid.

5. PROMPERÚ. (2017). Perfil del Turista Extranjero 2016. *PromPeru.* Lima: PROMPERÚ.

6. Ibid. p. 75.

7. Azusa Miyashita. (2009). *Killing the Snake of Poverty: Local Perceptions of Poverty and Well-Being and People's Capabilities to Improve Their Lives in the Southern Andes of Peru.* Amsterdam: Dutch University Press. https://dare.uva.nl/search?identifier=99c80eb8-a60a-41b2-835a-77dbad35ed1c

8. UNESCO. (2020). "Historic Sanctuary of Machu Picchu." http://whc.unesco.org/EN/LIST/274.

9. United Nations Environment Programme and World Conservation Monitoring Centre. (2011). *Historic Sanctuary of Machu Picchu Peru.* https://conservation-development.net/Projekte/Nachhaltigkeit/DVD_12_WHS/Material/files/WCMC_Machu_Picchu.pdf; ParksWatch Peru. (2004). "Machupicchu Historical Sanctuary." http://www.parkswatch.org/parkprofile.php?l=eng&country=per&park=mphs&page=inf.

10. Pellegrino A. Luciano. (2011). "Where Are the Edges of a Protected Area? Political Dispossession in Machu Picchu, Peru." *Conservation and Society* 9(1): 35–41.

11. Hyung yu Park. (2014). *Heritage Tourism.* Abingdon: Routledge. p. 114.

12. Observatorio Turístico del Perú. (2019). http://www.observatorioturisticodel peru.com/mapas/impne.pdf.

13. Andrea Sachs. (2018). "Peru Devises New Rules to Tackle the Mounting Crowds on Machu Picchu." *Washington Post.* https://www.washingtonpost.com/.

14. Agence France-Presse. (January 25, 2020). "Peru Promises to Protect Machu Picchu as It Builds Airport." *The Jakarta Post.* https://www.thejakartapost.com/travel/2020/01/24/peru-promises-to-protect-machu-picchu-as-it-builds-airport.html.

15. Tyler Moss. (October 24, 2018). "No, We're Not Trampling Machu Picchu Out of Existence." *Conde Naste Traveler.* https://www.cntraveler.com/story/we-are-not-trampling-machu-picchu-out-of-existence.

16. Keely Maxwell. (2012). "Tourism, Environment, and Development on the Inca Trail." *Hispanic American Historical Review* 92(1): 143–71.

17. Tim Leffel. (2006). "The Life of a Peruvian Porter." *International Travel News.* https://www.intltravelnews.com/2006/07/the-life-of-a-peruvian-porter.

18. Keely Maxwell. (2012). "Tourism, Environment, and Development on the Inca Trail." *Hispanic American Historical Review* 92(1): 143–71.

19. UNWTO. (2016). *Tourism and Culture Partnership in Peru: Models for Collaboration between Tourism, Culture and Community.* Madrid, Spain.

20. Alexandra Arrelano. (2011). "Tourism in Poor Regions and Social Inclusion: The Porters of the Inca Trail to Machu Picchu." *World Leisure Journal* 53(2): 104–18.

21. Keely Maxwell. (2012). "Tourism, Environment, and Development on the Inca Trail." *Hispanic American Historical Review* 92(1): 143–71.

22. Peruvian Government. (2001). "Reglamento de Uso Turistico Sostenible de la Red de Caminos Inka Del Santuario Histórico de Machupicchu." http://extwprlegs1.fao.org/docs/pdf/per167816anx.pdf.

23. Keely Maxwell. (2012). "Tourism, Environment, and Development on the Inca Trail." *Hispanic American Historical Review* 92(1): 143–71.

24. Ibid.

25. Gongora Meza and Miguel Angel. (October 23, 2017). "Are Working Conditions for Porters on the Inca Trail the Next Thailand Elephant Case?" *Living in Peru.* https://www.livinginperu.com/working-conditions-porters-inca-trail-next-thailand-elephant-case/.

26. Miguel, Inca Trail cook. (February 19, 2020). Personal communication with author.

27. Marinel de Jesus. (2020). "KM 82. A Documentary on the Porter Voices of Peru's Camino Inca." Indiegogo. https://www.indiegogo.com/projects/km-82#/.

28. Flor Quispe. (January 5, 2017). Resident of Historical Sanctuary of Machu Picchu. Personal communication with author.

29. Rumi Cevallos. (April 16, 2019). "Machu Picchu Sostenible." *La República.* https://larepublica.pe/la-contra/1450578-rumi-cevallos-machupicchu-sostenible/.

30. Keely Maxwell. (2012). "Tourism, Environment, and Development on the Inca Trail." *Hispanic American Historical Review* 92(1): 143-171; Machu Picchu Trek Guide. (2020). "Alternative Inca Trail Treks That Rock." https://www.machupicchutrek.net/alternative-inca-trail-treks/.

31. Ministerio de Cultura. (2017). *Resolución Ministerial.* http://transparencia.cultura.gob.pe/sites/default/files/transparencia/2017/02/resoluciones-ministeriales/rm070.pdf.

32. The complimentary measures for implementation of the regulations limit morning numbers to a maximum of 3,267 and afternoon numbers to 2,673 visitors.

33. Louise Norton. (2019). Unpublished Master's Project. *Guiding Machu Picchu:*

Tour Guides' Perceptions of Regulatory Change for Visiting the Site. Leeds: Leeds Beckett University.

34. Ministerio de Cultura. (2017). "Medidas Complementarias Para La Implementación De La Resolución Ministerial No. 070-2017-Mc Que Aprueba El Reglamento De La Llaqta De Machupicchu." Cusco: Direccion Desconcentrada de Cultura Cusco.

35. Andean Air Mail and *Peruvian Times*. (June 27, 2017). "Selfie Stick Ban among Several New Prohibitions at Machu Picchu." https://www.peruviantimes.com/27/selfie-stick-ban-among-several-new-prohibitions-at-machu-picchu/29847/

36. Louise Norton. (2019). Unpublished Master's Project. *Guiding Machu Picchu: Tour Guides' Perceptions of Regulatory Change for Visiting the Site*. Leeds: Leeds Beckett University.

37. Ibid.

38. Ibid.

39. Ibid.

40. "Machu Picchu: Management." (2020). *Barcelona Field Studies Centre*. https://geographyfieldwork.com/MachuTourismManagement.htm

41. Ministerio de Justicia y Derechos Humanos. (April 22, 2006). *Ley del Boleto Turístico*. http://spij.minjus.gob.pe/Normas/textos/220406T.pdf.

42. Boleto Tusitcio Delcusco. (January 2020). "Cuadro de Distribución Porcentual Mensualizado del Año 2019 a Participantes del BTC, de acuerdo a Ley Del Boteo Turístico No. 28719." *COSITUC*. http://cosituc.gob.pe/wp-content/uploads/2020/01/cuadro-de-distribucion-porcentual-a-diciembre-del-2019.pdf.

43. Rebecca Stone. (January 14, 2019). "Peru's Challenge to Build Tourism Outside the Shadows of Machu Picchu." *Skift*. https://skift.com/2019/01/14/perus-challenge-to-build-tourism-outside-the-shadows-of-machu-picchu/.

44. Annabel Pinker and Penny Harvey. (2015). "Negotiating Uncertainty Neo-Liberal Statecraft in Contemporary Peru." ~~~I~~~Social Analysis %%%I%%%9(4): 1531.

45. RPP Noticias. (September 9, 2014). "Cusco: destruyen muros incas en Ollantaytambo por construir vía." https://rpp.pe/peru/actualidad/cusco-destruyen-muros-incas-en-ollantaytambo-por-construir-via-noticia-723991.

46. Natalia Majluf. (2020). "Salvemos Chinchero, patrimonio cultural de la humanidad." Change.org. https://www.change.org/p/presidente-de-la-rep%C3%BAblica-del-per%C3%BA-salvemos-chinchero-patrimonio-cultural-de-la-humanidad.

47. Adele Berti. (October 31, 2019). "Saving Machu Picchu: Will a New Airport Create Problems in Peru?" *Airport Technology*. https://www.airport-technology.com/features/machu-picchu-airport-plans/

48. Fabian Weber, Juerg Stettler, Julianna Priskin, Barbara Rosenberg-Taufer, Sindhuri Ponnapureddy, Sarah Fux,Marc-Antoine Camp, and Martin Barth. (2017). *Tourism Destinations under Pressure*. Working Paper, Institute of Tourism ITW, Lucerne. https://static1.squarespace.com/static/56dacbc6d210b821510cf939/t/5909cb282e69cf1c85253749/1493814.

49. Dominic Medway, Gary Warnaby, and Sheetal Dharni. (2011). "Demarketing Places: Rationales and Strategies." *Journal of Marketing Management* 27(1–2): 124–42; C. Michael Hall. (2014). *Tourism and Social Marketing*. Abingdon: Routledge.

Chapter 4.3

Luang Prabang, Laos

By Robyn Bushell

The recorded history of Luang Prabang, in northern Lao People's Democratic Republic (Lao PDR, also known as Laos), dates to the fourteenth century when the first Lao Kingdom was established.[1] Luang Prabang was the former royal capital of the kingdom, perched on a peninsula at the confluence of two rivers, the mighty Mekong and the Nam Khan, and surrounded by lush mountains. The history of Laos is complex. A tiny landlocked country, sharing borders with China, Vietnam, Cambodia, Thailand, and Burma, it has experienced many forces shaping its history. As a former French colonial town, it was extensively rebuilt in brick and mortar in the late nineteenth and early twentieth centuries.

The political and economic capital moved to Vientiane in 1946, and Luang Prabang remained the royal capital and center of Buddhism until the revolution of 1975, when the Pathet Lao installed a communist regime with the establishment of the Lao PDR. The expulsion of the monarchy caused an exodus of the most wealthy. Then, the ravages of the US-Vietnam war left the nation in deep social conflict, with many residents deprived of access to high-quality education and food. The countryside was littered with unexploded landmines, devastating the community and food production. In 1986, the government made the pragmatic decision to introduce market reforms and reopen the country to foreign investment.[2] The population of Laos has grown from just 2.1 million in 1960 to 7 million in 2018.[3] Both improved standards of living and returning exiled Lao has driven this growth.

In 1995, Luang Prabang was inscribed as a World Heritage Site primarily because of its well-preserved traditional Lao and French-Lao colonial architecture.[4] The economic contribution of tourism in World Heritage destinations throughout Southeast Asia, to both local and national development, is well documented. The Asia-Pacific region had become the second most visited tourist region in the world (behind Europe), receiving 24 percent of the global total. Growth in arrivals to Southeast Asia, at 7 percent, is among the highest worldwide.[5] Likewise, World Heritage designation has ensured that tourism has been the driving force behind economic growth in Luang Prabang.[6]

The spike in tourists had put significant pressure on the historic town, however. As one blogger observed, "Luang Prabang has become a tourism focal point well beyond the capacity of the town's natural footprint."[7] In addition, in what anthropologist David Berliner calls "Unescoization," the city's coveted religious and cultural traditions "are being staged, sometimes inaccurately." Although UNESCO designation is designed specifically to protect sites and traditions, the designation can lead to overtourism, as witnessed in many World Heritage designated places.[8] Invaluable cultural resources are being compromised as tourism businesses and tourism authorities seek to capitalize on the marketability of these special places.

Causes of Overtourism

Small historic cities and towns, such as Luang Prabang in Laos, Siem Reap in Cambodia, and Hoi An in Vietnam, are key to national development agendas in countries working hard to move from "least developed" conditions. As UNESCO-listed World Heritage Sites, the governments of these locales are using the prestige of the designation to promote them as tourist destinations, encouraging investment and driving economic growth. With an average gross domestic product (GDP) growth of 7 percent and a gross national income per capita of US$2,460 in 2018 compared to only US$190 in 1990, Laos is on track to rise from least developed country status to a lower-middle income economy.[9] It is now one of the fastest growing economies in the East Asia and Pacific region. The tourism sector is a key focus in bolstering economic growth while other sectors have recently slowed, such as the agricultural sector in the face of natural disasters.[10]

The French government initially provided aid and expertise to develop tourism in Luang Prabang. A program to protect and develop the site was established in 1996 with the support of the City of Chinon and the Région Centre (France) and several partners, including the European Union and, as part of the France-UNESCO Cooperation Agreement, the French Ministry of Culture, the Ministry of Infrastructure, and the Ministry of Foreign Affairs. From this program came the creation of the Heritage Office, in which a Laotian operational team was made responsible for the ongoing protection and enhancement of the town. The program brought together UNESCO, French ministries, and local authorities. Intensive training and technical cooperation established a regulatory framework, with governance at the local, regional, and national levels, for the restoration of monuments, the built environment, and the landscape.[11] After initial successes, the program attracted funding from a wider range of donors with ongoing support from Japan, Korea, the Asian Development Bank, the Netherlands, Germany, and Australia. Luang Prabang became a tourism destination with strong appeal to Western aesthetics, being "enigmatic, seductive, lost in time, enchanting, rooted in tradition, spiritual and romantic."[12]

Ironically, even though World Heritage designation is intended to ensure the conservation and protection of listed sites, it frequently becomes a marketing tool for the "things to see before you die," contributing to issues of overtourism in many places.[13] Research indicates that residents see UNESCO recognition as a positive, adding to their sense of pride, and many have been able to achieve great improvements in their living standards because of tourism. But they are also feeling the impacts of the influx of tourism that World Heritage status has brought, for better and for worse,[14] especially in the historic core.

Central to this growth has been the continued rise in intraregional tourism, particularly from China. Expanded flight networks are catalytic as more tourists have been traveling in an "era of hyper mobility" and exceptional affluence.[15] For example, Vietnam Airlines began flying from many more locations in Vietnam, with more direct routes to numerous locations in neighboring countries belonging to the Association of Southeast Asian Nations. New flight schedules of this kind make it much easier to fly to Luang Prabang from Hanoi, Ho Chi Minh City, or Da Nang in Vietnam; Phnom Penh or Siem Reap (Angkor) in Cambodia; Bangkok and Chaing Mai in Thailand; Yangon in Myanmar; eight hubs in China; and multiple locations in Taiwan, South Korea, and Japan.

Impacts of Overtourism

Visitor numbers to Luang Prabang had grown exponentially from 30,769 in 1997 to 755,019 in 2018, a stark difference compared to the 24,000 residents within the World Heritage core.[16] Tourism has become the second major component of the economy, contributing 13.7 percent of GDP in 2017.[17]

Accommodation Development

Due to UNESCO restrictions on the height and style of buildings, there are no large hotels in the historic core of the town. To accommodate the demand and comply with the building regulations in the core, tourist guesthouses are being renovated and built in traditional style.[18] In 2017, there were 349 guesthouses, 81 hotels and resorts, 290 restaurants, and 93 tour companies within the historic core, in comparison to two decades prior when there were just 19 guest houses, 10 hotels, 22 restaurants, and 7 tour companies.[19] These numbers continue to grow, and pressures of visitor growth have not only shaped the physical landscape within the historic core, but the social one as well.

Outmigration from Historic Core

The Heritage Office of Luang Prabang, under the Ministry of Information, Culture, and Tourism, is responsible for the management of the heritage-listed items (buildings, monuments, temples, and a series of interlinking ponds) in the historic core. An extensive audit of the condition of the heritage-listed items in Luang Prabang identified major social repercussions due to tourism in the historic area. The audit found a huge conversion from residential to commercial use of buildings in the core, dropping from 47 percent residential in 2001 to 29 percent in 2018. Almost 43 percent of land use is now tourism-related.[20,21]

Although the regulations on hotel development and design style have preserved the streetscapes, this conversion from resident housing to tourist accommodations or other touristic uses is leading to a reduction in the local population. Property prices have skyrocketed; small plots of land in the core that would have sold for around USD$8,000 a few years earlier have increased in value to more than $100,000. The escalating surge in land value since the early 2000s led to many families selling or leasing property. This level of income was previously unheard of in a country with 77.2 percent of the population still working poor (earning less than USD$3.10 per day).[22] Many residents have leased their homes and shops to tourism businesses and investors (either foreigners or Lao returned from exile) who have the financial resources and expertise needed both to renovate these heritage-listed buildings and run a profitable tourism business. This loss of residents threatens the intangible heritage of the historic core and the future of many *wats* (temples) in the core.[23] Indicative of the high pressures of increased tourism, life in the World Heritage core has now become unsustainable (socially, culturally and/or financially) for many multigenerational local families.

Sai Bat

In Luang Prabang, this outmigration of locals from the core has not only changed the social landscape, but the spiritual and intellectual one as well. Here, the monastic community has traditionally been the source of spiritual and intellectual nourishment for each village. In return, the residents of the community care for the well-being of the monks and temples, evident in the morning ritual of almsgiving (*sai bat*) where monks receive their first meal of the day. Traditionally, families rise early and prepare fresh food to offer the monks in return for endowing merit on the families.

Home for lease in Luang Prabang. Source: Robyn Bushell.

Today, the *wats*, numbering fifty-eight in the city and thirty-four in the core area, have lost many of the residents who for generations lived in the surrounding thirty *bans* (villages). These residents are being replaced by tourists staying in converted houses. The visitors flock to observe and participate along the processional routes of the monks through town. Instead of fresh home-cooked local food, visitors offer prepacked sweets and snacks or rice. Monks can be observed at times dumping out the offerings, often now considered inappropriate and unhealthy. Instead, the very poorest are the recipients of this surplus, which is often high in salt and sugar or has a high potential for food poisoning, especially in children.

Sai bat is at risk of becoming a staged performance rather than a customary sacred ritual. The monks, on their morning procession, also have to suffer the constancy of the photographic/touristic gaze. The abbots of the temples, the Heritage Office, and local residents worry about this trajectory. As leaders in Luang Prabang's heritage preservation efforts have noted, "This is a religious procession, not Disneyland."[24]

The watchful eye of tourists over sai bat. Source: Robyn Bushell.

Visitor Dissatisfaction

Luang Prabang has been a very high yielding destination, with many five-star properties, high occupancies, and long lengths of stay. The visitor experience is now suffering as the growth in tourism dilutes the authentic heritage of Luang Prabang, however. Very few discerning travelers wish to observe other tourists participating in *sai bat,* a ritual for centuries centered on the sacred relationship between the village and its temple. Adverse comments on Tripadvisor highlight visitor dissatisfaction with the behavior of many tourists: "I witnessed what I can only describe as a complete circus."[25] Although the various authorities, community representatives, and some businesses provide advice on appropriate tourist behavior, the problem has gone beyond attempts to remedy it.

Solutions to Overtourism

Despite the growing prosperity, residents of Luang Prabang face many challenges due to the ever-increasing impositions of tourism: demand

Monks carry large bags of commercial produce given to them by tourists to hand out. Source: Robyn Bushell.

for resources, escalating prices, and dislocation from the family home and *ban*, along with the erosion of the most sacred aspect of their intangible cultural heritage. The effects of overtourism on communities are a powerful reminder to relevant government authorities to consider what constitutes good tourism planning and to monitor outcomes. The goal is not just increasing visitor numbers, but ensuring satisfaction for visitors and residents alike. As an expatriate tourism operator concedes, "On one hand, it's a positive thing that those families benefitted from a major injection of funds into their households. . . . On the other hand, now that there are fewer and fewer Lao families living in the heart of Luang Prabang, a certain amount of the charm has been lost."[26]

Tourism Dispersal

As part of its commitment to the United Nations' Sustainable Development agenda, and with concern for the integrity of the World Heritage system, UNESCO, along with the assistance of other agencies, has been proactively seeking to better understand the costs of development on intangible heritage and everyday life for residents. Among various strategies for handling rising visitor numbers, one UNESCO suggestion to local authorities is to draw visitors away from the old town by promoting attractions further afield.[27] Currently, the majority of Luang Prabang's accommodations and attractions are in the core or neighboring areas, limiting tourists' desire to venture elsewhere. Effective dispersal will require strategic tourism planning to improve infrastructure and accessibility, as well as good development and promotion of tourism attractions and accommodations outside of the historic core. For example, at a workshop in Luang Prabang, heritage and tourism officials agreed that crowding at key sites within the core should not be addressed by increasing parking or making seating available for people in long queues.[28] Rather, they recommended limiting the number of minibuses allowed at each site in a given period and working with tour companies to diversify their itineraries, thus taking visitors to a range of alternate sites outside the core. This approach will disperse the crowds and also the benefits, bringing visitors to other communities that have sites and products of interest and who are eager to receive tourists and tourism income.

Encouraging Longer Stays

Paired with increasing visitor dispersal, encouraging longer stays has the potential to alleviate many of the impacts of overtourism. By encouraging longer stays, Luang Prabang may be able to increase the economic benefit and reduce the need for more infrastructure such as accommodations, thus slowing the loss of locals from the core area. Again, the Heritage Office and the Department of Tourism plan to work with tour companies, accommodation providers, and river cruise operators to look at strategies and incentives to encourage longer stays.

Tourist Tax

The Heritage Office of Luang Prabang would also like to introduce a tourist tax, which would be used for conservation work and help support locals to be able to afford to maintain and stay in their homes.[29] A small amount per visitor would equate to a large total sum for conservation and the protection of the heritage of the site. The Heritage Office is championing the importance of keeping local residents living in the core, as they maintain the heritage values, they make the site authentic and interesting, and most importantly, they maintain its integrity.

Reevaluation of Tourism Metrics

To track the success of tourism in Luang Prabang, macroeconomic data—visitor numbers, contribution to GDP, employment, investment, and so forth—are routinely collected, as in most places. Harold Goodwin of the Responsible Tourism Partnership, however, considers such information "the wrong metric."[30] To ensure that tourism supports the local community and protects both the tangible and intangible heritage, great care is needed to design indicators that are site-specific and relevant to track progress against strategic goals. For example, are visitors staying long enough? Do the entry fees collected at heritage sites cover the costs of maintaining these sites? Are the most marginalized who lack education and experience provided with training opportunities to benefit from tourism? Unless the evaluation of tourism asks the right questions, many important aspects are overlooked.

In addition, tourism-related threats to Luang Prabang challenge the basis of World Heritage listing and the site's outstanding universal value (OUV). By UNESCO's definition, "Outstanding Universal Value means cultural and/or natural significance which is so exceptional as to transcend national boundaries and to be of common importance for present and future generations of all humanity."[31] The monitoring of and reporting on aspects of the site that are essential to maintain World Heritage listing can help combat the impacts of overtourism. For example, the Heritage Office is working to involve the hotels and guest houses in finding solutions to the problems with morning almsgiving so that it continues to align with the OUV. Other questions include how visitors might contribute to the conservation of the key heritage sites and how the right tourist market might support and sustain the traditional crafts of the ethnic communities.

This approach requires capacity-building for tourism staff and understanding the local values that are most affected and most important to local people, together with a reevaluation of tourism guidelines and policies. These aspects are all necessary so that development, strategic planning, and monitoring are consistent with the objectives of the

World Heritage Convention. This approach is sorely needed in small historic cities such as Luang Prabang. Visitor numbers and GDP are not the only measures of success. In fact, they might actually point to problems. The Heritage Office of Luang Prabang welcomes such an approach.

Success to Date

Although the concept of carrying capacity is well defined, the reality of determining how many is "too many" is much more complicated, and what constitutes "overtourism" is very subjective. If you ask local residents about the status of overtourism in their community, their perception is colored by their relationship to tourism, such as how close they live to the tourist area and whether or not their household income is directly related to tourism. Those with economic responsibility for seeing tourism grow or who have an investment linked to tourism are far more resilient and tolerant than those who gain little and stand to lose the traditions and ambience of their neighborhood. As tourism academic George Doxey noted many years ago, the "irritation index" of tourism varies from place to place and from person to person.[32]

What is clear, however, is that when local people are unable to stay in their homes due to tourism's impacts, we have exceeded the social carrying capacity of the place. Displacement of local people goes beyond discontent and changes the very fabric of the destinations that bring tourists in the first place, which raises questions of sustainability. The focus on effective management should encompass what needs to be protected, including local heritage, local people, the intangible aspects of everyday life, and the effects on the most vulnerable in the community. There is an important message for the tourism businesses, who have an ethical responsibility to care for the people and the place where they operate. They also need to understand that this approach is important to protect their own investments.

Notes

1. Martin Stuart-Fox. (1997). *A History of Laos.* Cambridge: Cambridge University Press.
2. Ibid.
3. World Bank. (2016). *Working for a World Free of Poverty: Lao PDR Country Report.*

http://www.worldbank.org/en/country/lao/overview; United Nations Development Program. (2019). "Lao People's Democratic Republic Human Development Index." http://hdr.undp.org/en/countries/profiles/LAO.

4. Russell Staiff and Robyn Bushell. (2017). "The 'Old' and the 'New': Events and Place Making in Luang Prabang, Laos." *International Journal of Event and Festival Management.* 8(1): 55–65.

5. United Nations World Tourism Organization. (2019). "World Tourism Highlights 2018." https://www.e unwto.org/doi/pdf/10.18111/9789284419876.

6. Russell Staiff and Robyn Bushell. (2012). "Mobility and Modernity in Luang Prabang, Laos: Re-Thinking Heritage and Tourism." *International Journal of Heritage Studies* 19(1): 98–113.

7. Tiffany Funk. (2019). "Visiting Luang Prabang, Laos: Social Responsibility in a Tourist Mecca." *One Mile at a Time.* https://onemileatatime.com/luang-prabang/.

8. Simon Usborne. (April 29, 2009). "Is UNESCO Damaging the World's Treasures?" *The Independent.* https://www.independent.co.uk/travel/news-and-advice/is-unesco-damaging-the-worlds-treasures-1675637.html.

9. World Bank. (2018). "World Development Indicators— Lao PDR Country Report." http://www.worldbank.org/en/country/lao/overview.

10. Buavanh Vilavong and Sitthiroth Rasphone. (January 1, 2020). "Laos on Course to Graduate from Least Developed Country Status." *East Asia Forum.* https://www.eastasiaforum.org/2020/01/01/laos-is-on-course-to-graduate-from-least-developed-country-status/; World Bank. (2016). *Working for a World Free of Poverty: Lao PDR Country Report.* http://www.worldbank.org/en/country/lao/overview; United Nations Development Program. (2019). "Lao People's Democratic Republic Human Development Index." http://hdr.undp.org/en/countries/profiles/LAO.

11. UNESCO World Heritage Centre. (March 2020). "Technical Co-operation for the Enhancement, Development and Protection of the Town of Luang Prabang, Lao PDR." https://whc.unesco.org/en/activities/29/.

12. Russell Staiff. (2012). "The Somatic and the Aesthetic: Embodied Heritage Tourism Experiences of Luang Prabang, Laos." In *The Cultural Moment in Tourism.* Eds. Laurajane Smith, Emma Waterton, and Steve Watson. pp. 3855. Abingdon and New York: Routledge.

13. Simon Usborne. (April 29, 2009). "Is UNESCO Damaging the World's Treasures?" *The Independent.* https://www.independent.co.uk/travel/news-and-advice/is-unesco-damaging-the-worlds-treasures-1675637.html.

14. Russell Staiff and Robyn Bushell. (2012). "Mobility and Modernity in Luang Prabang, Laos: Re-Thinking Heritage and Tourism." *International Journal of Heritage Studies* 19(1): 98–113.

15. Claudio Milano, Jospeh M. Cheer, and Marina Novelli. (2018). "Overtourism: A Growing Global Problem." *The Conversation.* http://theconversation.com/overtourism-a-growing-global-problem-100029.

16. Robyn Bushell. (2020). *Study of ASEAN World Heritage Sites and ASEAN Heritage Parks.* Report to ASEAN Tourism Committee, GIZ, and Ministry of Information, Culture & Tourism, Laos.

17. World Travel and Tourism Council (WTTC). (2018). "Travel and Tourism Economic Impact: Laos 2018." https://www.wttc.org/economic-impact/country-analysis/country-reports/.

18. Ceelia Leong, Jun-ichi Takada, and Shinobu Yamaguichi. (2016). "Analysis of the Changing Landscape of a World Heritage Site: Case of Luang Prabang, Lao PDR." *Sustainability* 8(8): 747.

19. Department of Information, Culture & Tourism. (2018). "The Luang Prabang Development and Marketing Strategy 20112020." With support from AFD and ADB and updated DICT Tourism Statistics.

20. Robyn Bushell. (2020). *Study of ASEAN World Heritage Sites and ASEAN Heritage Parks.* Report to ASEAN Tourism Committee, GIZ, and Ministry of Information, Culture & Tourism, Laos.

21. Ceelia Leong, Jun-ichi Takada, and Shinobu Yamaguichi. (2016). "Analysis of the Changing Landscape of a World Heritage Site: Case of Luang Prabang, Lao PDR." *Sustainability*, 8(8): 747.

22. United Nations Development Program. (2019). "Lao People's Democratic Republic Human Development Index." http://hdr.undp.org/en/countries/profiles/LAO.

23. Robyn Bushell and Russell Staiff. (2012). "Rethinking Relationships: World Heritage, Communities and Tourism." In: Patrick Daly and Tim Winter. (eds). *Routledge Handbook of Heritage in Asia.* Abingdon and New York: Routledge. pp. 247–65.

24. Denis D. Gray. (2016). "UNESCO World Heritage Sites and the Downside of Cultural Tourism." *Skift.* https://skift.com/2016/01/28/unesco-world-heritage-sites-and-the-downside-of-cultural-tourism/.

25. Keelin R. (September 10, 2018). "Ruined by Ignorant Tourists." Tripadvisor. https://www.tripadvisor.com/ShowUserReviews-g295415-d2229605-r615365151-Alms_Giving_Ceremony-Luang_Prabang_Luang_Prabang_Province.html?m=19905.

26. Sebastian Strangio. (May 2016). "Tourist Hordes Put Strains on Luang Prabang's Heritage." *Southeast Asia Globe.* https://southeastasiaglobe.com/19355-2-luang-prabang-tourism/.

27. Ibid.

28. Robyn Bushell. (2020). *Study of ASEAN World Heritage Sites and ASEAN Heritage Parks.* Report to ASEAN Tourism Committee, GIZ, and Ministry of Information, Culture & Tourism, Laos.

29. Ibid.

30. Harold Goodwin. (2017). *The Challenge of Overtourism.* Responsible Tourism Partnerships. Working Paper 4.

31. Georgia State University. (2020). "Outstanding Universal Value." https://worldheritage.gsu.edu/outstanding-universal-value/.

32. George V. Doxey. (1975). "A Causation Theory of Visitor-Resident Irritants: Methodology and Research Inferences in the Impact of Tourism." Sixth Annual Conference Proceedings of the Travel Research Association. San Diego.

Chapter 4.4

Mount Everest

By Birendra KC with Kelsey Frenkiel

Growing up in Nepal, I was fascinated by tourism. There was an ingrained perception that the tourism industry could do no wrong. Only the positive aspects of tourism were highlighted, acknowledging that it generates jobs for local people and increases foreign exchange earnings, as well as strengthens Nepal's international reputation. As Nepal is known for its majestic mountains, tourism in the mountain regions seemed related to those positive tourism impacts, even though ripple effects of tourism were occurring in different regions of Nepal. With the growing popularity of certain destinations and the rising number of tourists around the globe, however, it is clear that tourism can go awry if not managed properly.

In May 2019, one image of dozens of mountaineers climbing toe to toe on Mount Everest hit headlines. This human traffic jam is likely not the view that most climbers imagine when making their ascent, nor was it the view experienced by the first ones to reach the summit, Edmund Hillary and Tenzing Norgay, on May 29, 1953.[1] How is it that one of the most difficult and dangerous climbs in the world has become victim to overtourism? How can we better manage the site, one so important to our global natural heritage, so that the Nepalese perception of tourism's good blessings becomes a reality?

Causes of Overtourism

Nepal welcomed more than 1.1 million international tourists in 2018, a record high to date.[2] After Chitwan National Park, Sagarmatha National Park (the home of Mount Everest in Nepal) is the second most visited national park in the country. Chosen for its incredible beauty and valuable contribution to the preservation of exceptional natural phenomena, Sagarmatha National Park became a UNESCO World Heritage Site in 1979.[3] Regardless of the national park's isolated location, tourism to Mount Everest has been steadily on the rise. In 2017, for example, 50,381 international tourists visited Sagarmatha National Park, more than double the number of tourists who visited in 1998.[4] These numbers do not even include Nepalese travelers; citizens were not required to pay an entrance fee upon visiting the park until 2017, so records of domestic visitation are not available. (Although domestic visitation records do exist from 2018 onward, they have not yet been made public.)[5]

May is considered the safest month to summit Mount Everest due to the weather conditions, and the vast majority of ascents occur then. According to records from the Nepal side, in May 2019, approximately 450 climbers summited Mount Everest.[6] This figure is on par with recent years; since 2008, there have been more than 300 summiteers each year (other than 2014 and then again in 2015, when tourism was impacted by the earthquake). Sixty-three percent of all Everest summits, beginning in 1953, have occurred between 2008 and 2017. In 2013, there were a record 578 summiteers. And on a single day in 1993, a record 40 people reached the summit.[7]

Reports of overcrowding have more often targeted the Nepal side of Mount Everest, which has traditionally been more popular for climbers and where there is no cap to the number of permits that can be distributed. The number of climbers on the Tibetan side (the north side) has been increasing relative to Nepal (the south side) in recent years, however.[8] Climbers' preferences often shift along with natural disasters, political changes, worker strikes, and changes to regulations and permitting.

To fully capture the extent of overcrowding, though, one must also consider the vast unaccounted for number of climbers who failed to

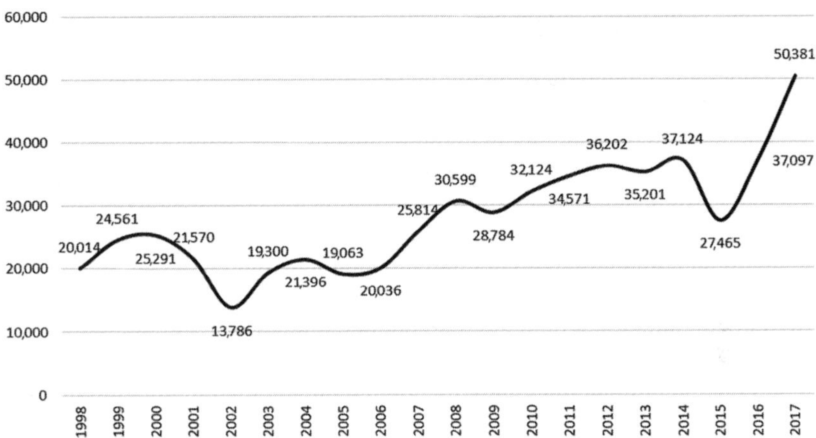

International visitation to Sagarmatha National Park, 1998–2017. Source: Prepared by Birendra KC and CREST with data from Sagarmatha National Park.

summit, as well as those whose destination was merely the base camp. The national park estimates that about 35,000 visitors trek to the base camp each year.[9]

Short Peak Season

Experts in the region report that one cause of overtourism is due to the short peak season, during which the climb is safest. Treks to the summit are normally scheduled within a tight window of time due to the ideal weather conditions, increasing the likelihood of a successful summit. That's why you will see lines of climbers all attempting the summit on the same day.

As a national park, one concern is that visitation is, geographically, not fairly distributed. Even though there are multiple seasons when the park is equally mesmerizing to visit and enjoy the natural beauty, there are only a few months out of the year when visitation remains high. Meanwhile, during the peak season, only a few locations within the park are popular, even though many parts of the park remain unexplored.[10]

Inexperienced Climbers, Guiding Companies, and Loose Government Regulations

Permits are required for foreigners to climb to the summit on both the Tibetan and Nepalese sides of Mount Everest. On the Tibetan side, permits are issued by the China Tibet Mountaineering Association and cost $11,450 per person in a group of four or more and $19,500 per person in a group of three or fewer.[11] For many years, Nepal had charged $25,000 per person during the peak climbing season, or $70,000 total for a group of seven or more (effectively $10,000 per person). In 2014, they cut the permits to $11,000, no matter the number of climbers in the party, to make the trek more accessible and removed the discounted permitting structure for groups. Experts were apprehensive even then; said managing director of Jagged Globe, a British expedition operator, "I find that terrifying frankly; safety comes from being in a team. This will open the floodgates for anyone to say 'I'm an expert mountaineer,' get a client and away they go. If something goes wrong they'll have to reach out to other teams. Help is always given, but it's frustrating when you end up having to help people who shouldn't be on the mountain."[12]

The percentage of relatively inexperienced climbers attempting this grueling journey is said to be increasing, due in part to these loose regulations. In addition to the cost of the permits, the price offered by tour companies is becoming more accessible too: "While most will pay about US$45,000, low-cost Nepalese operators have popped up in recent years offering to take people to the peak for US$30,000," said one experienced guide. "For many people, the only obstacle standing between them and conquering the world's most famous mountain is financial."[13] Depending on the tour operators, price negotiations take place indicating commercialization in mountain climbing. There is a perception that many new inexperienced climbers attempt to summit as a bragging point.[14]

According to one report:

> Fly-by-night adventure companies are taking up untrained climbers who pose a risk to everyone on the mountain. And the Nepalese government, hungry for every climbing dollar it can get, has issued more permits than Everest can safely handle, some experienced mountaineers say.... And the fact that Nepal, one of Asia's poor-

> est nations and the site of most Everest climbs, has a long record of shoddy regulations, mismanagement and corruption. . . . Nepal has no strict rules about who can climb Everest, and veteran climbers say that is a recipe for disaster.[15]

Impacts

The allure of Mount Everest, for some, lies in the sense of solitude—of being at the top of the world, "like the astronauts in their lunar module."[16] Many are attempting to accomplish what few others have, but that becomes a glorified fantasy when compared to the images described by the *New York Times* in 2019: "Climbers were pushing and shoving to take selfies. The flat part of the summit . . . was packed with 15 or 20 people. To get up there, [climber Ed Dohring] had to wait hours in line, chest to chest, one puffy jacket after the next . . . he said. . . . 'It was like a zoo.'"[17] Overtourism on Mount Everest has had some alarming consequences, beyond the degradation of the visitor experience.

Safety

An exponential growth in the number for mountain climbers demonstrates the issue of overtourism on Mount Everest.[18] Media scrutiny for overtourism reached new levels after eleven climbers died in 2019, some reportedly correlated with the tourist congestion.[19] Climbers are already at high risk of frostbite, fatigue, and hypoxemia/hypoxia from the lack of oxygen. Each of these factors is a by-product of time and can be exacerbated by delays. In numerous accounts of climber deaths, Sherpa guides have credited the difficulty and time involved in getting them to safety: "[A tour group manager] . . . said it took the group an additional three hours to return to camp, a wait that he believes contributed to [one woman's] death. Several of the climbers with Ms. Kulkarni returned to their camp with frostbite and other injuries."[20] In May 2019, climber Robin Fisher posted a video of his trek to Instagram with the caption: "I am hopeful to avoid crowds on summit day . . . With a single route to the summit delays caused by overcrowding could provide fatal so I am hopeful my decision to go for the 25th [of May] will mean fewer people. Unless of course everyone else plays the same waiting game."[21] Fisher later died during his descent, but it was unclear whether a delay due to overcrowding was a direct cause.[22]

Some have debated the assertion that overcrowding is the direct cause of these deaths on Mount Everest, attributing it to climbers' inexperience and poor decision-making by Sherpas.[23] One thing is certain: "If the crowds aren't directly culpable for killing people, they are unquestionably responsible for increasing the risks by necessitating longer summit days—indelibly changing the dynamic of climbing Everest."[24]

Environment

If summiting Mount Everest alone is a glorified fantasy, so too is the idea that the mountain remains a pristine, untouched landscape. Reports of overcrowding have also lamented the huge amounts of trash left by climbers. Estimates place the amount of waste left behind since about 1970 between 10 and 50 tons, which includes oxygen cylinders, human excrement, plastic bottles, food and supplies, and other trash.[25] More than two hundred human bodies also leave their mark.[26] Organic waste and the bodies will not biodegrade quickly due to the temperatures.

The waste left behind on Mount Everest is not simply a result of negligence; often, there is a danger involved in carrying extraneous weight to the summit or back down. Says climber Mark Jenkins, "Even under the best conditions, climbing the tallest mountain in the world is exhausting, dangerous work. Dropping used supplies on the mountain rather than carrying it with them can save vital energy and weight."[27] There is also a cost associated with returning human bodies to the base, and laws can prevent guides from burying or cremating them.[28]

In recent years, warming due to climate change has caused glaciers around the Himalayas to degrade. Popular routes have changed or become unsafe due to the risk of ice fall.[29] Warming has also had the unfortunate consequence of revealing human bodies. Climate change, combined with the waste issue, will surely leave Everest looking different than it did when Norgay and Hillary first made their ascent.

Solutions

As an internationally recognized site, the management of tourism on Mount Everest is the responsibility of the public, private, and civil sectors. Local government park authorities, tour operators, environmental organizations, and others all play a part in ensuring that Mount Everest is maintained. (As above, these sections focus on initiatives from the Nepal side of Mount Everest.)

Existing Management Framework

Before overtourism on Mount Everest was widely recognized, policies and management protocols were in place to conserve the park as a whole. Sagarmatha National Park was established in 1976 and quickly designated as a World Heritage Site in 1979.[30] It is managed by a government body, the National Park and Wildlife Conservation Office. The first management plan was not implemented until the period 2007–2012, and not all activities under the plan were carried out due to funding setbacks. A management plan for 2016–2020 was established, which acknowledged "overcrowding" and a "lack of clear tourism policy and dedicated institutional setup to deal with the increasing number of tourists in the Park."[31] It laid out a wide range of activities to protect the landscape and wildlife, as well as enhancing support for local communities, such as initiating a ticketing system at entry points to the park, developing a management body specifically to address tourism, designating special "zero carbon tourism" areas, and initiating studies of the impact of tourism on the environment. In addition, two of their special programs are generally recommending initiatives to protect the park's World Heritage status in the context of growing and unmanaged tourism and waste management. It is unclear whether all these activities have yet been implemented, however. It must also be noted that these management plans apply to the park more generally, rather than Mount Everest specifically.

According to UNESCO, one of the main characteristics protecting the integrity of Sagarmatha National Park is the buffer zone program, which "[enhances] conservation in combination with improved socioeconomic status of the local communities through a revenue plough back system."[32] Since 2002, local buffer zone committees have received up to 50 percent of the revenue from the park and use the funding for conservation, community development, income generation and skill development, and conservation education. The influx of tourists has no doubt had positive implications for the recipients of this funding.

Waste Management

The Sagarmatha Pollution Control Committee (SPCC), a Nepalese nonprofit organization that partners with the national park, is the main entity spearheading waste management initiatives in Sagarmatha National Park, in the main buffer zone, and on Mount Everest.[33] The SPCC is responsible for waste management on all peaks where a permit is required to trek. Climbers are required to register with the SPCC, which includes submitting a list of food and supplies, and then must return more than 8 kilograms (17.5 pounds) of waste after their trek to receive a refund of their deposit. The SPCC handles both "burnable" waste and recyclable waste and promotes a strict code of conduct for climbers that prohibits burning or burying garbage. It also requires climbers to bring down all human waste.

Although guides and nonprofit organizations have organized small cleanup projects in the past, the SPCC has now started to organize longer and more robust cleanup campaigns.[34] The SPCC reports that 775 kilograms (1,709 pounds) of waste were recovered from Mount Everest in 2018 (a two-week campaign) and 4,692 kilograms (10,356 pounds) in 2019 (a month-long campaign). The campaigns involved the Nepalese army, Sherpas, and other government entities and was funded by large organizations such as the World Wildlife Fund and the Coca-Cola Foundation. Multiple mountain closures to allow for cleanup projects have also occurred on the Tibetan side of the mountain, with one clearing, for example, 8.5 tons of waste in June 2018.[35]

Permitting and Climbing Regulations

After the deadly climbing season in the spring of 2019, many people pressured Nepal to increase its permit fees, but that did not happen. Instead in August 2019, Nepalese officials proposed new rules to restrict permits only to climbers who have summited mountains higher than 21,300 feet, as well as to tourism companies with a minimum of three years' experience.[36] To date, there is no limit on the number of climbers who can attempt to summit Mount Everest at one time.

In 2019, Nepal also released a more stringent set of regulations, including a $5,000 security deposit, a garbage collection fee of $1,500, clauses regarding the reputation of tour companies allowed on the mountain, and a stipulation that expenses incurred by rescue missions will be paid for by climbers.[37] Beyond just the cost of the permits, these regulations also included other price increases, such as the fee for Sherpas and transportation packages. Time will tell whether these new regulations will have an effect on overcrowding and the number of fatalities.

Lessons Learned from Other World Heritage Sites

Expedition companies are not the only responsible parties in ensuring Mount Everest is safe for climbers in light of the crowding. The government of Nepal (including national park authorities), UNESCO, and travelers themselves each play a role. It is up to authorities to mandate certain rules and regulations to prevent overtourism and the serious consequences that arise from it. An ongoing debate exists on how national and local authorities should address overtourism as Nepal continues to aggressively try to increase the number of international tourists.[38] Most recently, the Visit Nepal 2020 campaign was designed to increase international tourist arrivals to two million.[39]

The lack of a determined carrying capacity or quota system allows for an unlimited number of tourists to climb the mountain at any given time. Even though Everest climbers get permission from the Department of Tourism to enter Sagarmatha National Park and their visit is incorporated into park visitation data system, their trip schedule is beyond the park management team's control. To improve this system, the Department of Tourism and Mount Everest National Park authorities need to collaborate to achieve better efficiency and oversight. In addition, regulations need to be implemented to restrict the number of mountain climbers, even during the peak season, so that overcrowding is no longer a problem.

Around the world, several national parks and World Heritage Sites are taking small steps to curb overtourism. Some national parks in the United States, for example, are increasing their fees to generate the revenue needed to better manage their parks.[40] Machu Picchu in Peru (see chapter 4.2) is limiting the number of people and usable walking paths.[41]

As overtourism causes environmental degradation, authorities should also look to others for ideas on more generalized sustainability policies. One example is Chitwan National Park in Nepal, which recently implemented a plastic-free movement to promote environmental sustainability within the park.[42] A movement of this kind is needed in the Mount Everest region, which is also experiencing the impacts of climate change (so much so that Friends of the Earth and other environmental activists, including Edmund Hillary, petitioned to move the site to the UNESCO World Heritage in Danger List in 2005).[43] For a national sustainability movement throughout Nepalese tourism sites, broader collaborative effort is required.

Conclusion

Mount Everest is a matter of national pride for the Nepalese. Nepal's reputation and its microeconomies are directly and indirectly affected by the trajectory of overtourism on Mount Everest. Proper management and implementation of specific rules and regulations are critical. Sustainability-focused practices must be prioritized by the government, along with collaborative efforts from fellow government bodies, operating sectors of tourism industry, and nongovernmental organizations.

At the same time, it is unfair to put the blame for overtourism solely on local governments and management authorities. Locals should be held accountable, but it is important to remember that Nepal is heavily reliant on tourism for economic development, and management authorities may be reluctant to create regulations that in their minds may deter income. Outside influence from UNESCO and even climbers themselves needs to be leveraged to protect this internationally significant natural site together with local management authorities, addressing the barriers that they face. On an individual level, potential climbers should assess their level of expertise and experience before choosing to climb Mount Everest and potentially seek out alternative treks on other mountains in the region. While planning their trip, they need to thoroughly research their options and make conscious choices, which will have a positive impact on the destination.

In 1923, climber George Mallory famously responded this way when asked why he wanted to climb Mount Everest: "Because it's there."[44] The following year, Mallory joined the ranks of climbers who lost their lives while attempting to reach the peak. For each ascent that is made, tourism authorities, operators, and travelers must ask themselves whether "Because it's there" is a sufficient reason.

Notes

1. Ministry of Culture, Tourism & Civil Aviation. (2019). Mountaineering in Nepal: Facts and Figures 2018. https://nepalindata.com/resource/mountaineering-nepal-facts-and-figures-2018/.

2. Ministry of Culture, Tourism & Civil Aviation. (2019). Nepal Tourism Statistics 2018. https://www.tourism.gov.np//files/publication_files/287.pdf.

3. UNESCO World Heritage Center. (2020). World Heritage List. https://whc.unesco.org/en/list/.

4. Ministry of Forests and Environment. (2017). "Sagarmatha National Park Annual Report." http://www.sagarmathanationalpark.gov.np/index.php/documents/cat_view/5-annual-report.

5. Pramod Bhattarai. (July 22, 2019). Chief Conservation Officer, Sagarmatha National Park. Personal communication with author.

6. Ministry of Culture, Tourism & Civil Aviation. (2019). Mountaineering in Nepal: Facts and Figures 2018. https://nepalindata.com/resource/mountaineering-nepal-facts-and-figures-2018/.

7. The Daily Beast. (September 28, 2018). "Mount Everest by the Numbers: Deaths, Cost to Climb, and More Mountain Records." https://www.thedailybeast.com/mount-everest-by-the-numbers-deaths-cost-to-climb-and-more-mountain-records?ref=scroll.

8. Alan Arnette. (2019). "Everest by the Numbers: 2018 Edition." https://www.alanarnette.com/blog/2017/12/17/everest-by-the-numbers-2018-edition/.

9. Basecamp Trek Nepal. (2020). "How Many Tourists Visit Mount Everest Each Year." https://basecamptreknepal.com/how-many-tourists-visit-mount-everest-each-year.

10. Pramod Bhattarai. (July 22, 2019). Chief Conservation Officer, Sagarmatha National Park. Personal communication with author.

11. Alan Arnette. (June 21, 2019). "Everest: Nepal Looking at Huge Permit Fee Increase." https://www.alanarnette.com/blog/2019/06/21/everest-nepal-looking-at-huge-permit-fee-increase/.

12. Will Coldwell and agencies. (February 14, 2014.) "Nepal Slashes Cost of

Climbing Everest." *The Guardian.* https://www.theguardian.com/world/2014/feb/14/nepal-slashes-cost-climbing-mount-everest.

13. Mercedes Hutton. (May 29, 2019). "When Overtourism Turns Deadly—why Everest Should Be Off-Limits to Casual Climbers." *Post Magazine.* https://www.scmp.com/magazines/post-magazine/travel/article/3012085/when-overtourism-turns-deadly-why-everest-should-be.

14. Ibid.

15. Kai Schultz, Jeffrey Gettleman, Mujib Mashal, and Bhadra Sharma. (May 26, 2019). "'It Was Like a Zoo': Death on an Unruly, Overcrowded Everest." *New York Times.* https://www.nytimes.com/2019/05/26/world/asia/mount-everest-deaths.html.

16. Stephen Venables. (May 29, 2013). "Mount Everest: The View from the Top." *The Guardian.* https://www.theguardian.com/world/2013/may/29/mount-everest-view-from-top.

17. Kai Schultz, Jeffrey Gettleman, Mujib Mashal, and Bhadra Sharma. (May 26, 2019). "'It Was Like a Zoo': Death on an Unruly, Overcrowded Everest." *New York Times.* https://www.nytimes.com/2019/05/26/world/asia/mount-everest-deaths.html.

18. Ministry of Culture, Tourism & Civil Aviation. (2019). Mountaineering in Nepal: Facts and Figures 2018. https://nepalindata.com/resource/mountaineering-nepal-facts-and-figures-2018/.

19. Karen Zraick and Derrick Bryson Taylor. (May 29, 2019). "These Are the Victims of a Deadly Climbing Season on Mount Everest." *The New York Times.* https://www.nytimes.com/2019/05/29/world/asia/everest-deaths.html.

20. Megan Specia. (May 23, 2019). "On Everest, Traffic Isn't Just Inconvenient. It Can Be Deadly." *New York Times.* https://www.nytimes.com/2019/05/23/world/asia/deadly-everest-traffic-jam.html.

21. Robin Fisher. (May 19, 2019). Instagram caption. https://www.instagram.com/p/BxofaW7ljpj/.

22. CBS News. (May 27, 2019.) "Everest Climber Who Warned of Overcrowding Becomes 8th Person to Die on Peak This Year." https://www.cbsnews.com/news/everest-climber-robin-fisher-who-warned-of-overcrowding-becomes-8th-person-to-die-on-peak/.

23. Freddie Wilkinson. (May 29, 2019). "Traffic Jams Are Just One of the Problems Facing Climbers on Everest." *National Geographic.* https://www.nationalgeographic.com/adventure/2019/05/everest-season-deaths-controversy-crowding/.

24. Ibid.

25. TStreet Media. (June 27, 2017). "The Environmental Impact of Everest Mountain Climbing." *Drifter.* https://driftermagazine.ca/the-environmental-impact-of-everest-mountain-climbing-50fc9bdb100.

26. Rachel Nuwer. (October 8, 2015). "The Tragic Tale of Mt. Everest's Most Famous Dead Body." BBC Future. https://www.bbc.com/future/article/20151008-the-tragic-story-of-mt-everests-most-famous-dead-body?referer=https%3A%2F%2Fsearch.yahoo.com%2F.

27. Harold Maass. (May 29, 2013). "Mt. Everest's Filthy Secret: It's a Dump." *The Week.* https://theweek.com/articles/463825/mt-everests-filthy-secret-dump.

28. Bhadra Sharma and Kai Schultz. (March 20, 2018). "How Do You Get 200,000 Pounds of Trash Off Everest? Recruit Yaks." *New York Times.* https://www.nytimes.com/2018/03/20/world/asia/mount-everest-trash-nepal.html.

29. Bennett Slavsky. (May 29, 2019). "How Climate Change Is Making Mount Everest More Dangerous." *Climbing.* https://www.climbing.com/news/climate-change-on-mount-everest-old-bodies-and-new/.

30. Ministry of Forests and Environment. (2016). "Sagarmatha National Park and Its Buffer Zone Management Plan 2016–2020." http://www.sagarmathanationalpark.gov.np/index.php/documents.

31. Ibid.

32. UNESCO. (2020). "Sagarmatha National Park." https://whc.unesco.org/en/list/120.

33. Sagarmatha Pollution Control Committee. (2020). https://spcc.org.np/.

34. Sagarmatha Pollution Control Committee. (2020). "Cleanup Campaigns." https://spcc.org.np/cleanup-campaigns/ourwork.

35. Lily Kuo Beijing and Peter Beaumont. (January 21, 2019). "Mount Everest Climber Numbers Face Major Cut as China Starts Cleanup." *The Guardian.* https://www.theguardian.com/world/2019/jan/21/mount-everest-climbers-face-major-clampdown-as-china-begins-cleanup; Kelsey Cheng. (February 15, 2019). "China Closes 'Core Port' of Mount Everest Including a Base Camp to Tourists to Deal with Escalating Waste Problems." *Daily Mail.* https://www.dailymail.co.uk/news/article-6708351/China-closes-Mount-Everest-base-camp-Tibet-deal-escalating-waste-problems.html.

36. Bhadra Sharma and Kai Schultz. (August 14, 2019). "New Everest Rules Could Signficiantly Limit Who Gets to Climb." *New York Times.* https://www.nytimes.com/2019/08/14/world/asia/everest-climbing-rules.html.

37. Alan Arnette. (June 21, 2019). "Everest: Nepal Looking at Huge Permit Fee Increase." https://www.alanarnette.com/blog/2019/06/21/everest-nepal-looking-at-huge-permit-fee-increase/.

38. Pramod Bhattarai. (July 22, 2019). Chief Conservation Officer, Sagarmatha National Park. Personal communication with author.

39. Times of India. "Nepal Tourism to Strengthen Its Tourism Game in 2020 with 'Visit Nepal Year.'" https://timesofindia.indiatimes.com/travel/destinations/nepal-tourism-to-strengthen-its-tourism-game-in-2020-with-visit-nepal-year/as69789635.cms.

40. Brian Maffly and Erin Alberty. (October 25, 2017). "How's $70 to Get in to One of Your Favorite National Parks?" *Salt Lake Tribune.* https://www.sltrib.com/news/2017/10/24/national-park-service-proposes-massive-fee-hike-at-utah-parks/.

41. Johhny Jet. (August 20, 2018). "How Is Overtourism Impacting Travel to Popular Destinations?" *Forbes.* https://www.forbes.com/sites/johnnyjet/2018/08/20/how-is-overtourism-impacting-travel-to-popular-destinations/#cafb4a635b84.

42. World Wildlife Fund. (January 1, 2019). "Chitwan National Park Becomes Nepal's First Plastic-Free Protected Area." https://wwf.panda.org/knowledge_hub/where_we_work/eastern_himalaya/news_stories/?340910/Chitwan-National-Park-becomes-Nepals-first-plastic-free-Protected-Area.

43. CIPRA. (December 2, 2004). "Mount Everest Put Forward for UNESCO

World Heritage in Danger List." https://www.cipra.org/en/news/1510; Friends of the Earth International. (July 11, 2005). "Everest Must Be Put on a Danger List." https://www.foei.org/press_releases/everest-must-be-put-on-an-danger-list.

44. Lyndsey Matthews. (May 28, 2019). "Why Do People Climb Mount Everest? These Riveting Stories Explain the Fascination." *Afar.* https://www.afar.com/magazine/why-do-people-climb-mount-everest-these-riveting-stories-explain-the-fascination.

Chapter 5.1

Beaches and Coastlines

By Andrea Sachs with Martha Honey*

In sixth grade, Isabelle Ramage studied droplets from ponds on Cape Cod, where she grew up. After seeing what lives in the water, she never swam in ponds again. In high school, she took an environmental science class with a teacher who dedicated several months to the community's derelict septic system, which has been buckling under the pressure of too many users, including tourists. As a teenager, she knew more about how human waste ends up in the bays than most adults visiting the Massachusetts shoreline. Now finished with high school, the older and wiser graduate has the same reaction to the start of summer as her younger self. While thousands of suntan lotion-slicked tourists stampede the public beaches, she retreats. "I don't go to the public beaches. They are extremely crowded and noisy. We can reclaim them after Labor Day but never, ever, during the summer. We don't have that option," the then-eighteen-year-old said. "Being right on the ocean is an amazing thing, but we have to be cautious. We need to educate the people coming here."[2]

Living on Cape Cod, or any coastal region for that matter, can be

* Martha Honey contributed to the research and writing of the sections on cruise tourism.

idyllic. The ocean, the sun, the leaping dolphins all cue the sunset and sighs. Yet as Isabelle can attest, residing in a coastal community overrun with tourists has its downsides, and Cape Cod is not unique. "There are seven billion people in the world, and probably half like to go the beach," said Jonathan Tourtellot, chief executive officer of the Destination Stewardship Center and founder of the National Geographic Center for Sustainable Destinations. "But the beaches aren't getting larger, except maybe in Dubai."[3]

Overtourism in coastal destinations shares several traits with overtourism in land-locked locations. Residents in both geographies are finding themselves marginalized as rents skyrocket, cultural traditions vanish, and traffic grows even knottier. Isabelle calls these challenges "normal." As in, it's normal that the twenty-minute drive to dance class takes twice as long in the summer. That her volunteer group, in Hyannis, requires extra garbage bags to pick up the trash left by tourists. That she can't drink from the unfiltered water fountain at school because of the presence of toxic chemicals—the septic system strikes again. As Andrew Gottlieb, executive director of the Association to Preserve Cape Cod, put it, "Half-a-million people visit over three months, more than doubling the year-round population, and all those people pee and poop."[4]

"People in local communities begin to turn against tourists because they feel that their needs are being pushed to the side in favor of the tourists' needs," said Rochelle Turner, research director of the World Travel and Tourism Council (WTTC). "Higher property rents, the inability of shops to sell products that residents need, and crowded streets are all issues that cause local people to feel alienated." She added that tourists also "pick plants, tread on areas outside of boundaries, leave behind waste and rubbish, trample on paths, and impact the natural environment of a destination."[5]

Even at her age, Isabelle understands the need for tourist dollars, although she sweetly admitted, "The ongoing joke is that we don't like tourists. They're always in the way."

Residents aren't the only ones feeling the squeeze; tourists also complain of too many tourists. Researchers at George Washington University's International Institute of Tourism Studies explored this subject in a spring 2019 project. Using Cruise Critic as their data

A typical beach scene on Cape Cod, Massachusetts. Source: Henrik Driesler (Flickr).

pool, the group evaluated 450 consumer reviews of nine attractions in Cozumel, the Bahamas, and the Cayman Islands, the three busiest cruise ports in the Caribbean region. Focusing on comments posted between February 2012 and April 2019, they discovered that overcrowding was a frequent gripe, along with value and cost. In February 2019, a cruise passenger in Grand Cayman reported having "a very fun experience. However, there were so many boats and people in the area, that stingrays were in limited supply, so to speak. Our boat captain had to go catch one and bring it over to our group." Another cruise passenger shared her disappointment after a May 2015 excursion to Grand Cayman's Seven Mile Beach, saying that the trip was "OK, if you like to feel like a sardine on a dirty beach, surrounded by drunks!"[6]

At the same time, in a number of coastal and beach communities, excessive numbers of cruise passengers are degrading experiences for both stayover vacationers and local residents. A review of Tripadvisor finds overnight visitors voicing concerns about cruise tourism. Take, for instance, Key West, Florida. A February 2020 post reads: "Disappointment. Read so much about sunset parties at Mallory Square. Arrived and found a huge cruise ship taking over the entire view of the ocean." Another in January 2019 notes: "Watch the sunset, that is, if a giant cruise ship is not blocking your view. Years ago this spot like most of Key West was charming. Now it's become more of a cruise ship port—tourist trap."[7]

Key West residents have in recent years also sought to draw a line in the sand (or sea) to control cruise tourism. In October 2013, a citizen's coalition, the Key West Committee for Responsible Tourism, mounted a Vote No campaign on a referendum to study plans to dredge a deeper channel to accommodate larger cruise ships. Voters turned out in large numbers, and nearly three-fourths (74 percent) cast ballots against the dredging plan. The channel runs directly across the Florida Keys National Marine Sanctuary, which includes the only barrier coral reef in North America. According to the Army Corps of Engineers, deepening the channel would "have a significant negative effect on environmental quality."[8]

Then in the wake of the 2020 pandemic and revelations of cruise ships as "superspreaders" of COVID-19,[9] a new coalition, the Key West Committee for Safer Cleaner Ships, succeeded in getting three referenda on the November ballot. The ballots, which called for limiting the size of ships and number of disembarking passengers and giving preference to those ships with better environmental practices, had wide public support.[10]

Key West is not alone. In other cruise destinations, residents have taken to the streets or the sea, as well as the media and the ballot box, to decry the impacts of cruise tourism on their quality of life. Campaigns in US and European port cities, including Charleston (South Carolina), Juneau (Alaska), Seattle (Washington), Venice (Italy), Barcelona (Spain), and Dubrovnik (Croatia), have sought to limit the number of cruise ships and passengers, control environmental impacts, and increase the earnings from cruise tourism retained by the coastal destinations.[11] In response to the pandemic, a global coalition of activists in some dozen port cities around the world mounted the first-of-its-kind virtual "Rally to Clean Up Cruising" on World Oceans Day. Under the umbrella of "Clean Up Carnival," the coalition continued to challenge the resumption of cruise tourism until significant reforms had been put in place.[12]

Causes of Beach Overtourism

Myriad factors cause overtourism, including one that is hard to fault: we simply love to swim, snorkel, surf, stand-up paddle, splash in waves, and loaf on the beach like a sea lion after a big pescatarian lunch. Water comprises 60 percent of the human body, so maybe our draw to the ocean is biological or maybe it's spiritual. "The earth has guilt, the earth has care, Unquiet are its graves; But peaceful sleep is ever there, Beneath the dark blue waves," wrote poet Nathaniel Hawthorne, who spent his childhood in Maine.[13]

Overdevelopment

Real estate developers and hoteliers contribute to the seaside crush by building accommodations that jockey for water views and compete for the shortest distance to the beach. By offering a range of rates, from budget to expensive, the hotels, motels, and rental properties create a destination for the people—all 23.3 million (Miami), 10 million (Venice Beach, California), and 8 million (Ocean City, Maryland) of them. Visitors to Ocean City, for instance, can choose from among 10,000 hotel rooms, 25,000 condo units, and more than 300 Airbnb properties, including a bungalow near a crabbing pier and a yurt.[14] The greater Miami area has 57,415 hotel rooms and sold 15.6 million hotel room nights in 2018.[15] Equal access to the beach is not the issue; controlling overdevelopment is. Ideally zoning regulations should provide for a range of residential and nonresidential accommodations along with other facilities that are both affordable and luxurious.

Poor Planning

If any locale is the poster child for the lure of the sea and the challenges of tourism, Hawai'i is it (see chapter 5.4). In the 1980s, Hawai'i became a pioneer in sustainable tourism, passing a management plan that promoted tourism without compromising the "aina" spirit. "We were ahead of the ball," said James Mak, professor emeritus at the University of Hawai'i at Manoa, "but things really slid."[16] Mak has been voicing his concerns about overtourism and the state's lack of preparedness for years. In 2008, his book, *Developing a Dream Destination: Tourism and Tourism Policy Planning in Hawaii*, was published. Today, he is banging the same drum, a slow and steady beat on a snail's shell. "Nothing moves quickly here in Hawaii," he said—except tourism.[17]

In 2019, nearly 10,425,000 travelers visited Hawai'i, a 5.4 percent increase from the previous year.[18] On average, the islands hosted about 249,000 tourists per paradisiacal day, a 3 percent spike from previous years. In 2017, *Hawaii News Now* reported that "over the last five years, about 37,000 more people have left Hawaii for the mainland than moved in" because of congestion, traffic, and the high cost of living.[19]

Expansion of Cruise Tourism

The cruise industry is the fastest growing and most profitable sector of the leisure tourism industry, with numbers of passengers increasing forty-fold since 1970 and earnings doubling between 2008 and 2018.[20] It is dominated by three megalines—Carnival, Royal Caribbean, and Norwegian—which together control 90 percent of the US market and make cruising the most consolidated sector of the tourism industry. These three corporate conglomerates are headquartered in Florida but have registered, using a legal loophole, in Panama, Liberia, or elsewhere, which provide "flags of convenience" to avoid US taxes and compliance with environmental and labor regulations.

More than 35 percent of cruise tourism takes place in the Caribbean, which is the leading cruise region in the world, but despite its importance, the Caribbean captures only 6 percent of the global cruise earnings. Caribbean ports and island nations often invest tens of millions of dollars in cruise port infrastructure, while the return is often inequitable. Low head (or passenger) and other taxes paid by cruise lines, steep commissions paid to the lines by "preferred" onshore tour operators and shops, and low spending from passengers equate to exceedingly modest earnings from cruise tourism, especially in contrast with stayover tourism. In 2015, for instance, the number of cruise passenger and stayover arrivals in the Caribbean were almost the same. According to the Caribbean Tourism Organization, total onshore spending by tourists was $30 billion, and of this amount, $2.4 billion was by cruise passengers and $27.6 billion, or 11.5 times more, was from stayover visitors. The unmistakable conclusion of this economic analysis is that stayover tourism is far more beneficial to coastal ports and island economies in the Caribbean than is cruise tourism.[21]

At the same time, cruise ships have increasingly become destinations themselves, with ever greater attractions and amenities to entice passengers to spend more on board than on shore. According to the UN World Tourism Organization, "Onboard spending is an increasingly important part of cruise line income, ranging between 25% and 35% of total income," with gambling and beverages accounting for half the total.[22]

Furthermore, planning and development for cruise tourism do not often address the societal and infrastructure challenges associated with accepting cruise ships. "Inadequate planning and infrastructure development in popular ports of call have meant that many destinations are unable to handle the concentrated influx of cruise passengers during their short periods of stay, in a sustainable manner," said Seleni Matus, executive director of George Washington University's International Institute of Tourism Studies. "This is disrupting the daily life of residents, e.g., the vehicular and pedestrian traffic in these ports of call."[23]

Cruise lines also pose a range of environmental challenges for ports of call and local residents, including sewage, gray water, and solid waste dumped near shorelines; damage to coral reefs, mangroves, and other sensitive natural sites; and air and noise pollution, to name a few. Historically, maintaining environmental oversight of cruise ships has been difficult, and illegal dumping has proliferated, resulting in backlash, including a 2016 lawsuit against Princess Cruise Lines for using a "magic pipe" to illegally dump raw bilge waste directly into the ocean. The environmental degradation is not only the result of illegal acts, however; according to international regulations, cruise lines are allowed to dump untreated sewage when they are more than 12 miles from any shore.

Moreover, the direct impacts to destinations and their surrounding coastal and marine areas listed above do not account for the cruise sector's contributions to climate change. One cruise ship's emissions equals that of five million cars over the same distance, and by one estimate, its greenhouse gas emissions are 7.6 times that of an airplane over the same distance. These impacts are almost completely unregulated. Likewise, the increase in natural disasters, most notably hurricanes, has disproportionately impacted coastal and island communities. According to the Caribbean Maritime Institute in Jamaica: "[Since around 1980], the intensity of hurricanes in the Caribbean has been increasing. [Since around 2000], extreme events resulting from climate change have produced widespread damage to infrastructure, housing, and hotel and recreational facilities, including the shoreline for accommodating cruise activities."[24] The message is that not only does overtourism's environmental impacts hurt residents and travelers, but it harms the tourism industry itself.

Overpublicity: Social Media, Selfies, and Films

Social media can never keep a secret, and all the sharing and liking has plucked some destinations from obscurity and thrown them onto a global stage. Pinterest and Instagram accounts are the glossy new vacation brochures. Several recent studies have determined that travelers scour social media for trip ideas and may pick a locale because of its photogenic qualities. For example, a 2017 survey by Schofields, a UK-based holiday rental home insurance company, polled one thousand millennials about how they choose a vacation spot, and more than 40 percent said the place must be Instagrammable; sampling the local cuisine and sightseeing ranked a distant fourth and fifth.[25]

Travel influencers wield a lot of power. Chanel Cartell and Stevo Dirnberger, who share their adventures on the Instagram account #HowFarFromHome, have 164,000 followers. Nearly 10,500 fans liked their photo of Bondi Beach in Australia, despite the warning in the caption: "The entire golden curve was packed."[26]

"We've become victims of the hit-and-run selfie," said Tourtellot of the Destination Stewardship Center. "People only go to prove that they've been there. It's high quantity versus high quality."[27] Alastair Danter, project manager of SkyeConnect, the Isle of Skye's tourism management organization, has witnessed the sheep behavior on the Scottish island, particularly at sundown. "There are in excess of 150 great locations on Skye to see the sunset," he said, "but the selfie/social media mentality means that everyone has to go and see it at Neist Point."[28]

Danter, who prefers the term *volume tourism* to *overtourism*, blames Tripadvisor reviewers who hype up certain attractions at the expense of others. He also disapproves of media outlets that churn out listicles to feed the insatiable content beast. To demonstrate his point, he highlighted two CNN roundups. In 2018, the news organization mentioned Isle of Skye as a place to avoid, due to crowds. A year later, the island appeared on CNN's top ten list of destinations to visit.

Television and movie location tours also contribute to the Isle of Skye's overtourism problems, with many filmmakers gravitating to the same cinematic sites, such as the Quiraing and the Old Man of Storr. "The real question is figuring out what this opportunity is and how it can be met," Danter said when referring to the growing interest in film site tours, "and how it can be harnessed to enhance social, economic, and cultural development."[29]

Convergence of Climate Change and Overtourism

In addition to the dangers of overtourism, coastal communities are witnessing another menace: the destruction of their natural environments, further enhanced by climate change. Seaside towns are struggling with sand erosion, the loss of coral reefs, a plunge in marine life, algae blooms, and beached animals. Although climate change is a primary cause, overtourism is no friend to the fragile ecosystem either. In fact, among overtouristed areas, the combined forces of these perils seem to be hitting beachfront destinations the hardest. To handle the overlapping threats, conservationists, legislators, and other problem solvers must address a multitude of challenges and pursue multipronged strategies.

Solutions

Unlike national parks and protected areas, many beaches have failed to address overtourism and overdevelopment through carefully crafted and implemented management strategies. Rather, efforts around the world to tackle overtourism on beaches and coastline have varied widely and have been carried out by a range of local or national governments, destination marketing organizations, nonprofit organizations, and even, in some instances, hotels and tour operators. Although no clear model or formula has yet emerged for how to effectively protect and manage beaches and coastlines, a range of experiments, taken together, may offer tools for ensuring long-term sustainability.

Temporary Closures

In 2018, Thailand's Department of National Park, Wildlife and Plant Conservation sounded the alarm on its dying coral reef at Maya Bay. On average, five thousand tourists and two hundred boats a day were swarming the cove on Ko Phi Phi Leh Island, jeopardizing the health of the reef. Park officials decided to close the beach for four months, and then a year, and now possibly until June 2021. When the government reopens Maya Bay, it may cap the number of arrivals at twelve hundred a day.[30] For now, tourists can only visit the endangered paradise through Leonardo DiCaprio's movie *The Beach*, which portrayed the once-pristine island before its downfall. Oh, the irony— the celebrity environmentalist indirectly triggered an overtourism eco-crisis.

Thailand is not the only country in the region to give its beaches a break by barring tourists. Boracay Island in the Philippines, which the country's President Rodrigo Duterte once described as a "cesspool," shuttered for six months in 2018 for repairs and reconceptualizing. The former bacchanal on the beach reopened as an exemplar of sustainable practices. Visitors must follow a lengthy list of rules, including no single-use plastics, beach beds, umbrellas, or deck chairs and no barbecuing, pets, partying, or throwing up. Watersport enthusiasts can no longer zoom around in swimming zones but must stay at least one hundred yards offshore. Not even the legendary fire dancers escaped regulation: the performers must trade in their fiery kerosene torches for eco-sensitive LED lights.[31]

"The president took the very drastic step of closing the island, demolishing illegal buildings, and investing in new waste infrastructure to create a destination that is better managed and controlled for the longer term," WTTC's Turner said. "This is clearly not a step that many destinations would take, but the impact from poor management, over the years, had made it necessary."[32]

Managing Transportation and Timing

Most locales are not ready to shut down tourism altogether, but some are trying to change how tourists travel to their destinations, and when and where they visit. On the Isle of Skye, official Danter said that building a public transportation system with paid parking lots (versus stashing your car on the side of the road) could alleviate some of the pressure.[33] To encourage nonmotorized transport, a nonprofit organization called the Broadford and Strath Community Company has been working, for years, on creating the Skye Cycle Way for hikers and bikers.

For a more immediate solution, Danter is encouraging travelers to visit during off-peak months and venture into lesser-known parts: "Go two miles down the road past the Fairy Pools, walk a quarter mile up the glen on your left, and instead of being in a line of people walking along a straight footpath," he said, "you will be alone with a few sheep, the eagles from the Cuillin, some deer, views over to Rum and Minch, rock pools and waterfalls—and not a person in sight.[34]"

In March 2019, the Isle of Skye launched a campaign that urges visitors to simplify their itineraries and slow down, way down. The philosophy of #skyetime contains three simple directives: "Stay Longer. See Less. Experience More."[35]

Conserving Resources

Catalina Island, a bucolic speck off the coast of Los Angeles, is also reaching out to its one million annual visitors, although its message is more pragmatic than poetic. The desert island with a full-time population of about four thousand residents struggles with drought conditions. Several times a year, its desalination plant has not been able to meet its citizens' daily needs, and officials have to tap into its freshwater reserves. As if that wasn't enough, the saltwater brine is deteriorating the island's wastewater-processing system. Moreover, its landfill could reach maximum capacity any year now. So far, though, islanders have successfully avoided the 2021 doomsday deadline thanks to a campaign that led to less waste and more recycling, which the island barges to the mainland.

"It's amazing what you can do when you need to do it," said Jim Luttjohann, president and chief executive officer of Love Catalina Island, formerly the Catalina Island Chamber of Commerce and Visitors Bureau.[36] In 2016, the chamber delivered door hangers to about twenty hotels with more than eight hundred guest rooms and two hundred vacation rental units. The knob accessories urged guests to skip housekeeping in exchange for a $5 Island Green credit redeemable at local retailers. This one act helped the island conserve seventeen gallons of water per person.[37]

Educating Visitors

Educating visitors is key to curtailing the ill effects of overtourism. Love Catalina Island has a Care for Catalina guide that it disperses through several outlets, such as its website, local access TV channels in hotels and rentals, and supermarket monitors. A link to the site is included on public maps posted around downtown and at boat terminals, as well as on the confirmation emails of ferry bookings. Care for Catalina preaches the holy commandments advanced by many overtouristed destinations, such as to veer off the beaten path and avoid peak travel times (Saturdays and July for Catalina). The good-behavior manual also offers tips on water and fuel conservation, plastic reduction, trail usage, wildlife protection, and cell phone etiquette: "Get some JOMO (Joy of Missing Out) by silencing your cell phone before stepping into nature and speaking softly without using the speaker function."[38] According to Luttjohann, "On a community level, [overtourism is] coming up because of Barcelona and Machu Picchu. We need to be thoughtful and proactive and get ahead of the problem."[39]

Limiting Visitation

In Hawai'i, Mak points to a few incremental—and hopeful—advancements in curbing overtourism. At Hanauma Bay Nature Preserve, about ten miles east of Waikiki, park officials raised the tourist entry fee from $3 to $7.50, plus $1 for parking in the three-hundred-spot lot. Visitors must watch an educational video about Hanauma Bay and other reef environments before bouncing off to the beach.[40] The Department of

Parks and Recreation for the city and county of Honolulu isn't ready to break out the cake and champagne yet, but it does have reason to celebrate.

"We believe these mitigation measures are working to improve the amount of people entering the nature preserve and reduce the negative impacts on the environment," said Nathan Serota, a public information officer with the department. "Most people visit Hanauma Bay to experience nature, so its preservation should enhance the enjoyment of this environmental treasure."[41]

Mak said that more parks—such as Diamond Head State Monument on Oahu, which draws more than a million hikers each year—should follow suit. Currently, walk-in visitors pay $1, a figure set by the state. "Diamond Head needs-two million dollars in maintenance," he said. "But to raise the fee, you will have to change state law."[42]

In Kauai, Haena State Park and Napali Coast State Wilderness Park embraced an opportunity to update their management plan when heavy rains, flooding, and landslides forced officials to close in April 2018. The parks reopened more than a year later with new and improved amenities, including a parking lot, boardwalk, and walkways to Kē'ē Beach and the Kalalau Trailhead, plus entry requirements for out-of-state visitors. To control numbers, guests who arrive by car, shuttle bus, bike, or foot must secure a reservation from one to thirty days in advance. In addition, drivers wishing to park all day must purchase morning and afternoon passes; for stays stretching to sunset, they must buy all three time slots. One of the best ways of protecting these Kauai state parks "is to limit visitation," Alan Carpenter, assistant administrator for the Department of Land and Natural Resource's Division of State Parks, said in a statement, "and any mainland or world travelers have likely already experienced limited-park entrance at certain National Parks and international destinations like Machu Picchu or Nepal."[43]

Controlling Short-Term Rentals

Hawai'i legislators are also tackling one of their biggest headaches, short-term rentals. The rules, which vary by county, range from limiting where short-term rentals can operate (for instance, not in nonresort areas) to requiring property owners to provide on-site guest parking and information on neighborhood quiet hours. Tourists can also do their part by sleeping in legal accommodations.

Shifting from Marketing to Management

For the long-term health and prosperity of the islands, Mak said that Hawai'i needs to shift its focus more aggressively from marketing to tourism management. Even the governor agrees. In response to Southwest Airlines' new daily service to Hawai'i, which kicked off in May 2019, Governor David Ige said, "We definitely are concerned about managing the number of visitors we have here. We are definitely working to do that."[44]

Several government agencies can rescue Hawai'i from the masses, if they choose to act; the Hawai'i Tourism Authority (HTA) is one of them. Marisa Yamane, director of communications with the state-funded tourism office, described HTA's four areas of focus as natural resources, Hawaiian culture, community, and brand marketing.[45] Since 2019, the organization has tied its success to several key performance indicators, including resident and visitor satisfaction.[46] In the foreword to its Five-Year Strategic Plan 2020–2025, HTA demonstrates its dedication to balancing the needs of both groups and the demand for moving forward without leaving anyone behind: "With HTA now in our third decade, this plan responds to new levels of tourism shaped by new technologies with new opportunities and new challenges. What remains unchanged is HTA's mission: to strategically manage Hawai'i tourism in a sustainable manner consistent with economic goals, cultural values, preservation of natural resources, community desires, and visitor industry needs."[47]

To achieve this objective, HTA has helped the Hawaii Department of Land and Natural Resources with repairs and improvements to hiking paths, such as the Manoa Falls Trail on Oahu, and has assisted with workforce development programs and visitor assistance initiatives. "HTA has shifted its priorities away from increasing visitor arrivals and is focusing its efforts on enhancing the quality of life for residents and the visitor experience," Yamane said. "The measures of success are visitor satisfaction, resident sentiment, per person per day spend, and total visitor expenditures."[48]

The 10.25 percent transient accommodations tax, which the state established in 1987, helps subsidize projects that preserve the Hawaiian culture, safeguard natural resources, and benefit the community, such as the Hawaii Food and Wine Festival, the Pan Pacific Festival, and the Okinawan Festival.[49] "In 2020, we will work diligently to continue growing the number of events and programs," HTA president and chief executive officer Chris Tatum stated in the 2019 annual report to the Hawai'i state legislature.[50] In 2019, the state funded nearly 130 projects; in 2020, the number rose to more than 170.

Industry Stewardship

Hawai'i is hardly alone in its management struggles; oceanfront destinations around the world are looking toward their stewards for solutions. "We need to think about the sustainability of these coastal communities," said Paula Vlamings, chief impact officer of Tourism Cares. "Coral reefs are collapsing. Changing hotel towels is not enough."[51] Vlamings espouses an all-hands-on-deck approach to help overcrowded coastal destinations. Developers and government officials should stop pursuing unfettered "growth, growth, growth" and fix what is broken or ailing.[52]

As an example, Vlamings points to Chris Blackwell, the boutique hotelier in Jamaica who established the Oracabessa Foundation in 1995. Working with local partners, the organization supports several programs on the Caribbean island. One such program trains students who wish to work in the hotel industry—several hundred students have completed the curriculum and secured jobs in hospitality. The group also created a fish sanctuary with St. Mary Fisherman's Cooperative. Since 2010, the coral and fish have rebounded by nearly 50 percent.[53] "I think that if all hotels got their own fish sanctuary," said Vlamings, "then they could bring the reef back."[54]

As vice president of the Grupo Puntacana Foundation, Jake Kheel is dedicated to thoughtful sustainable growth in Punta Cana, a popular tourist area in the Dominican Republic. One of the group's approaches is to build resorts horizontally, not vertically. The Punta Cana Resort and Club, for instance, stops at five stories. "By not building up and limiting the number of rooms by square meters of the property, you limit the density," Kheel said. "You don't cram as many people in as possible."[55] Another is the Ojos Indígenas Ecological Park and Reserve, a fifteen-hundred-acre private reserve that the organization owns and protects through a payment plan. Guests of Tortuga Bay, the Westin Puntacana Resort & Club, and Puntacana Resort & Club can explore the reserve for free; nonguests, however, must pay $50. "Solutions in destinations are only going to be effective when a destination thinks strategically over the long term about the type of place it wants to be," said Turner of the WTTC. "It also requires leadership to make tough and, at times, regulatory decisions."[56]

Community-wide Engagement

Back on Cape Cod, CARE for the Cape and Islands has been working since 2012 to engage tourists, businesses, local organizations, and residents in raising funds and awareness for environmental stewardship programs. Working with travel industry businesses, CARE creates opportunities for travelers to contribute to local efforts that promote long-term environmental conservation and cultural heritage preservation. Community projects cover a wide range of interests and needs. They have improved bicycle trails, historical walking maps, and interpretive signage; promoted public murals, trash-to-treasure art, and art by those with disabilities; supported local waste and recycling stations; and provided content for a "learning book" that offers visitors information about the local fishing industry and ocean stewardship. CARE has also been active in local campaigns to stop the use of single-use plastic straws and bottles, funded water bottle filling stations and marine wildlife entanglement projects, and sponsored annual beach cleanup days for visitors and residents.[57]

To modernize its antiquated septic system, Cape Cod needs residents, town officials, and the government to work together on a common cause and not ignite a turf war. Installing the necessary wastewater infrastructure in the fifteen incorporated towns will cost $4 billion. In December 2018, state legislators created the Cape Cod and Islands Water Protection Fund and established a new nightly 2.75 percent excise tax on lodging and short-term rentals. The surcharge, which officials earmarked for the construction of a new sewage system, should raise $1 billion to help Cape Cod better handle the busy toilet flushing season. "It's complicated. It's expensive. It's not saving cute, fuzzy animals," said Gottlieb of the Association to Preserve Cape Cod. "It's a slow-going grind."[58] By 2019, six towns had started to build the new infrastructure. The others, said Gottlieb, are "talking about it. The majority of the region has not come up with a way to implement a comprehensive plan."[59]

As for the mobbed public beaches, Isabelle has a solution: limit the number of people allowed on the beach or set a legal distance between towels. "If they did that," the teen said, "I might possibly want to go to the public beaches."[60]

Notes

1. Isabelle Ramage. (June 7, 2019). Personal communication with author.

2. Jonathan Tourtellot. (April 25, 2019). Destination Stewardship Center. Personal communication with author.

3. Andrew Gottlieb. (April 25, 2019). Association to Preserve Cape Cod. Personal communication with author.

4. Rochelle Turner. (April 18, 2019). World Travel and Tourism Council. Personal communication with author.

5. Seleni Matus. (June 3, 2019). "Overtourism in Coastal Communities in the Caribbean." [Draft.] Washington, DC: George Washington University.

6. Tripadvisor. Key West. (October 3, 2020). Reviewed by Martha Honey and Claire Campbell. https://www.tripadvisor.com/Attraction_Review-g34345-d104133-Reviews-Mallory_Square-Key_West_Florida_Keys_Florida.html.

7. Lynn-Marie Smith. (March 30, 2014). "Key West Votes NO to Sanctuary Dredging." *Ecology*. https://www.ecologyflorida.org/2014/03/feature-key-west-votes-no-to-sanctuary-dredging/.

8. Rosalind S. Helderman, Hannah Sampson, Dalton Bennett and Andrew Ba Tran. "The Pandemic at Sea." (April 25, 2020). *Washington Post*. https://www.washingtonpost.com/graphics/2020/politics/cruise-ships-coronavirus.

9. Key West Committee for Safer Cleaner Ships. https://www.safercleanerships.com/.

10. Martha Honey, editor. (2019) *Cruise Tourism in the Caribbean: Selling Sunshine*. Routledge. London and New York.

11. Clean Up Carnival. Stand.earth. https://www.stand.earth/content/clean-carnival.

12. Nathaniel Hawthorne. (1834). "The Ocean." Eds. Lily, Walt, Coleman & Holt. *The Mariners Library of Voyagers Companion: Containing Narratives of the Most Popular Voyages*. Boston: C. Gaylord.

13. Ocean City Maryland. (2019). https://ococean.com/media.

14. Greater Miami Convention and Visitors Bureau. (April 30, 2019). *Record Tourism Industry Performance in 2018.* https://www.miamiandbeaches.com/press-room/miami-press-releases/record-tourism-industry-performance-in-2018.

15. James Mak. (April 4, 2019). University of Hawai'i at Manoa. Personal communication with author.

16. Ibid

17. Hawai'i Tourism Authority. (January 29, 2020). *Hawai'i Visitor Statistics Released for 2019.* https://hawaiitourismauthority.org/media/4120/december-2019-visitor-stats-press-release.pdf.

18. HNN Staff. (December 21, 2017). "People Are Leaving Hawaii in Droves—and This Year Was No Exception." *Hawaii News Now.* https://www.hawaiinewsnow.com/story/37122464/people-keep-leaving-hawaii-in-droves-for-the-mainland-and-this-year-was-no-exception/.

19. Martha Honey. (2019). *Cruise Tourism in the Caribbean: Selling Sunshine.* Routledge. London. pp. 4–31.

20. Martha Honey, ed. (2019) *Cruise Tourism in the Caribbean: Selling Sunshine.* Routledge. London and New York. pp. 4-31.

21. UN World Tourism Organization. (2010). *Cruise Tourism—Current Situation and Trends.* Madrid, Spain: World Tourism Organization. p. 16.

22. Seleni Matus. (June 3, 2019). Washington University's International Institute of Tourism Studies. Personal communication with author.

23. Ibrahim Ajagunna and Fritz Pinnock. (2016). "The Impact of Climate Change on Cruise Tourism: A Case Study from the Caribbean." Unpublished paper prepared for the Center for Responsible Travel.

24. Philip Schofields. (April 3, 2017). "Two Fifths of Millennials Choose Their Holiday Destination Based on How 'Instagrammable' the Holiday Pics Will Be." *Schofields Insurance.* https://www.schofields.ltd.uk/blog/5123/two-fifths-of-millennials-choose-their-holiday-destination-based-on-how-instagrammable-the-holiday-pics-will-be/.

25. Chanel Dirbergy and Stevo Dirberger. (2019). "How Far from Home." *Instagram.* https://www.instagram.com/HowFarFromHome/?hl=en.

26. Jonathan Tourtellot. (April 25, 2019). Destination Stewardship Center. Personal communication with author.

27. Alistair Danter. (April 19, 2019). SkyeConnect. Personal communication with author.

28. Ibid.

29. Karla Cripps and Kocha Olarn. (May 10, 2019). "Thailand Bay Made Popular by 'The Beach' to Remain Closed for Two More Years." *CNN Travel.* https://www.cnn.com/travel/article/thailand-maya-bay-reopening-date/index.html.

30. Euan McKirdy. (October 26, 2018). "Boracay Reopens to Tourism, but Its Party Days Are Over." CNN Travel. https://www.cnn.com/travel/article/boracay-reopening-restrictions-intl/index.html.

31. Rochelle Turner. (April 18, 2019). World Travel & Tourism Council. Personal communication with author.

32. Alistair Danter. (April 19, 2019). SkyeConnect. Personal communication with author.

33. Ibid.

34. Skye-Time. (2019). https://www.skye-time.com.

35. Jim Luttjohann. (March 28, 2019). Catalina Island Chamber of Commerce and Visitors Bureau. Personal Communication with author.

36. Ibid.

37. Catalina Island Chamber of Commerce & Visitors Bureau. (2019). *Care for Catalina Guidelines.* https://www.catalinachamber.com/community-information/care-for-catalina/.

38. Jim Luttjohann. (March 28, 2019). Catalina Island Chamber of Commerce and Visitors Bureau. Personal Communication with author.

39. James Mak. (April 4, 2019). University of Hawai'i at Manoa. Personal communication with author.

40. Nathan Serota. (July 31, 2019). Department of Parks and Recreation, City and County of Honolulu. Personal communication with author.

41. James Mak. (April 4, 2019). University of Hawai'i at Manoa. Personal communication with author.

42. Department of Land and Natural Resources. (February 28, 2019). *Ha'ena State Park Provides Silver Lining to Historic April '18 Flooding.* https://dlnr.hawaii.gov/blog/2019/02/28/nr19-042/.

43. Alex Temblador. (March 20, 2019). "Hawaii Tourism Is at a 'Tipping Point' of Overtourism." *Travel Pulse.* https://www.travelpulse.com/news/destinations/hawaii-tourism-is-at-a-tipping-point-of-overtourism.html.

44. Marisa Yamane. (September 3, 2019). Hawai'i Tourism Authority. Personal communication with author.

45. Hawai'i Tourism Authority. (December 9, 2019). *2019 Annual Report to the Hawai'i State Legislature.* https://hawaiitourismauthority.org/media/4052/2019-annual-report-to-the-hawaii-state-legislature.pd.f

46. Hawai'i Tourism Authority. (2016). *Five-Year Strategic Plan.* https://www.hawaiitourismauthority.org/media/1849/hta15001-strategic-plan_web.pdf.

47. Marisa Yamane. (September 3, 2019). Hawai'i Tourism Authority. Personal communication with author.

48. State of Hawaii: Department of Taxation. (January 12, 1987). *Transient Accommodations Tax Technical Memorandum No.1.* http://files.hawaii.gov/tax/legal/memorandums/tatmem-01.pdf.

49. Hawai'i Tourism Authority. (December 9, 2019). *2019 Annual Report to the Hawai'i State Legislature.* https://hawaiitourismauthority.org/media/4052/2019-annual-report-to-the-hawaii-state-legislature.pdf.

50. Paula Vlamings. (May 3, 2019). Tourism Cares. Personal communication with author.

51. Ibid.

52. Oracabessa Foundation. (Accessed March 2020). http://www.oracabessafoundation.org/a-healthy-bay.

53. Ibid.

54. Jake Kheel. (May 31, 2019). Grupo Puntacana Foundation. Personal communication with author.

55. Rochelle Turner. (April 18, 2019). World Travel and Tourism Council. Personal communication with author.

56. CARE for the Cape and Islands. (2020). https://careforthecapeandislands.org/about-us/.

57. Andrew Gottlieb. (May 9, 2019). Association to Preserve Cape Cod. Personal communication with author.

58. Ibid.

59. Isabelle Ramage. (June 7, 2019). Personal communication with author.

Chapter 5.2

The Galapagos Islands, Ecuador

By Carter A. Hunt

"The isolation of Galapagos has been broken. . . . The resident population jumped from 6,200 in 1982 to 9,800 in 1990, and is growing at the unprecedented rate of 6.3 percent per annum. . . . Recommended quotas on the number of tourists that should be allowed to visit the islands each year, [originally set at] 12,000 in 1973 and 25,000 in 1981, have been repeatedly surpassed. . . . Tourism . . . is the driving force which, directly and indirectly, dictates the pace and types of changes that are occurring in the islands."[1]

These words were written in 1993 by Bruce Epler, a North American scientist who for decades chronicled the Galapagos Islands. At that time, domestic and international visitation to the islands totaled less than fifty thousand per year. Eller's passage illustrates that concerns for the Galapagos' human carrying capacity—or what is today referred to as overtourism—have been raised since organized tourism began to harm the fragile archipelago in the early 1970s.

Located roughly six hundred miles off the coast of Ecuador and declared UNESCO's first World Heritage Site in 1978, visitor arrivals have continued to increase dramatically despite decades of efforts to set annual limits. By 2007, visitor arrivals exceeded 160,000 a year, and by 2018 arrivals had increased to a whopping 275,000, a twenty-three-fold increase over 1973, when a quota to curb overtourism was first proposed.[2]

In addition, the local resident population of farmers, fishers, conservationists, and others catering directly and indirectly to tourism in the Galapagos has now grown from around fourteen hundred in the 1950s to more than thirty-five thousand people in 2019.[3] The place described by Charles Darwin as a "little world unto itself" clearly no longer is.

Roots of Overtourism in the Galapagos

The beginning of the contemporary history of Galapagos is often pegged to the building of a US military base, including an airstrip, on Baltra Island during World War II. These facilities were transferred to the Ecuador government in 1946.[4] Little population growth or demographic change has occurred in the small agricultural communities established by the pioneers who were sent to the islands throughout the mid-twentieth century. In 1959, the Charles Darwin Foundation and the Galapagos National Park—Ecuador's first national park—were established, both headquartered in Puerto Ayora on Santa Cruz Island.

Tourism-related change began a decade later, in 1969, when Metropolitan Touring, an Ecuadorian company headquartered in Quito, initiated the first "floating hotel" in the Galapagos. This innovation grew to be the dominant tourism model in the Galapagos, with some one hundred small boats running weeklong tours of the islands. Visitors ate and slept on the boats as they followed a carefully controlled itinerary accompanied by naturalist guides. The focus on nature-based tourism was intended to keep a conservation ethic squarely at the center of economic development in the islands.[5] Santa Cruz Island became the population and decision-making center and placed the archipelago on a course of economic development and human growth centered around ecotourism that continues to this day. As Epler noted, however, even in these early days of tourism, the islands were already wrestling with the question of how many visitors were too many.[6]

Throughout the subsequent decades, tourism development stimulated population growth as immigrants were lured to the islands by the higher standard of living compared to other provinces in Ecuador. In addition, the 1990s saw a global boom in the lobster and sea cucumber markets (which fetched higher prices than gold), causing an even faster influx of fisherman seeking to exploit the Galapagos' marine reserve. During this decade, concerns about overfishing trumped concerns about overtourism.

In 1998, the Ecuadorian government passed the Special Law for Galapagos in an effort to alleviate rising local conflicts over restrictions on use of marine resources and concerns about the impacts of tourism on the islands' fragile environment. The law set forth new quarantine controls on invasive species, resource management policies, and residency rules to limit immigration. In an effort to ensure more tourism benefits flowed to local communities, the law also specified that 10 percent of park entrance fees went to the Provincial Council of Galapagos and an additional 20 percent to municipalities in the Galapagos. The law also gave preferential treatment to new tourism businesses operated by permanent Galapagos residents.[7]

Because residency restrictions and work permit requirements were loosely enforced, however, migrants seeking opportunities in tourism continued to arrive.[8] In addition, given the reporting difficulties, the official 2019 estimate of thirty-five thousand local residents is likely conservative.

In the Galapagos today, the influx of diverse groups from Europe, mainland Ecuador, and elsewhere has led to clashes that have influenced decision-making and governance of the islands.[9] Bitter environmental conflicts erupted during the lobster and sea cucumber fishing booms of the 1990s, culminating in physical violence as bands of resident and recent immigrant fishermen burned the national park headquarters, took staff hostage, and hung several giant tortoises in the street to protest newly imposed fishing restrictions.[10] These conflicts, which were rooted in incompatible cultural worldviews, led to complex coalitions of residents whose actions failed, at times, to recognize and respect long-standing conservation values in the Galapagos. They included the careful use of natural resources, appropriate forms of economic development, and the need to protect endemic species and the islands' fragile ecosystems.[11] These differences helped precipitate a revolving door of leadership at key institutions like the national park, the Charles Darwin Foundation, and local government offices. Taken together, the divisions have also hindered broader collective action built upon shared conservation values, including the need for policies to effectively address overtourism in the archipelago.

The rise in local conflicts led UNESCO to put the Galapagos on the World Heritage in Danger List between 2007 and 2010.[12] As UNESCO officials stated:

> The principal factor leading to the inscription of the property [as a] World Heritage in Danger arises from the breakdown of its ecological isolation due to the increasing movement of people and goods between the islands and the continent, facilitating the introduction of alien species which threaten species native to the Galápagos. This is fueled by poor governance leading to inadequate regional planning and unsustainable tourism development. Illegal and excessive fishing pressures in the Galápagos Marine Reserve (GMR) were another factor contributing to the Danger listing.[13]

In response, the government improved its regional planning and expanded local participation in fishery management, while the Ministry of Environment placed tighter controls on migration and the introduction of invasive species.[14] Visitation to the islands continued to grow, however, even though UNESCO listed "unsustainable tourism" as one of the reasons for putting the Galapagos on the World Heritage in Danger list.

Accelerating Overtourism

The floating hotel tourism model that dominated in the Galapagos from the early 1970s until the early 2000s is widely cited as an ecotourism success story that kept environmental impacts to a minimum while providing visitors with high-quality nature experiences. Tourists were required to travel in the company of guides at all times, their activities were restricted to designated areas, and quotas for visitation to different park sites were carefully managed by the national park, both to control for ecological impacts and to preserve the visitors' immersive and uncrowded nature experience.

Under this model, however, there were considerable leakages of tourism earnings and jobs from the islands. Most of the boats were owned by non-Galapagos companies, most supplies and equipment were imported, and many of the top guides were not locals. Eventually, the leakages from the floating hotel model began to generate resentment. With the backing of the 1998 Special Law for Galapagos, local community leaders and nongovernmental organizations began to advocate for "a new model of ecotourism" that placed the priority on the maximization of economic benefits for local communities, not just for the external tour operators in charge of the floating hotels.[15] By 2010, the tourism industry was undergoing several dynamic shifts. These changes are increasing the magnitude of tourism-related impacts on local residents, on the visitor experience, and on the islands' unique ecosystems. In particular, three key trends are exacerbating concern about overtourism.

The Changing Model of Island Visitation

Although the long-established model of small cruise visitation on a preset itinerary of islands within the Galapagos National Park continues to grow in popularity, the historic model has been overtaken by a model of island-based visitation. By 2011, more visitors were staying in hotels on the islands than aboard cruise boats.[16] Airlines operating flights from the mainland began making visitation more affordable for both international and national visitors, and park entrance fees of $100 for foreign tourists and $6 for Ecuadorian nationals have, incredibly, not been raised since 1993.[17] Rather than a slow-paced week-long tour aboard luxury boats, the majority of visitors now stay in more modest land-based accommodations and do shorter island-hopping excursions to local park sites. This increase in on-island tourism has led to extensive construction of new infrastructure, including accommodations, dining and drinking establishments, and improved airports on Santa Cruz and San Cristobal Islands. Shipments of food and supplies from the mainland that used to arrive every couple of weeks now arrive daily to support this new model of tourism. The growth of on-island tourism puts further stress on the sources of finite freshwater, requires additional energy resources, and generates a massive amount of new waste and trash.[18]

The last fifty years of development in the Galapagos Islands. Source: Carter Hunt.

The Changing Nature of Tourists to the Galapagos

Tourism in the Galapagos has long been dominated by US visitors, although since 2017, Ecuador itself accounts for the largest contingent (31 percent) of annual visitors.[19] US visitors now comprise 29 percent of total arrivals, and no other country provides more than 5 percent of the total.[20] With the boat tour model of visitation dominated by international visitors, the growth of the Ecuadorian market that dominates the on-island model represents the bulk of all new growth in visitation in recent years. Although international visitation to the islands will remain extremely important, especially given the higher per capita revenue it generates, foreign visitors are far more likely to spend their time on boats and to have more limited interactions with the local population. In contrast, growth of the Ecuadorian market corresponds to a further intensification of the on-island model of tourism.

For some, the growth of the national market represents a positive trend toward "democratization" of tourism to Galapagos, with more Ecuadorians visiting their own natural heritage than ever before. The visitor experience of land-based tourism, however, is now less explicitly linked to nature-based activities and the conservation of the islands' unique habitat that has been the hallmark of the boat-based model. National park site quotas to visit specific islands are filled by the long-standing agreements with the operators of larger, forty- to one-hundred-passenger boats, leaving little opportunity for shorter-term, on-island visitors to tour the farther reaches of the park. As a result, on-island visitation is becoming more akin to coastal mass tourism, characterized by beach visits, surf and scuba lessons, upscale cafés selling coffee grown on the islands, rented electric scooters zipping up and down the streets of Puerto Ayora, and numerous craft breweries. This erosion of the nature-based visitor experience and the decoupling land-based tourism from the conservation ethic embodied in small cruise tourism are undermining the fundamental character of the Galapagos, a concern central to all discussions of overtourism.

The Changing Nature of the Resident Population

The Galapagos Islands had no native human population. After serving as a penal colony in the nineteenth century, subsequent waves of immigrants were drawn by the archipelago's economic opportunities, first in agriculture, then fishing followed by conservation, then floating hotel tourism, and, most recently, on-island tourism.[21]

Under the 1998 Special Law, local residents who had lived in the islands for more than five years were declared "permanent residents." The law also granted residence rights to all future descendants of Galapagos residents, whether or not they had ever lived in the archipelago, thereby opening a large pool of potential immigrants.[22] In addition, the absence of effective controls on migration as well as various loopholes in temporary work-permitting policies, has almost certainly led to underreporting of the resident population. Indeed, despite the Special Law's stated attempt to control immigration, in-migration has dwarfed endogenous population growth.[23] As of 2015, just 36 percent of the 35,000+ local residents were born in the islands.[24] Tourism continues to grow and foment both formal and informal migration to the Galapagos, furthering overdevelopment and other undesirable outcomes.

Solutions to Overtourism

There is an urgent need to address these issues to preserve the Galapagos, one of the natural scientific community's most iconic sites. The following recommendations include better tourism management and improved, and more cooperative, research.

Controlling Visitor Numbers, Raising Entrance Fees

One obvious measure that the Ecuador government could take to address overtourism is to place restrictions on the number of visitors that can enter the Galapagos. Doing so would have to proceed by simultaneously revisiting the Galapagos National Park entrance fee structure, which has remained unchanged since 1993. An international adult visitor to Galapagos pays $100 for up to sixty days in the islands; in comparison, an international visitor pays $95 daily to visit Tanzania's Serengeti National Park, also a World Heritage Site. There is an urgent need for research that provides simulations of how changes to this fee structure might influence overall visitor numbers, as well as overall revenue for the Galapagos National Park, the Galapagos Marine Reserve, and the local provincial and municipal governments.

Research Focused on Human-Environment Relations

Historically, research in the Galapagos has focused on its unique natural history, which has created an incredible repository of scientific knowledge about the islands' ecological systems and processes. Essential to implementing policies that effectively address long-standing overtourism concerns is a better understanding of the dynamic milieu of cultural worldviews among the islands' population and the ways that these differing attitudes influence conservation, development governance, and decision-making.

The 2016 renewal of cooperative agreements between the Charles Darwin Foundation and the Ecuadorian Government call for human-environment relations to be a specific research focus of the foundation going forward. Although little such research has yet gotten underway, promising lines of investigation involve better understanding of tourism activities and other human drivers of invasive species introduction, facilitation of more inclusive management of fisheries and other natural resources, efforts to promote and reward organic agriculture, deeper exploration of the persistent influence of imported cultural worldviews on local natural resource decision-making, and more sustainable management of tourists and the tourist experience.

Research Partnerships

In undertaking overtourism-related research, institutional partnerships remain crucial. The Charles Darwin Foundation has a history of collaboration dating back to its cocreation with the Galapagos National Park in 1959. In addition, the Universidad San Francisco de Quito and its partner, the University of North Carolina at Chapel Hill, have greatly expanded their joint tourism-related social science research, especially through their comanaged Galapagos Science Center on San Cristobal island.

These partnerships often encounter challenges, including a sense of "territoriality" among competing research institutions that must be overcome. As with all major conservation and development challenges, greater cooperation among academic and research institutions, nongovernmental organizations, civic organizations, and government institutions will be essential for addressing the long-expressed concern for overtourism in the Galapagos.

Conclusion

Today, tourism is recognized as one of the primary global drivers of human-caused socioeconomic and environmental disturbances, and in few places is this truer than in the Galapagos. Indeed, the archipelago has a lengthy history of unheeded warnings about too much tourism, including, most notably, UNESCO's declaration of the Galapagos as a World Heritage Site in Danger between 2007 and 2010. It remains unclear at what point the institutional and political forces will act to set limits on tourism visitation, reform the Galapagos National Park entrance fees, place effective restrictions on in-migration, and enact other measures to offset the overtourism pressures on the islands. Despite the well-documented disruptions overtourism creates in local communities and environments, however, well-managed tourism remains a powerful tool for conservation and socioeconomic well-being, given the scale of threats to the marine and terrestrial ecosystems of the Galapagos.[25] From shark finning to the influx of invasive species and from overfishing of lobster and sea cucumber fisheries to the impacts of global climate change, tourism, responsibly done, is likely to remain the best hope for both conservation and local communities in UNESCO's first World Heritage Site, the Galapagos Islands.

Notes

1. Bruce Epler. (1993). *An Economic and Social Analysis of Tourism in the Galapagos Islands*. Kingston: Coastal Resources Center, University of Rhode Island.

2. Bruce Epler. (2007). *Tourism, the Economy, Population Growth, and Conservation in Galapagos*. Puerto Ayora, Ecuador: Charles Darwin Foundation; Graham Watkins and Felipe Cruz. (2007). *Galapagos at Risk: A Socioeconomic Analysis of the Situation in the Archipelago*. Puerto Ayora, Ecuador: Charles Darwin Foundation; Observatorio de Turismo. (2018). *Informe Annual de Visitantes a la Areas Protegidas de Galapagos 2018*. Puerto Ayora, Ecuador: Observatorio de Turismo.

3. Carlos Carrion. (2007). "Migración y crecimiento poblacional en las islas Galapagos." Eds. Pablo Ospina and Cecilia Falconí. *Galápagos: Migraciones, economía, cultura, conflictos y acuerdos*. Quito: Corporación Editora Nacional. pp. 93–113; Carlos Mena. (June 25, 2019). *El suministro de alimentos en galápagos: enlazando agricultura, importaciones y turismo para crear escenarios futuros*. Presentation, Puerto Ayora, Ecuador: Galapagos National Park and the Charles Darwin Foundation.

4. Pablo Ospina. (2001). *Migraciones, actores e identidades en Galápagos*. Quito: Consejo Latinoamericano de Ciencias Sociales.

5. Bruce Epler. (1993). *An Economic and Social Analysis of Tourism in the Galapagos Islands*. Kingston: Coastal Resources Center, University of Rhode Island; Bruce Epler. (2007). *Tourism, the Economy, Population Growth, and Conservation in Galapagos*. Puerto Ayora, Ecuador: Charles Darwin Foundation.

6. Bruce Epler. (1993). *An Economic And Social Analysis of Tourism in the Galapagos Islands*. Kingston: Coastal Resources Center, University of Rhode Island.

7. William Durham. (2008). "Fishing for Solutions: Ecotourism and Conservation in Galapagos National Park." Eds. Amanda Stronza and William Durham. *Ecotourism and Conservation in the Americas*. Wallingford: CABI. pp. 66–89.

8. Byron Villacis and Daniela Carrillo. (2012). "The Socioeconomic Paradox of Galapagos." Eds. Stephen Walsh and Carlos Mena. *Science and Conservation in the Galapagos Islands: Frameworks and Perspectives*. New York: Springer. pp. 69–85.

9. Christophe Grenier. (2007). *Conservación contra natura. Las islas Galápagos*. Quito: Editorial Abya Yala.

10. Flora Lu, Gabriela Valdivia, and Wendy Wolford. (2013). "Social Dimensions of 'Nature at Risk' in the Galápagos Islands, Ecuador." *Conservation and Society*. 11(1): 83; William Durham. (2008). "Fishing for Solutions: Ecotourism and Conservation in

Galapagos National Park." Eds. Amanda Stronza and William Durham. *Ecotourism and Conservation in the Americas*. Wallingford: CABI, pps. 66-89.

11. Pablo Ospina. (2001). *Migraciones, actores e identidades en Galápagos*. Quito: Consejo Latinoamericano de Ciencias Sociales; Christophe Grenier. (2007). *Conservación contra natura. Las islas Galápagos*. Quito: Editorial Abya Yala.

12. UNESCO. (2010). *State of Conservation of Properties Inscribed on the List of World Heritage in Danger: Reactive Monitoring Mission Report Galapagos Islands*. Brasilia: UNESCO World Heritage Committee; Martha Honey. (2008). *Ecotourism and Sustainable Development: Who Owns Paradise?* Washington, DC: Island Press.

13. Wendy Strahm and Marc Patry. (June 17, 2010). *Reactive Monitoring Mission Report, Galapagos Islands 27 April–6 May 2010*. Paris: World Heritage Center UNESCO.

14. Rory Carroll. (July 29, 2010). "UN Withdraws Galápagos from World Heritage Danger List." *The Guardian*. https://www.theguardian.com/world/2010/jul/29/galapagos-withdrawn-heritage-danger-list.

15. Juan Carlos García, Daniel Orellana, and Eddy Araujo. (2013). The New Model of Tourism: Definition and Implementation of the Principles of Ecotourism in Galapagos. *Galápagos Report 2011–2012*. Puerto Ayora, Ecuador: Galapagos National Park, Governing Council of Galapagos, Charles Darwin Foundation, and Galapagos Conservancy. pp. 95–99.

16. Observatorio de Turismo. (2011). *Boletín No. 1 Edición Trimestral Junio–Septiembre 2011*. Puerto Ayora, Ecuador: Observatorio de Turismo.

17. Bruce Epler. (2007). *Tourism, the Economy, Population Growth, and Conservation in Galapagos*. Puerto Ayora, Ecuador: Charles Darwin Foundation; Parque Nacional Galapagos. (2020). *Tribute of Entry Fee for Galapagos National Park*. http://www.galapagos.gob.ec/en/tribute-of-entry/.

18. Stephen Walsh, Amy McCleary, Benjamin Heumann, Laura Brewington, Evan Raczkowski, and Carlos Mena. (2010). "Community Expansion and Infrastructure Development: Implications for Human Health and Environmental Quality in the Galápagos Islands of Ecuador." *Journal of Latin American Geography*. pp. 137–59; Kerstin Zander, Angelica Saeteros, Daniel Orellana, Veronica Toral Granda, Aggie Wegner, Arturo Izurietah, and Stephen Garnett. (2016). "Determinants of Tourist Satisfaction with National Park Guides and Facilities in the Galápagos." *International Journal of Tourism Sciences*, 16(2): 60–82.

19. Observatorio de Turismo (2017). *Estadística Turismo Galapagos 2017*. Puerto Ayora, Ecuador: Observatorio de Turismo.

20. Ibid.

21. Kerstin Zander, Angelica Saeteros, Daniel Orellana, Veronica Toral Granda, Aggie Wegner, Arturo Izurietah, and Stephen Garnett. (2016). "Determinants of Tourist Satisfaction with National Park Guides and Facilities in the Galápagos." *International Journal of Tourism Sciences*, 16(2). pp. 60–82.

22. Martha Honey. (2008). *Ecotourism and Sustainable Development: Who Owns Paradise?* Washington, DC: Island Press.

23. Byron Villacis and Daniela Carrillo. (2012). "The Socioeconomic Paradox of Galapagos." Eds. Stephen Walsh and Carlos Mena. *Science and Conservation in the Galapagos Islands: Frameworks and Perspectives*. New York: Springer. pp. 69–85.

24. Instituto Nacional de Estadística y Censos. (2015). *Análisis de resultados definitivos Censo de Población y Vivienda: Galápagos 2015*. https://www.ecuadorencifras

.gob.ec/documentos/web-inec/Poblacion_y_Demografia/CPV_Galapagos_2015 /Analisis_Galapagos%202015.pdf.

25. Amanda Stronza, Carter Hunt, and Lee Fitzgerald. (2019). "Ecotourism for Conservation?" *Annual Review of Environment and Resources.* https://doi.org/10.1146 /annurev-environ-101718-033046.

Chapter 5.3

Big Sur, California

By Roberta Atzori and Laura Kasa

Tourists to the stunning coastal area of Big Sur in Monterey County, California, historically had come as "spiritual seekers": counterculture worshippers of natural beauty. In the 1970s, people had heard of a beautiful area on the coast of central California, where a majestic highway hugged the cliffs hanging over the Pacific Ocean. At that time, visitors would perhaps send postcards with an iconic picture of a 1930s bridge spanning the hills with the ocean in the background and a note that might say, "Find your way to this amazing place and your soul will be touched by the magic, peacefulness, and breathtaking beauty of nature in Big Sur." The massive influx of tourists in Big Sur, which has mainly reflected the rise of social media, has, however, detracted from those amazing aspects that originally led people to come experience it. Now, postcards of the scenery are replaced with Instagram posts of visitors looking at the scenery.

Big Sur has long been settled by individuals looking to immerse themselves in natural beauty and get away from it all. It has been a bohemian haven for artists, writers, and photographers, including Jack Keroauc, Hunter S. Thompson, and Henry Miller. It is the home of the Esalen Institute, a new age retreat famous for its hot springs and counterculture workshops. Understandably, the rise of overtourism in

Selfie seekers in Big Sur, California. Source: Roberta Atzori.

this region has been met by a plethora of grassroots initiatives and resident protests.

Causes of Overtourism

Big Sur is a breathtaking scenic area along California's central coast. Although it has no set boundaries, it is generally considered to include more than seventy miles from Carmel-by-the-Sea in Monterey County on the north to the San Luis Obispo County line near San Simeon on the south.

When the section of Highway 1 that runs along the Big Sur coast opened in 1937, the main purpose of the new scenic route was for motorists' enjoyment and recreation.[1] Still today, sightseeing and scenic driving are the main recreational activities along one of the most spectacular roads in the world, parts of which are designated "All-American Roads" by the Federal Highway Administration's scenic byways program.[2]

Since the latter half of the twentieth century, visitation to Big Sur has skyrocketed. An article in *Outside* magazine from 2019 placed the number of one-way road trips on Highway 1 around Big Sur at an estimated 4.6 million per year and visitation to Big Sur at 5.8 million visitors per year.[3] Due to its proximity to the Highway 1 route that leads to Big Sur, Monterey city and surrounding areas can be considered gateway communities for Big Sur; in 2017, the Monterey County Convention and Visitors Bureau (MCCVB) reported in 2017 that almost 20 percent of survey participants visited Big Sur.[4] In 2018, travel spending in the Big Sur gateway county had increased almost 6 percent ($2.9 billion) from 2017 ($2.8 billion), whereas the average annual change between 1992 and 2018 was only 3.5 percent.[5]

A variety of factors has likely led to this explosion, and they are reflective of the global increase in tourism, including a growing middle class, more cost-effective methods of transportation and lodging vis-à-vis the sharing economy, and an aging group of millennial travelers with expendable income. The most prominent cause of overtourism and unsafe tourist behavior in Big Sur, though, is the region's popularity on social media and in film.

In 2020, Instagram searches for "#bixbybridge," "#highway1," and "#bigsur" returned 98,200, 428,000, and 1.2 million posts, respectively. All depict stunning scenes of cliffs, aquamarine waters, and couples or solo visitors gazing out over the scenery. On Bixby Creek Bridge, a stunning arched bridge that connects northern Monterey County to Big Sur, these images do not show the frequent pileup of cars stopped at one small, precipitous pull-off, creating a dangerous situation for drivers.

Some reports have also attributed recent increases in visitation to *Big Little Lies*, an HBO series based in Monterey that, in its opening sequence, follows the main characters as they drive Highway 1.[6] Monterey tourism authorities have called the show's impact on marketing the area a "bonanza," said Rob O'Keefe, chief marketing officer and vice president of the MCCVB. "Our destination is the unedited star of the show."[7]

The Big Sur Coast Land Use Plan (LUP) of 1986 was intended to limit all private, public, and commercial development to a calculated maximum sustainable level supporting both public and private priorities.[8] It is often considered one of the most restrictive development policies in the United States. By dictating specific land use and development restrictions, the plan makes adding infrastructure to accommodate the increased number of tourists challenging. Although the LUP anticipated increasing visitor pressure, visitation when it was created was approximately half today's levels. There are very few gas stations, public restrooms, visitor information centers, and grocery stores and a limited number of accommodations. In addition, rather than being governed by a single entity, Big Sur is a mix of private, state, and federal properties. As a result, the area is now faced with serious infrastructure and safety challenges—and no easy way to address them.

Impacts of Overtourism

In the summer of 2019, photos of a banner that read "Overtourism Is Killing Big Sur," hung over Bixby Bridge, flooded headlines.[9] The protest was organized by a group of locals called Take Back Big Sur. The same message was illegally spray-painted onto the road a week after the banner was taken down.[10] That same summer, an Instagram account with the handle "Big Sur Hates You" circulated photos of bad behavior by tourists.[11] The account was quickly changed to "Big Sur Educates You," but the original phrasing is still used as a hashtag by some locals.

The photos show a plethora of traffic jams, illegally parked cars, and trash, as well as selfie seekers positioned in the middle of the road or making contact with sensitive wildlife. Human nature is not entirely to blame; due to limited infrastructure, visitors are competing for the few resources that are available, often resulting in disrespectful, illegal, and unsafe behavior. The lack of sufficient restroom and trash facilities along the coast means that visitors often use the roadsides and trails for these purposes, polluting the area that leads into the Monterey Bay National Marine Sanctuary.[12] The lack of sufficient parking in pull-outs and parks results in lines at park entrances, vehicles double-parked, and vehicles left in the middle of the road while visitors jump out to take a photo.[13]

There has been an increase in traffic congestion along Highway 1, which is already a risky drive along a winding cliffside. Dangers on the

Traffic jam at Bixby Creek Bridge. Source: Kodiak Greenwood.

roadway are exacerbated by the threat of landslides; the May 2017 Mud Creek landslide caused damage that cost the state $54 million and a year to fix.[14] Referred to by some as the "mother" of all landslides, it was one of the largest the state had ever seen and is thought to have been brought on by extreme drought followed by heavy rainfall. The Pfeiffer Canyon Bridge, near a popular tourist beach, also closed during this time due to cracking from the heavy rains.[15] Road closures and congestion can be detrimental for tourism and for locals; Highway 1 is the only access point into and out of the area, and the nature of the route makes it difficult for emergency services to intervene.

Tourism has also abetted other climate change–related threats. In 2016, an extremely costly wildfire, one of the costliest in the state's history, was caused by a campfire in an illegal campsite near Garrapata State Park and resulted in more than $250 million in suppression costs, seventy homes lost, and one fatality.[16] Illegal camping and campfires in the backcountry are commonplace, and due to a lack of enforcement, locals often take reprimanding the lawbreakers into their own hands.[17]

Massive wildfires, landslides, and damage to infrastructure along Highway 1 and in Big Sur can be devastating to the tourism economy

Ecosystem destruction and fire hazard caused by visitors driving off-road in Big Sur. Source: Kate Novoa.

and the operators who depend on it. Furthermore, a lack of affordable housing in the area makes hiring challenging; a chef at a Big Sur hotel confirmed that finding new staff in the area was difficult and that he commuted from Monterey (a highly variable thirty-mile drive depending on the congestion and impassability in many areas).[18] Some high-end hotels, such as the Post Ranch Inn and Ventana Big Sur, with staff housing on property, have had to undergo expansion projects to meet this need.[19] How can local government, businesses, citizens, and other stakeholders ensure that there is a balance between tourism's needed revenue and the effects it has on the community and this unique landscape?

Solutions

Operating in the confines of the challenges outlined above, Big Sur is in need of innovative solutions so as to preserve the natural environment, community, and tranquility of the area. There are a limited number of options to improve congestion on Highway 1; more infrastructure cannot easily be built at this time because of the LUP. Furthermore, implementing tolls on state highways is uncommon, as the criteria is highly restrictive. There are, however, a variety of ways to reduce tourism's impacts, including outreach to help educate the public to be more aware of the limited services, safety issues, and ways they can be respectful of the area during their visits. Ways to improve traffic flow and reduce trash that do not require major infrastructure changes have also been tested, and these efforts can continue to be expanded.

Cross-Sector Partnerships

To address these problems and craft potential solutions, the policy framework included in the LUP resulted in the creation of the Big Sur Multi-Agency Advisory Council (BSMAAC) in 1986.[20] Members include representatives of the California state legislature, the California Department of Parks and Recreation, the California Department of Transportation, the Property Owners' Association of Big Sur, the Big Sur Chamber of Commerce, and the Monterey Bay National Marine Sanctuary. Quarterly meetings provide a public forum where citizens can offer ideas and concerns, which in 2019 included traffic control, fire prevention and illegal camping, the need for public restrooms for visitors, cell phone coverage in the area, signage gaps, and infrastructure improvements.[21] The strict nature of the LUP, paired with the many public agencies involved and their different missions, as well as insufficient funding and resources, pose a challenge to finding effective and timely solutions, however.

Grassroots Initiatives

A wide array of nonprofit organizations and associations have sprung to action to address tourism issues with more dexterity. The most prevalent of them is the Community Association of Big Sur (CABS), formed in the 1960s by property owners to thwart efforts to turn Big Sur into a national park.[22] Now, CABS has partnered with Beyond Green Travel to call for a comprehensive destination stewardship plan. It also created the Big Sur Pledge, which prompts visitors to exhibit environmentally friendly behavior while in Big Sur, and CABS members greet tourists at local hotspots to ask them to take the pledge.[23] CABS also created an article for an airline in-flight magazine, available in multiple languages, that addresses the unwritten behavioral rules to observe while visiting Big Sur and highlights area services available.[24]

In addition, a variety of nonprofits are indirectly mitigating the impacts of tourism in Big Sur through environmental education initiatives and cleanups, such as Friends of Garrapata State Park and Big Sur Advocates for a Green Environment (BSAGE). The Ventana Wilderness Alliance has a volunteer park ranger program to patrol backcountry and trail areas for illegal camping and campfires.[25]

Using Technology

Successful efforts have also been made to reach tourists via social media. Some state parks, for example, have produced videos to show how to properly view wildflowers.[26] In addition, the MCCVB has developed a "Sustainable Moments" campaign, communicated through the MCCVB website and social media, that uses humorous videos such as "Avoid a #TravelFail" to instruct travelers on how to have a safe, low-impact visit [27]

Residents have also worked to distribute helpful information via blogs and the internet. Kate Woods Novoa, or BigSurKate, keeps an independent blog containing up-to-date information to keep residents and visitors informed on the latest news in Big Sur, including road conditions, events announcements, and fires and landslides.[28] Through the blog, Novoa also provides updates on the work of CABS and the formation of a Big Bay Stewardship Council. In addition, the Big Sur Visitor Guide website provides information to visitors on what services the area offers and does not offer (such as "hospitals, lifeguards, fast-food restaurants, and grocery stores") along with a detailed review of local rules and regulations.[29] The purpose is to help visitors familiarize with the area and plan their trip accordingly before they visit.

Controlling Traffic

The Sycamore Canyon Shuttle Project, led by CABS, provided trial shuttle runs in the summer of 2018 for visitors to Pfeiffer Beach.[30]

The trial was deemed successful, but CABS is still waiting for beach management authorities to approve a plan to decrease visitor traffic through parking limits and fees. CABS argues that this step must be taken in tandem with the shuttle system to have a significant impact.

Innovative Recommendations

The above techniques to address the problems of tourist education, traffic congestion, and litter have been successful. Due to the unique constraints in Big Sur and the lack of more overarching interventions, however, there is still a long way to go, and innovative ideas are welcome.

One recommendation came from retired California State Highway patrol officer Ken Wright. "Rental car agencies could play a part in providing rules of the road in several different languages for people driving in the US and on the Big Sur Highway, to reduce traffic and safety issues," he said.[31]

"An online reservation system is being considered for neighboring Point Lobos State Reserve," stated Pat Clark-Grey, regional interpretation specialist for California State Parks. This approach may also be considered for Julia Pfeiffer Burns State Park to reduce congestion at the picturesque tourist spot.[32]

And yet another idea came from Butch Kronlund, executive director of CABS. "Webcams could be installed at choke points and hotspots along Highway 1 so visitors could view real-time traffic before they make their trip," he remarked.[33]

According to former congressman and Big Sur property owner Sam Farr, "We need to get people out of cars and into buses that can provide a value-added experience of seeing Big Sur. Narration of what you are seeing on California's first Scenic Highway. Every twist and turn has a story. . . . Congestion is caused by stopping to see something. That something could be narrated on a bus."[34]

Visitor centers at both ends of Big Sur, which would not be under the jurisdiction of the LUP, could be installed and could offer tourist information, refreshments, and public restrooms. LaVerne McLeod, leader of BSAGE, suggests, "This could even be a pilot program where a pop-up tent is manned by a volunteer who could provide maps, tourist services information such as where cell phone signals are available, and a code of conduct (available in several languages)."[35]

Finally, incentives could be offered to tourists who agree to act responsibly, such as coupons to a local store, attraction, or restaurant in the area. It would be an excellent way to partner with local businesses to encourage people to spend money in gateway communities.

Improving the Tourist Impact

With social media increasingly dictating the way travel is experienced, beautiful natural areas around the world are attracting more tourists than the location can sustain. Instead of looking toward permanent infrastructure changes, which would be costly and topographically challenging within Big Sur, the solutions presented focus on improving the tourist impact on the area.

There is, however, a clear need for concerted multisector efforts to address the sustainability of tourism in Big Sur. Factors that are putting the region at risk, such as increased visitation and landslides and wildfires brought on by climate change, are all predicted to increase. Management authorities in Big Sur, as well as in the entry points to the area to the north and south, need to collaborate to develop a stewardship plan (as has been suggested by CABS) and recognize the collective impact on seventy miles of California's coastline.

For the time being, the goal among grassroots organizers is to improve the situation in Big Sur by focusing on solutions that can be put in place quickly and that do not require large investments or coordination among multiple government agencies. The hope is that once the successes of these innovative approaches can be showcased, the many government agencies involved with Big Sur will be inspired to work cooperatively to dedicate focus and resources to address more complicated issues.

Notes

1. Monterey Peninsula Herald. (June 26, 1937). "Highway along Monterey Coast Will Be Open Tomorrow." *Monterey Peninsula Herald.*

2. US Department of Transportation Federal Highway Administration. (2020). "America's Byways." https://www.fhwa.dot.gov/byways/.

3. Josh Marcus. (October 3, 2019). "Has Overtourism Killed Big Sur?" *Outside Magazine*. https://www.outsideonline.com/2401685/big-sur-overtourism.

4. Monterey County Convention and Visitors Bureau. (January 2017). "2016 Monterey County Visitor Profile Final Report of Findings." https://assets.simpleviewinc.com/simpleview/image/upload/v1/clients/montereycounty/Monterey_County_Visitor_Profile_2016_Final_Report_of_Findings_26af80ed-36b8-4570-845c-bfae665e0349.pdf

5. Dean Runyan Associates. (April 2019). "Monterey County Travel Impacts, 1992–2018." https://assets.simpleviewinc.com/simpleview/image/upload/v1/clients/montereycounty/Dean_Runyan_2018_c28ef92d-b189-492c-ad5c-0404bed745c9.pdf.

6. Kari Paul. (July 9, 2019). "Overtourism Is Killing Big Sur: Activists Raise Banner in California Vacation Spot." *The Guardian*. https://www.theguardian.com/us-news/2019/jul/09/big-sur-banner-bixby-bridge-tourism.

7. Carly Matberry. (December 14, 2017). "HBO's 'Big Little Lies' a Financial Boon for Monterey Peninsula." *East Bay Times*. https://www.eastbaytimes.com/2017/12/14/hbos-big-little-lies-a-financial-boon-for-monterey-peninsula/.

8. Monterey County Planning. (April 10, 1986). "Big Sur Coast Land Use Plan." http://www.co.monterey.ca.us/planning/docs/plans/Big_Sur_LUP_complete.PDF; Martha Diehl. (2006). "Land Use in Big Sur: In Search of Sustainable Balance between Community Needs and Resource Protection." Master's thesis. California State University, Monterey Bay.

9. Kari Paul. (July 9, 2019). "Overtourism Is Killing Big Sur: Activists Raise Banner in California Vacation Spot." *The Guardian*. https://www.theguardian.com/us-news/2019/jul/09/big-sur-banner-bixby-bridge-tourism.

10. Chelcey Adami and Kate Cimini. (July 16, 2019). "'Overtourism Is Killing Big Sur' Spray Painted on Big Sur's Bixby Bridge Parking Lot." *The Californian*. https://www.thecalifornian.com/story/news/2019/07/16/california-chp-search-suspects-big-sur-bixby-bridge-vandalism-overtourism-tourism/1751508001/

11. Caitlin Conrad. (May 14, 2019). "An Instagram Account Is Shaming Bad Tourist Behavior at Big Sur. But is it Actually Helpful?" *SF Gate*. https://www.sfgate.com/travel/article/Big-Sur-Hates-You-shames-bad-tourist-behavior-13842140.php.

12. Kate Woods Novoa. (March 4, 2019). BigSurKate. Personal communication with author

13. Tammy Blount Canavan. (March 4, 2019). Monterey County Convention & Visitors Bureau. Personal communication with author

14. Pam Wright. (February 19, 2019). "California's Big Sur's $54 Million 'Catastrophic Landslide' a Result of Drought Followed by Deluge, Scientists Say." The Weather Channel. https://weather.com/science/environment/news/2019-02-19-big-sur-catastrophic-landslide-drought-deluge; Laurel Wamsley. (May 25, 2017). "'Mother of All Landslides' in Big Sur Buries Section of California's Highway 1." *NPR*. https://www.npr.org/sections/thetwo-way/2017/05/25/530025850/-mother-of-all-landslides-closes-section-of-california-s-highway-1.

15. Amy Graff. (February 23, 2017). "Pfeiffer Canyon Bridge in Big Sur Cracked Beyond Repair by Rains, Must be Replaced." *SF Gate*. https://www.sfgate.com/news/article/See-the-Big-Sur-bridge-that-s-cracked-and-10951909.php

16. Big Sur Fire. (2019). "Soberanes Fire." Bigsurfire.org. http://bigsurfire.org/?page_id=2579.

17. Alani Letang. (September 3, 2019). "Big Sur Local Putting Out Illegal Campfires." KSBW8. https://www.ksbw.com/article/big-sur-local-putting-out-illegal-campfires/28906172#

18. Anonymous source. (February 2020). Personal communication with the editors.

19. Community Association of Big Sur. (2019). "Community Housing and Visitor-Serving Infrastructure." https://www.cabigsur.org/community-housing/

20. County of Monterey. (2020). Big Sur Multi-Agency Advisory Council. https://www.co.monterey.ca.us/government/departments-a-h/clerk-of-the-board/boards-committees-and-commissions/big-sur-multi-agency-advisory-council.

21. Big Sure Multi-Agency Advisory Council. (August 23, 2019). "Big Sur Multi-Agency Advisory Council Meeting Minutes for August 23, 2019." Pfeiffer Big Sur Lodge Conference Center, Pfeiffer Big Sur State Park, Big Sur. https://www.co.monterey.ca.us/home/showdocument?id=84427

22. Community Association of Big Sur. (2019). "Welcome." https://www.cabigsur.org/.

23. Big Sur Pledge. (2020). "Big Sur Pledge." https://bigsurpledge.org/.

24. Butch Kronlund. (March 8, 2019). Community Association of Big Sur. Personal communication with author

25. Ventana Wilderness Alliance. (2020). "Home." https://www.ventanawild.org/index.php.

26. Patricia Clark-Grey. (March 8, 2019). California State Parks. Personal communication with author

27. See Monterey. (April 27, 2017). "Avoid a #TravelFail on Your Next Adventure to Monterey County, California." Monterey County Convention & Visitor Bureau. https://youtu.be/Yc1KueI4adY

28. BigSurKate. (2020). "Big Sur Information." https://bigsurkate.blog/

29. Big Sur Visitor Guide. (2020). "Big Sur Visitor Guide." http://www.bigsurvisitorguide.com/

30. Community Association of Big Sur. (2020). "Sycamore Canyon Shuttle to Pfeiffer Beach Day Use Area (PBDUA)." https://www.cabigsur.org/sycamore-canyon-shuttle/

31. Ken Wright. (March 20, 2019). California Highway Patrol. Personal communication with author

32. Patricia Clark-Grey. (March 8, 2019). California State Parks. Personal communication with author

33. Butch Kronlund. (March 8, 2019). Community Association of Big Sur. Personal communication with author

34. Sam Farr. (March 13, 2019). Former US Congressman. Personal communication with author

35. LaVerne McLeod. (March 1, 2019). BSAGE. Personal communication with author.

Chapter 5.4

Hawai'i

By Frank Haas

When Hawai'i first welcomed outsiders, awestruck visitors described the islands as "paradise." After his visit in 1866, Mark Twain pronounced Hawai'i to be "the loveliest fleet of islands that lies anchored in any ocean."[1] That allure is the foundation of a tourism industry that attracted more than ten million visitors in 2019. Arrivals set records for most years in the 2010s. For residents who think that there are "too many tourists" in Hawai'i, consider that on an average day in 2019, there were nearly 250,000 visitors in a state with only about 1.4 million residents. On the tourist-heavy islands of Maui and Kaua'i, about 27 percent of the people on island on an average day are visitors.[2]

With all that growth, the islands are experiencing the strain of mass tourism, but things haven't reached a crisis stage yet. Despite the string of record arrivals and a large tourist population, visitor satisfaction with Hawai'i remains strong and hotels continue to command premium rates. But as tourism numbers increase, there are signs of trouble ahead.

Causes of Overtourism

The Hawaiian Islands are as fragile as they are alluring. Hawai'i's "anchorage" (as Twain put it) is relatively small geographically and isolated some two thousand miles from the nearest inhabited landmass. As a result, the islands developed a unique culture and fragile ecosystems that can been threatened by development and growth—including the growth of tourism. Intense development stresses resources, detracts from the character of the place, and encroaches on what visitors come to see.

Social Media

Hundreds—or even thousands—of visitors descending on fragile sites are symptomatic of the tourism management challenge in Hawai'i. The sheer numbers cause congestion and traffic jams and can degrade the quality of experience at these sites. New technologies and social media exacerbate the problem. Sites that once were "secret" hideaways are now often overrun with visitors. One example is Laniākea Beach on the island of O'ahu, which was "discovered" on social media as a spot where visitors could often see turtles. So many visitors now have Laniākea on their itinerary that long traffic jams plague the adjacent highway.[3]

Another hotspot is the trailhead for Maunawili Falls on O'ahu. Although the trailhead is in a secluded residential neighborhood, it can easily be found by GPS. There is no parking lot, so hikers park on the street. There are no restrooms. The trail is generally muddy, and returning hikers often leave a mess in the neighborhood. Noise and litter have become constant neighbors. "To describe what has happened to the peaceful Maunawili neighborhood around the trailhead as 'inconvenient' or 'disruptive' is like calling a bomb blast 'kinda noisy,'" Lee Cataluna, a columnist for the *Honolulu Star Advertiser*, complained.[4]

These once "secret" sites do not have the infrastructure to accommodate large numbers of visitors, and the invading crowds upset local residents. Many of these sites are culturally sensitive, on private land, or pose a danger to life and limb. In 2018, for example, the Honolulu Fire Department rescued 346 hikers on O'ahu. Although not all the rescues were visitors, visitors very often get into trouble because they aren't familiar with local conditions.[5]

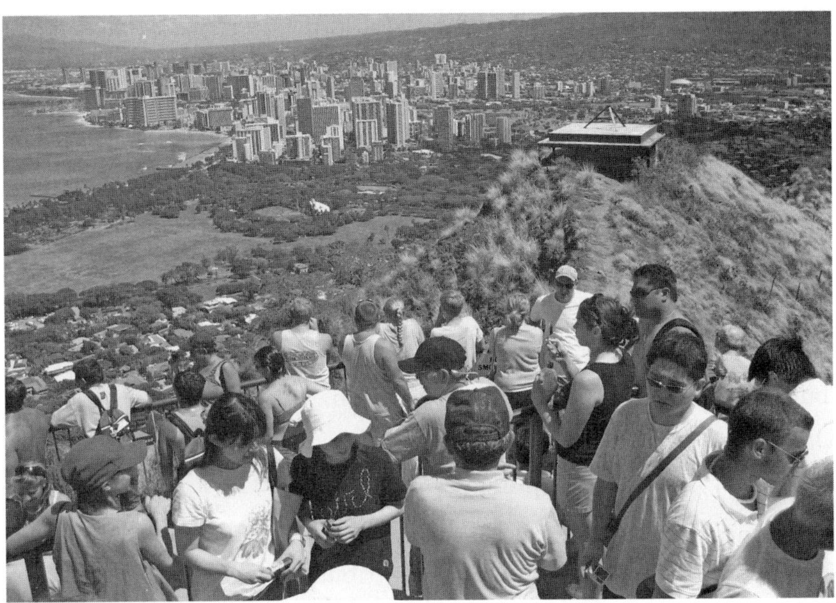

View from Diamond Head Lookout. Source: Pierre Omidyar (Flickr).

Alternative Accommodation Options

Technology and social media enable visitors to efficiently seek out lower-priced alternatives to traditional accommodations and experiences. Hawai'i has seen a dramatic rise in the use of alternative accommodations, including time-share units, condominiums, and vacation rentals. In 2019, the Hawai'i Tourism Authority (HTA) conducted a point-in-time analysis of three vacation rental booking platforms and found 33,118 advertised vacation rental units in the state, 85 percent of which were whole houses or condominium units (not shared rooms).[6] The rate of increase each year between 2014 and 2018 hovered around 20 percent.

The government has been slow in coming to grips with the rapid expansion of vacation rentals. Between the 1980s through 2019, the county of Honolulu had not issued any permits because of the controversy surrounding vacation rentals. As a result, as their numbers grew, most of the vacation rentals on the island have been operating illegally.

Proponents of vacation rentals argue that these rentals provide supplemental income to offset Hawai'i's high cost of living. Opponents contend that, to the extent that these rentals remove housing units from the residential pool, they drive up the high cost of housing and contribute to homelessness. Hawai'i has the fourth-highest percentage of long-term renters in the United States, and "vacationers are competing with long-term renters for property, especially since a homeowner can earn more than three times as much from renting short-term," as reporter Michelle Broder Van Dyke explained.[7] In addition, the HTA reported in 2016 that almost 70 percent of vacation rental properties available for short-term rentals are owned by nonresidents.[8] Hawai'i's counties have now begun to take steps in restricting vacation rentals, but enforcement will undoubtedly be a challenge.

Impacts of Overtourism

The amount of money visitors to Hawai'i spend per day has been declining, and tourism's economic contribution to Hawai'i's economy has not kept pace with arrivals growth. With these negative forces in play, resident attitudes toward tourism have shifted. Native Hawaiian culture and language are at risk of being overwhelmed and overcommercialized by mass tourism.

Economic Impacts

In 2019, Hawai'i welcomed 10.4 million visitors with total expenditures of $17.8 billion.[9] Although that number sounds impressive, it is less than the real (inflation-adjusted) $18 billion in total spending in the peak year of 1989, which saw 6.5 million visitors. Thus, the state has about 3.9 million more visitors than it did in 1989, causing significantly more congestion and stress, but with about the same real economic impact. This situation is likely due to a combination of factors; notably, visitors staying in short-term rentals in Hawai'i tend to spend less than those who stay in a hotel.[10]

Resident Attitudes

In a research series on resident attitudes toward tourism, support for the statement that tourism "brings more benefits than problems" declined from a high of 80 percent in 2010 to 58 percent in 2018. A local resident writing in Kaua'i's *Garden Island* newspaper summed up the problem this way: "Apart from the job opportunities, the problem is that only a small circle of residents or investors will enjoy the tourism-generated benefits and a large number of residents will pay for it by losing the once-preferred laid-back island lifestyle."[11]

Trivialization of Hawaiian Culture

The tourism industry has historically packaged the complex Native Hawaiian culture into digestible tidbits. Despite a significant renaissance in Hawaiian language and culture, the culture is often marginalized or misrepresented, and important, nuanced cultural concepts can become cartoonish. The word *aloha,* a central concept in the Hawaiian culture, is often a pep-rally-like shout-out at events for visitors. A cruise line once featured a photo of the statue of King Kamehameha I (a revered monarch who united the islands) with a photoshopped champagne flute in his outstretched hand.

Solutions

With the state and counties traditionally following a relatively laissez-faire, hands-off approach to tourism management, a more holistic, integrated plan for addressing these issues is needed. Tourism management and regulation of short-term rentals have been largely at the county or island-level and have been case-specific.

Individual Site Management

Tourists in Hawai'i have historically enjoyed free and easy access to sites and attractions. Beaches in Hawai'i are free and open to the public with mandated public access provided at regular intervals. Traditionally, parks and trails, too, have been open to visitors and residents alike, free of charge. With growing negative visitor impacts, restricted access has increasingly been implemented on a case-by-case basis as sites became overwhelmed. To date, implementation of restrictions and fees and other management programs have only been adopted when a site reaches a crisis point.

Perhaps the most significant move to restrict access to a popular site is Hanauma Bay on the island of O'ahu. As the number of visitors grew, this spectacular snorkeling site saw visitation increase from about half a million in 1975 to about 2.8 million in the late 1980s (about 7,500 per day).[12] The crush of visitors damaged live coral in the bay and upset the ecosystem as visitors fed fish bread and frozen peas. In 1990, Honolulu finally implemented an initial management plan for the bay, limiting access, educating visitors, and improving facilities. Once the three-hundred-spot parking lot filled, visitors were turned away. Over time, additional controls were adopted, including an entrance fee in 1995 and the complete closure of the park on Tuesdays for cleaning and maintenance. Today, conditions at Hanauma have improved due to the cap on the number of visitors and revenues from admissions contributing to the maintenance of park facilities. Visitation stabilized at 1.6 million visitors in 2017, more than 40 percent lower than its peak.[13]

At the federal level, the National Park Service has placed limits on the number of people allowed at the ten-thousand-foot summit of Haleakalā National Park on Maui at sunrise. Sunrise at Haleakalā is often described as a must-see event for Maui visitors. Before regulation, unrestricted access at sunrise attracted more than one thousand visitors, with three hundred cars jockeying for one hundred parking spaces.[14] In the predawn darkness, visitors parking outside of the paved lot often trampled seedlings and root systems of rare silversword plants. To address overcrowding, visitors coming to the park between 3 a.m. and 7 a.m. now need to apply for a limited number of permits in addition to the park entrance fee.[15]

A state-led management initiative addresses overcrowding issues at Hā'ena State Park on the island of Kaua'i. Hā'ena lies at the end of the road and at the beginning of the Napali Coast State Wilderness Park. This once-quiet site was drawing hundreds of thousands of visitors annually. Without controls, cars would sometimes illegally park as far as two miles from the entrance to gain access.[16] In response to overcrowding, deteriorating facilities, and community concerns about traffic, the state Department of Land and Natural Resources' Hā'ena Master Plan went into effect in 2019. The plan limits parking and allows access to only nine hundred people during peak hours. Fines for illegally parked cars increased from $30 to $200. The plan establishes a cultural advisory committee and a community advisory committee to provide advice on all aspects of park management and improvements.[17]

Short-Term Rental Regulations

Each of Hawai'i's counties has tightened regulations on short-term vacation rentals. In Honolulu, where many illegal vacation rentals have been operating, the county passed an ordinance in 2019 putting a cap of 1,715 units that can rent out individual rooms outside of resort areas.[18] Full home rentals are still allowed in resort areas but nowhere else. The new restrictions set a dramatic reduction in vacation rental inventory from the estimated 10,000 short-term rentals operating on the island (many of which were not permitted). To preserve the residential "feel" of neighborhoods, Maui capped the number of vacation rentals by district in 2019.[19] In 2018, Hawai'i county and the island of Kaua'i added restrictions to their short-term rental policies.[20] O'ahu has the most restrictive policies for "unhosted" short-term rentals, allowing them only in resort zones. Available rental nights for short term rentals on O'ahu dropped from 261,000 in December 2018 to 207,300 the following year.[21]

Remaining Challenges

With alarm bells ringing, Hawai'i is awakening to the issues brought on by unbridled growth. A new five-year strategic plan adopted by the Hawai'i Tourism Authority in 2020 speaks of balancing tourism marketing and management. The agency's budgets are being "rebalanced" to provide more support for natural resources, Hawaiian culture, and community programs. The definition of tourism success is shifting from gross measurements like visitor arrivals and nominal spending to measurements better aligned with sustainability, including both resident and visitor satisfaction. County governments are enforcing regulations to control vacation rentals. Different agencies responsible for tourism are responding to management issues, although they have only approached controversial issues once they reached a crisis point (such as government agencies beginning serious work on addressing traffic backups near "turtle beach" after a young visitor was injured crossing the highway and the community organized demonstrations).[22] Although an emerging focus on tourism management is a welcome change from a mindless focus on growth, the actions are still generally uncoordinated and underfunded.

As the state's lead agency for tourism, the HTA has a mission "to strategically manage Hawai'i tourism in a sustainable manner consistent with economic goals, community desires, and visitor industry needs."[23] In the years since its founding in 1998, achieving this mission on its own has been unrealistic because the agency lacks authority and funding to implement effective solutions. Despite more than half a billion dollars in receipts from the transient accommodations tax (Hawai'i's "hotel tax"), only about 13.2 percent went into HTA-funded programs in 2019, down from 34.1 percent in 2008. A growing visitor base with complex management issues requires more—not less—funding for visitor safety, site management, and other programs to maintain the visitor experience and resident quality of life. One example of inadequate funding for important tourism functions is the millions of dollars in deferred maintenance needed at the Hawai'i Convention Center.

Furthermore, oversight for tourism does not rest with a single agency like the HTA. Instead, it is spread over multiple state agencies, jurisdictions, nonprofit organizations, and industry groups. Until there is a coordinating mechanism and a statewide, adequately funded strategic plan in place, effective responses to mass tourism will remain a challenge. The management strategies discussed here have been ad hoc, site-specific exceptions to Hawai'i's hands-off management approach, and the response going forward begs for a long-term outlook, serious funding, and high-level planning.

The challenges that have cropped up with the rapid growth of tourism are not insurmountable, but they need to be addressed before Hawai'i faces a true crisis. The destination is at a proverbial tipping point, requiring new thinking, additional resources, and new management strategies to avoid descent from paradise to paradise lost. When Joni Mitchell wrote her 1970 hit "Big Yellow Taxi," she noted that its warning was written about Hawai'i: "Don't it always seem to go / that you don't know what you've got 'til it's gone."[24]

Notes

1. Mark Twain. (1990). "Mark Twain in Hawaii: Roughing It in the Sandwich Islands." Honolulu: Mutual Publishing
2. Hawaii Tourism Authority. (2020). http://www.hawaiitourismauthority.org/.
3. Nicole Tam. (August 4, 2019). "Concerned Residents Plead for the State to Take Action along a Busy North Shore Highway." KITV Island News. https://www.kitv.com/story/40874614/concerned-residents-plead-the-state-to-take-action-along-a-busy-north-shore-highway.
4. Lee Cataluna. (January 4, 2017). "At Last, Some Reprieve for Hike-Ravaged Maunawili." *Honolulu Star Advertiser.* https://www.staradvertiser.com/2017/01/04/hawaii-news/lee-cataluna/at-last-some-reprieve-for-hike-ravaged-maunawili/ .
5. Nina Wu. (September 30, 2019). "Diamond Head's Popularity with Visitors Keeps It in the Top Position on a List of Sites Requiring Rescues." *Honolulu Star Advertiser.* https://www.staradvertiser.com/2017/01/04/hawaii-news/lee-cataluna/at-last-some-reprieve-for-hike-ravaged-maunawili/
6. Hawaii Tourism Authority. (2020). https://hawaiitourismauthority.org/media/4085/2019-visitor-plant-inventory-report-final-rev.pdf. pp 59-60.
7. Michelle Broder Van Dyke. (March 30, 2018). "People in Hawaii Are Fed Up with Vacation Rentals." *Buzzfeed News.* https://www.buzzfeednews.com/article/mbvd/hawaii-airbnb-tourism-vacation-rentals.
8. SMS Research and Marketing Services, Inc. (November 2016). *The Impact of Vacation Rental Units in Hawai'i, 2016.* https://hawaiitourismauthority.org/media/2005/impact-of-vacation-rental-units-in-hawaii-2016.pdf
9. Hawai'i Tourism Authority. (2020). *Total Real Visitor Expenditures.* https://data.uhero.hawaii.edu/#/series?id=164719&sa=true&seriesCat=36&geo=HI&freq=A.
10. Carli Procell and Stewart Yerton. (July 10, 2019). "9 Charts That Show How Hawai'i Tourism Is Changing." *Honolulu Civil Beat.* https://www.civilbeat.org/2019/07/9-charts-that-show-how-hawaii-tourism-is-changing/
11. Janos Keoni Somu. (February 22, 2018). "Too Many Tourists—Blessing or a Curse." *The Garden Island.* https://www.thegardenisland.com/2018/02/22/opinion/too-many-tourists-a-blessing-or-a-curse/
12. James Mak. (2008). "Developing a Dream Destination: Tourism and Tourism Policy Planning in Hawaii." Honolulu: University of Hawai'i Press
13. Department of Business, Economic Development and Research. (2018). *2017 State of Hawaii Data Book Individual Tables.* http://dbedt.hawaii.gov/economic/databook/2017-individual/.
14. Allison Schaefers. (February 6, 2017). "New Rules Curb Haleakalā Sunrise Crowd." *Honolulu Star Advertiser* https://www.staradvertiser.com/2017/02/06/hawaii-news/new-rules-curb-haleakala-sunrise -crowd/.
15. Ibid
16. Allison Schaefers. (April 15, 2019). "Community Fears Reopening Kuhio Highway Will Create Flood of Tourists." *Honolulu Star Advertiser.* https://www.staradvertiser.com/2019/04/15/hawaii-news/community-fears-reopening-kuhio-highway-will-create-flood-of-tourists/.
17. PBR Hawai'i & Associates, Inc. (May 2018). *Haena State Park Master Plan.* https://dlnr.hawaii.gov/dsp/files/2018/05/FEIS-BLNR-with-Appendices.pdf.

18. State of Hawai'i: City Department of Planning and Permitting. (August 8, 2019). *New Ordinance on Short-Term Rentals.* http://www.honolulu.gov/rep/site/dpp/str/faqs.pdf.

19. Pete Jalbert. (March 2020). "Maui County Passes New Vacation Rental Bill." *Maui Real Estate Blog.* https://www.mauirealestate.com/maui-county-passes-new-vacation-rental-bill/; Maui County Code. (2020). *Chapter 19.65: Short-Term Rental Homes.* https://library.municode.com/hi/county_of_maui/codes/code_of_ordinances?nodeId=TIT19ZO_ARTIVREMIAR_CH19.65SHRMREHO_19.65.030REST.

20. Stephanie Vermillion. (August 2, 2019). "A Guide to Hawaii's Short-Term Rental Regulations." *Turnkey Blog.* https://blog.turnkeyvr.com/a-guide-to-hawaiis-short-term-rental-regulations/.

21. Hawai'i Tourism Authority. (2020). *December 2019 Hawai'i Vacation Rental Performance Report.* https://hawaiitourismauthority.org/media/4115/hta-december-2019-hawaii-vacation-rental-performance.pdf.

22. Ben Gutierrez. (August 11, 2019). "Demonstration at 'Turtle Beach' Underscores Tensions between Residents, Tour Busses." *Hawai'i News Now.* https://www.hawaiinewsnow.com/2019/08/12/protest-turtle-beach-underscores-tensions-between-residents-tour-buses/

23. Hawai'i Tourism Authority. (2020). https://hawaiitourismauthority.org/who-we-are/our-strategic-plan/.

24. Joni Mitchell. (1970). "Ladies of the Canyon: Big Yellow Taxi." *Songfacts.* https://www.songfacts.com/facts/joni-mitchell/big-yellow-taxi.

Chapter 6.1

Overtourism and Destination Governance

By Jonathan B. Tourtellot

The year was 1996, long before the term *overtourism* existed. I was visiting the Caribbean island nation of Dominica to write a travel story that included a stop at the Dominica Botanic Gardens. Its tiny zoo had a pair of Dominica's endemic national bird, the endangered sisserou parrot. The two or three hundred that survive in the wild are hard to see, so I was enjoying a close-up look at the parrots in their roomy outdoor cage. A minibus pulled up and disgorged a load of tourists. Then came another van, then another, and another, and more, all coming from a small cruise ship in port. Eventually, about a hundred people were crowded around the cage, trying to see the two parrots.

I'd long been concerned with the likely impact of population growth on visitation to appealing places, but in this moment the future crystallized. I thought, "We're going to have a problem."

Surely, crowding would never happen in one of my other favorite destinations—Iceland. It was too chilly, too rainy, too expensive. But I was wrong, as subsequent events have shown. When my then-girlfriend and I first went in 1973 for a weeklong trip, our US friends ribbed us: "Only you two would go to *Iceland* for a vacation!" But after that visit I was hooked, entranced by the wild, weird scenery and geologic wonders juxtaposed with the fascinating culture, deep heritage, and intriguing medieval language. I would return repeatedly over the years.

The Overtourism Paradox

By my seventh, most recent trip, in 2018, tourism was skyrocketing. Strings of tour buses were charging around the Golden Circle, the day-trip circuit out of Reykjavík. A sprawling commercial shopping mall had popped up across the road from the compact, very active Geysir hot springs area. The highway north to Akureyri had sprouted assembly-line restaurants for tour groups. Interminable lines fetched up against security and passport formalities at the airport. Lately, Iceland (see chapter 6.2) has been trying to get a handle on its runaway popularity, which shows every sign of weathering the March 2019 collapse of its discount carrier, WOW Air.

Now I constantly encounter Americans who have just been or are planning to go to Iceland. Although it's great that so many more people are experiencing other lands, cases of overtourism continue to grow, which leads, then, to the overtourism paradox: If it is good for everyone to be able to travel, and if all who can afford to travel do travel, the resulting volume may not be good at all.

Governance, or Lack of It

Previous chapters have looked at overtourism in specific types of destinations. Now we must look instead at "governance" (or lack of it) for any type of destination—at how poor governance has contributed to overtourism and why a new approach must be central to the solution. This examination is needed at national, regional, and local levels and in hard-to-administer transborder destinations, as in Lake Tahoe (see chapter 6.3).

As noted earlier, the rate of increase in tourism has even outpaced projections by the boosterish United Nations World Tourism Organization (UNWTO). By 2030, we may well exceed UNWTO's current forecast of 1.8 billion international arrivals.[1] This often-misinterpreted UNWTO statistic ("arrivals" means border crossings, not individuals) does not even reflect the near impossible-to-count domestic trips made

Tour buses on Iceland's Golden Circle, leaving Thingvellir. Source: Jonathan B. Tourtellot.

inside each country. One UNWTO estimate puts the global aggregate of internal trips at seven times international.[2]

Even as the number of tourists soar, the sites and places they visit remain the size they have always been. Five people gazing at the *Mona Lisa* at a given moment in 1955 have morphed into more than two hundred today. It's no wonder, then, that overtourism has cropped up in places far beyond the storied throngs in Venice and Barcelona. Crowds are creating problems on Scotland's Isle of Skye, in California's Big Sur, and even in hard-to-reach places like El Nido on Palawan, Philippines, and in the parks of Chilean Patagonia.

A Cancerous Addiction to Numbers

Despite the simple arithmetic of relentless growth, governments in most cases have done nothing about overtourism—and often, even worse than nothing. In the 2018 documentary *Crowded Out: The Story of Overtourism*, featuring Barcelona, Venice, and other crunch points, you could hear a constant refrain: "No one's in charge." Nor are those places exceptions. Even the most heavily visited destinations have

Chinese domestic travelers pack the walkways at Hongcun village, a World Heritage site in Anhui province. Source: Jonathan B. Tourtellot.

rarely if ever had a manager in charge of coping with the growth of tourism.[3]

Why not? Basically, governments haven't seen any reason for it. For decades, policy makers around the world have operated on the assumption that growth in tourism is good. That attitude relies on a chain of unstated—and perhaps unconscious—fallacious economic assumptions:

- The growth fallacy: Economic growth through tourism is necessary and good.
- The equivalency fallacy: Tourism is similar to any other industry.
- The accounting fallacy: Transactions by tourism businesses portray the total tourism picture.

These fallacies have led governments to measure success in tourism based on only two numbers: gross domestic product (GDP) and tourist arrivals, the second metric being impelled by the first. GDP measures an economy by toting up its transactions. If you, the tourist, rent a hotel room, that counts. If you photograph the appealing scenery outside the window, it doesn't—there's no transaction involved.

The first fallacy, the idea of mandatory economic growth, has been disputed by the very architect of GDP himself. Back in the 1930s, Nobel prize–winning economist Simon Kuznets expressed grave reservations about the way government leaders were embracing his statistical tool to the exclusion of all else. It's important, to be sure. Impoverished countries struggling toward affluence obviously need to grow GDP. Tourism's contribution to GDP has been enormously important there, even outstripping that of extractive industries in some places. But Kuznets warned that improving the well-being of a nation requires more than just boosting its transactional totals.[4]

That's doubly true for tourism because, unlike many other industries, it depends on a uniquely limited resource: the place. Policy makers who fall prey to the second fallacy—that tourism is like other industries—don't recognize this reality. They see increasing tourism revenues on the spreadsheet, and that's good enough. "More bums in seats, more heads in beds," as an industry catchphrase goes.

What don't appear on the spreadsheet are the things that attract tourists in the first place. The lakes, mountain views, city skylines, sea breezes, inviting street life, birds in the trees, historic architecture, the people themselves—all those come for free. GDP economics dismisses such things as "externalities"; because they cannot be assigned a dollar value, neither can their loss. Under the third fallacy, the accounting fallacy, externalities are literally not counted because there are costs invisible in "tourism industry" economics. Even toting up transactions is a complex job, so governments are prone to rely on the handiest indicator available, arrivals.

The Push for Infinite Growth in Finite Places

Thanks to point-of-entry formalities when entering a country, arrival totals are the simplest proxy for measuring international tourism growth. Policy targets are publicized in terms of growing the arrivals count, which is an incentive for quantity, not quality. One thousand tourists spending $50 each yields the same daily revenue as one hundred tourists spending $500 each, but the first group beats the arrival count of the second by a politically pleasing order of magnitude. It's irresistible. A Google news search limited just to the month of January 2020 yielded these quoted tourism goals:

- Japan: "Japan's inbound numbers are on track to meet the country's 2020 goal of 40 million international visitors."[5]
- Maldives: "The government has raised its target to 2 million visitors for the year 2020."[6]
- Greece: "Target for 2020 and the years to come is an annual one-digit rise in arrivals."[7]
- Nepal: "Should add 800 thousand more tourists this year to meet the target of 2 million."[8]
- Anguilla: "We have set ambitious targets of a 20% increase in tourist arrivals."[9]
- Kerala, India: "The Department of Tourism . . . has set a target of increasing the domestic tourist by ten percent."[10]
- Cyprus: "The Ministry aims to increase tourist flows to the island by 30% by the end of the decade, according to the national strategic plan."[11]
- Cambodia: "The Tourism Ministry has set the ambitious goal to attract at least 11 million of foreign tourists by 2025 and 25 million by 2030."[12]

Such thinking does not anticipate and deter overtourism; rather, it fans the flames. It is worse than doing nothing, indeed. This type of turnstile counting is deeply ingrained. Seleni Matus, executive director of the International Institute of Tourism Studies at the George Washington University, has been researching destination governance. Her conclusion: "Increased visitation is the primordial measurement of a tourism minister's success."[13]

So government leaders keep pushing growth—at least until constituents complain or the destination collapses. Rarely are short-term interests and political braggadocio so thoroughly at odds with wise long-term planning. "Growth for the sake of growth," warned conservationist Edward Abbey, "is the ideology of the cancer cell."[14]

To extend his analogy, some painful chemotherapy ought to help: The Philippines had to close severely overdeveloped Boracay island for half of 2018 just to restore it. Undeterred, the national Department of Tourism set a 2020 arrivals target of 9.2 million, up from the 8.2 million set the previous year. Let no tumor go unnourished.

Marketing versus Management

For much of the twentieth century and into the twenty-first, the go-to entity in most destinations for boosting tourism growth has been the destination marketing organization, or DMO. The "marketing" part of the name says it all. Often dominated by hoteliers, the DMO's main job has been to promote the place, casting the net far and wide to maximize heads in beds.

Changing the mindset will not be easy. As Richard Butler, originator in 1980 of the Tourism Area Life Cycle model, has pointed out, "DMOs have traditionally been, and mostly still are, promotion focused bodies charged with at least maintaining, if not increasing, visitor numbers."[15] Ask them to "reduce" numbers? You might as well ask an Olympic sprinter to run the race backwards.

Most governments also have a national tourism authority, often placed under the economic development department. In theory, the authority works to boost the economy through tourism, and the DMOs market the destinations. The missions may overlap or even reverse. They may not be coordinated. State, provincial, and local DMOs add layers of complexity. Significantly, Matus pointed out, almost none have roles in actually "managing" tourist sites and neighborhoods. Stewardship has not been their job.

The result can be a total policy disconnect, as demonstrated by cases from opposite ends of the world. In 2017, authorities in charge of Ireland's renowned Cliffs of Moher reportedly wanted to raise parking fees on tour coaches, fearing that the summer crowding was damaging their reputation. County Clare Council resisted, preferring more tourists, not fewer. In 2018, visitation at the Cliffs topped 1.5 million. A compromise has now been worked out, with such policies as favoring visitors who stay overnight in the county and charging a premium for day-tour visitors.[16]

In Tasmania, visitation to the Australian state's iconic Cradle Mountain had soared 25 percent in only three years, reaching 250,000 in 2017. Outfitters called for visitation caps, fearing loss of its selling point as an isolated wilderness. Fully bought into the first fallacy, government pushed back. National news quoted the state's premier, Will Hodgman, as saying a cap "will damage and hurt Tasmanian businesses right across the state. It'll cost jobs. We want to see growth in our tourism industry continue."[17] Until when? When is enough?

"Enough" may be now. Ever so slowly, the DMO's pure focus on marketing is beginning to change. According to a survey conducted in 2018 by the consulting firm NEXTFactor, 77 percent of North American DMOs claim to be "increasing their focus on sustainable tourism."[18] As missions broaden beyond simply pumping up tourist numbers, the question has arisen as to just what the "M" in DMO stands for—marketing or management? Some organizations go for both as a DMMO or, as in Europe, for the stuttering acronym DDMMO, for destination development, marketing, and management organizations.[19]

A US-based group, the former Destination Marketing Association International, wisely sidestepped the nomenclature trap in 2017 by renaming itself Destinations International (DI). Under its "DestinationNEXT" program, DI has been gently urging its membership to think outside of the hotelier box, calling for more community involvement and attention to sustainability.

Andreas Weissenborn, DI's senior director of research and advocacy, says that part of their challenge is to promote the principle of shared community value, arguing that for DMOs, "the true client is the citizens." To this sentiment I would add that (1) the true client is also the tourist, whose hotel bed tax often helps pay for the DMOs' existence, and (2) the true product is the place, whose quality and sustainability is essential. A few trailblazing DMOs are beginning to reflect this change in attitude, partly at the urging of their own citizens.

The DMO for British Columbia's Ireland-size Thompson-Okanagan region is one of them. Under CEO Glenn Mandziuk, the Thompson Okanagan Tourism Association (TOTA) adopted a sustainability stance after extensive consultation with residents, some of whom complained about visitor behavior in sensitive natural areas. After much back-and-forth, TOTA's new tourism management plan has been endorsed by all its fifty-plus communities. By then, TOTA had transformed itself from a stagnant, small DMO with two hundred to three hundred relatively inert members into a dynamic operation by simply dropping the yearly fee. Its stakeholder-based membership totals some forty-five hundred (although members must still be tourism-related enterprises).[20]

Most DMOs have yet to follow suit. Destination consultant Paul Ouimet, chief executive officer of NEXTFactor and the architect of DestinationNEXT, pointed out that "tourism and hospitality have outpaced [overall] economic growth in most destinations."[21] Governments and DMOs have trouble keeping up. Doing so requires planning, but in North America, Ouimet estimated that fewer than 10 percent of destinations even have a management plan. Community buy-in like TOTA's is essential for the plan actually to be adhered to over time. Ouimet said that places "that go through the effort to get a professional plan, done properly with a lot of engagement, those communities are moving ahead."[22] In Europe, Barcelona is widely viewed as a leader, having developed an ambitious, multisector Strategic Plan for Tourism 2020 to deal with overtourism (see chapter 2.2).

Colorado, also prompted by its citizenry, has begun to instruct visitors in proper behavior (see chapter 6.4). One well-established Rocky Mountain resort town, Breckenridge, has gone further. Being within day-trip range of Denver, its five thousand residents already faced an overtourism challenge: some one million visitors a year, of whom perhaps 20 percent are day-trippers, drawn especially to festivals and ski competitions.

The town's style is holistic. There's a sustainability officer, and membership in the DMO, the Breckenridge Tourism Office, includes all town businesses, not just tourism enterprises. Its chief executive officer, Lucy Kay, said that when her office invited Ouimet's company to help develop a tourism plan, hundreds of community residents provided feedback, complaining especially about the rise in "red flag" days—some forty days a year—when event-drawn tourist crowds were disrupting everything from street traffic to grocery shopping. The town's residents said that they wanted a destination "management" plan, not a growth plan: "No more growth, just manage what we've got."[23]

The vision statement Breckenridge subsequently adopted is striking: "Harmony of quality of life for residents and quality of place for visitors." Note the word *harmony*, an improvement on the more commonly seen term *balance*, a zero-sum assumption that portions of nature or scenery or peaceful living must be sacrificed at the altar of economic growth.

Breckenridge now mitigates red flag days with measures such as satellite parking, free shuttles, and—most important, said Kay—advance communication with residents to ensure local peace of mind and engagement in the process. Her advice for other DMOs reflects a needed tilt toward management: Deliver what your marketing promises, she said. "Have a hand in how you manage the experience."[24]

That includes freedom from overcrowding. Most DMOs have neither the political clout nor breadth of vision to handle destination management by themselves. They must collaborate with various government departments, have access to useful data from both residents and tourists, and be able to coordinate with stewardship organizations (conservation, historic preservation, beautification, and so forth), and the private sector.

So far, actions to cope with overtourism have typically focused less on systemic fixes and more on mitigations that may apply to any of the types of destinations throughout this book. What's needed next are a few short-term solutions and then a look at some longer-term approaches needed to address existing and impending overtourism. All require collaboration to ensure effective governance.

Short-Term Solutions: Mitigate the Problem

Some destinations where overtourism is already occurring are taking action to cope with it. Methods of mitigation include limitation, physical modifications, regulation, visitor education, and dispersion, often in combination. Such suites of action work reasonably well, at least for the time being. The following are some specific tactics that are known to be effective.

Increase Capacity

Probably no place has needed to rebuild infrastructure for crowds more than Mecca, which must accommodate two million to three million pilgrims "at one time" during the annual hajj. To avoid the deadly stampedes of the past, the Saudis have rebuilt the Jumaraat Bridge to provide five levels for as many as three thousand pilgrims per hour during the ritual Stoning of the Devil.[25]

For many places, though, expanding infrastructure with broader highways, more parking, or an added airport merely postpones trouble or even hastens it. Easier access can induce higher visitation. It would be better for infrastructure improvements to be part of a holistic management plan for optimum tourist traffic.

Modify Tourist Behavior

Regulations are one method, like the flood of new prohibitions in Rome and elsewhere in Italy (see chapter 2.1).[26] The overtourism paradox comes into play here: there's nothing wrong with one person sitting on the Spanish Steps, but if four hundred are sitting there, that's a problem. So now, no one is allowed.

Education and persuasion don't spoil as much fun as regulation. At the most basic level, regulation involves asking the tourist to behave and not be an idiot. Tahoe's humorous "Take Care" campaign (see chapter 6.3) has been successful enough to inspire Cape Cod's Care for the Cape and Islands group to adopt a similar approach.[27]

"Visitor education is becoming a critical function of DMOs," said Ouimet. A Colorado visitor's brochure asks, "Are You Colo-ready?" and offers tips interlaced with requests to pack out your trash and stay on the trails. Citing research by Leave No Trace, the state director of tourism, Cathy Ritter, said that "the best time to influence [visitor] behavior is the planning stage" (see chapter 6.4).

Tourist pledges are gaining favor. Overtouristed Palau—160,000 visitors a year visit this Micronesian country of 20,000 people—inaugurated one of the first in December 2017. Brilliantly, it's addressed to the next generation. As an independent island state, Palau has put the pledge right on the visa page. You can't enter the country without signing it:

> Children of Palau,
> I take this Pledge,
> To preserve and protect your beautiful and unique island home.

> I vow to tread lightly, act kindly, and explore mindfully.
> I shall not take what is not given.
> I shall not harm what does not harm me.
> The only footprints I shall leave are those that will wash away.[28]

Iceland, Colorado, Yellowstone, Hawai'i, and other places are adopting pledges of their own. Social pressure and coordinated governance will likely determine how well they work.

Modify Visitor Access

Dubrovnik, Croatia, Santorini, and Greece have capped per-day cruise-ship visitation.[29] Machu Picchu and the Taj Mahal limit visitors by day or hour.[30] Charleston, South Carolina, enacted a ban on tour buses in certain historic neighborhoods in favor of a set number of historically evocative horse-drawn carriages (see chapter 2.3). Timed admissions by the hour have worked well for the Alhambra in Spain, as have those by calendar date for rafting down the Colorado River in the Grand Canyon (see chapter 3.1). Italy's Cinque Terre villages have been threatening to limit visitors, even more swollen now by Chinese cruise passengers on day hikes.[31] Some destinations have put a ceiling on accommodation by curbing new hotel construction (Barcelona) and short-term rentals (New Orleans).

A growing number of others are requiring advance reservations. As Colorado's Ritter notes, visitors and even locals have welcomed the new reservation system for accessing overpopular Hanging Lake. Setting a tolerable social carrying capacity has made the lake once again a pleasant place to visit.

Disperse Tourist Flows

Dispersion has become a common response to overtourism: redistribute tourists to lessen or avoid crowding. At the national level, Ireland created a signed route, the Wild Atlantic Way, to send tourists along less visited parts of the west coast. Scotland designated a North Coast 500 route for the same reason. Inspired by Ireland, Iceland declared an Arctic Coast Way in June 2019, for which *Conde Nast Traveler* published a story under the obedient title, "Skip the Golden Circle—Take Iceland's New Arctic Coast Way Instead."[32] The roads were already there, of course; only the designation was new.

Will dispersion tactics work over the long term? They may not. Oregon suspended a program that was directing visitors to out-of-the-way places "because past campaigns had actually proved too successful at driving traffic to remote, fragile areas, and therefore contributed to degrading pristine environments."[33] More successful has been the state's Trailhead Ambassadors program whereby locals redirect hikers and bikers from overcrowded trails to less well-known alternatives[34] in a good example of community engagement.

Given ever-increasing tourist numbers, though, dispersion programs may amount to a temporary fix. How long will it be before the redirected places become overcrowded themselves? And does dispersion really relieve pressure on over trafficked sites? Research and monitoring are required to track the impacts over time.

Rebrand for Selected Markets

Under the "Pure Grenada" brand, the Caribbean island nation of Grenada is focusing more on its own endemic character than on the Caribbean's default recipe of beaches, time-shares, and golf courses. The approach is likely to capture the attention of people who value some natural and cultural Caribbean experiences along with their sand, rum, and sunburn.

The ever-growing supply of tourists makes it easier for DMOs to abandon "come one, come all" marketing. Figuring that day-trippers from Denver don't need more encouragement, Breckenridge's DMO now markets only to out-of-state visitors likely to stay for a while. Thus destinations can pick and choose what kind of tourism they want. For the DMO, it requires a mental marketing shift from "who have we left out?" to "who do we want to attract?"

Finally, there is a cluster of tactics aimed at disincentivizing travelers, often by raising the cost of visitation. Two ways of doing so are to discourage day-trippers altogether and to set a high price.

For example, conditions have become so dire in places such as Venice and Barcelona that proposals for raising day-tripper taxes and

Attractive architecture in St. George's, contributing to the "Pure Grenada" brand. Source: Jonathan B. Tourtellot.

implementing other controls are on the agenda.[35] In Ireland, Cliffs of Moher managers now plan to use pricing to disincentivize day-trip tour bus operations in favor of tours that carry a night or more in County Clare.

In many places, "hit-and-run" selfie-stick tourism accounts for a substantial portion of overcrowding. Legions of poorly informed visitors pile out of buses and ships, grab selfies and maybe buy a cheap souvenir, and then pile back in and leave. They take up lots of space, virtually tax-free. Worse, they discourage overnighters who do pay taxes and contribute far more per capita to the local economy.

One solution might be to incentivize beneficial length-of-stay with progressively lower taxes per night for longer-stay visitors and higher fees for day-trippers. Look for some places to try it in years to come.

Likewise, setting a high price can change the tourism dynamics. Do you want to visit Bhutan? If you are from overseas and hope to see this small, self-aware Himalayan country, you'll need to clear several hurdles, one of which is committing to a "minimum daily package" of US$200 to $250 daily in touring expenses, "per person."[36] That includes a $65 daily tax to benefit the Bhutanese.[37] When you fly in, you must use one of the two national airlines.[38] That puts a cap on fly-in visitation—some sixty-three thousand in 2018. Notably, Bhutan does not follow a GDP model of growth but rather the "nine domains" that make up its renowned "gross national happiness" yardstick of success.[39] Maximizing tourism arrivals was never on the table.

By world standards, Bhutan does not have an overtourism problem—well, almost not. Tourists from the Indian subcontinent have always been excused from the spending requirement. But Indian visitation quadrupled from fifty thousand in 2014 to two hundred thousand in 2018, overcrowding several sites. Now the Tourism Council of Bhutan is thinking of tightening rules for regional visitors.[40]

Another approach is the Botswana solution. The Okavango Delta and Botswana's northern parks and reserves have some of Africa's richest wildlife populations—and richest tourists. The standard way to tour the Okavango is by luxury safari camps on vast private reserves, where $2,000 per person per night (not per couple!) would not be unusual. There's no danger here of the crowding associated with places like Kenya's Masai Mara (see chapter 3.4). Many of the Okavango camps are models of conservation, extremely responsible, awarding-winning operations that make a point of local hiring.

The high-price model can be effective, but does it mean that Earth's last areas of charismatic wildlife and distinctive culture must be reserved for the rich?

Longer-Term Solutions: Restructure

Successful overtourism mitigation strategies—successful for now, that is—are playing out over the ominous subsonic of unrelenting tourism growth. This rising tide calls for more than a couple of fingers in the dike. A destination can enact most of the preceding mitigations without significantly restructuring governance—and that is their weakness. Conventional institutions are simply inadequate for destination management. Rare are the Bhutan-type cases that set up controls on visitation from the start. Once mass tourism has taken hold in a place, it can be extremely difficult to backtrack. Powerful interests that benefit from high-volume visitation will lobby to perpetuate it.

The challenge is how to manage a diffuse industry—tourism—that sells a holistic product: the place. If war is too important to leave to the generals, as French statesman Georges Clemençeau allegedly said, tourism is too important to leave to tourism ministries or DMOs, although both have their roles. Coordinated, collaborative governance is needed to keep government departments from working at cross-purposes, as they often do—economic development versus environment, for instance, or city versus port authority. All should have a role in crafting reforms to prevent overtourism before it happens.

The Global Sustainable Tourism Council (GSTC) has specified that well-run destinations should have a broad-based, multisector structure. Originally affiliated with UNWTO and the UN Foundation, the now-independent GSTC is tasked with setting sustainability criteria for certification of tourism businesses and for destinations.[41] GSTC's destination criteria were revised and updated in 2019. Criterion A1 (formerly A2) reads in part:

> The destination has an effective organization, department, group, or committee responsible for a coordinated approach to sustainable tourism, with involvement by the private sector, public sector and civil society. This group . . . has defined responsibilities, oversight, and implementation capability for the management of environmental, economic, social, and cultural issues.[42]

Such groups can be referred to generically as destination stewardship councils. Rather than a single monolithic entity, says George Washington University's Matus, they need to be structured network-style so as to formulate policy that takes into account all pertinent actors and points of view. Councils would include the DMO—perhaps even be led by the DMO—but would perform far broader functions. Councils might be advisory, might have governing authority, or some combination of the two.

Unfortunately, few such arrangements yet exist. Barcelona is working on a council-type approach, and Teton County, Wyoming, keystone of what's called the greater Yellowstone region, has been trying. It's tough going.

"We do have a challenge in uniting the land management agencies and destination management organizations, given the number and variety of governmental, nongovernmental, and business stakeholders," said Timothy O'Donoghue, executive director of the Riverwind Foundation and leader of the collaborative multiyear effort called the Jackson Hole and Yellowstone Sustainable Destination Program. "We are still in the process of developing the scope and structure of a community stewardship council that will coordinate these stakeholders."[43]

A Model Council

My own years of experience while promoting the geotourism concept[44] via National Geographic suggest that stewardship councils be proportioned roughly one-third sustainable private sector, one-third stewardship-related civil society (nongovernmental organizations/community-based organizations and academia), and one-third government. No one sector should have total control. That ensures continuity even after a change in government.

In this conception, the core group does most of the planning, coordination, and policy-making. A much larger pool of affiliate members can help with specific projects or expertise as needed. Projects and events can engage both residents and tourists as appropriate.

Councils need geographical representation that fits tourist behavior. A visitor doesn't care that the pyramids are actually in Giza, not Cairo, nor that the town of Banff isn't actually part of the surrounding Canadian park. A popular protected site and its separate gateway community are effectively the same destination. Stewardship council membership should represent the entire local tourism ecosystem, regardless of jurisdiction.

Composition of the ideal council should be able to address issues arising from overtourism. The council should include the following:

- Environmental representatives and land managers to prevent degradation of natural environments

- DMOs for rebranding and selectively messaging to the traveling public as well as expanded management roles
- Infrastructure agencies for proactive traffic management, traffic taming, and parking capacities, along with restrictions, transportation alternatives, and information technology hardware
- Site managers to avoid wear and tear on historic and archaeological sites
- Representatives for other unique sites, routes, and destination activities, existing or potential
- Methods for public engagement by both tourist and residents

In the case of the "traveling public"—the destination's guests—the council needs a monitoring process for collecting tourist perceptions of the destination and the quality of its management. A method for soliciting feedback from travel journalists, researchers, academics, and online tourism influencers would also give a valuable perspective.

In the case of the "resident public"—the destination's hosts—two-way communication is essential. We've seen that local people have proven to be excellent whistleblowers when tourism is getting out of hand. They may not, however, be so excellent at recognizing their own benefits from tourism, such as a reduced tax bill, beneficial business creation, and sustainable tourism's economic value as a viable alternative to more destructive types of development.

Enlisting citizens in telling the story of their destination is one way to promote such understanding. Resident tourism volunteer programs, such as Oregon's Trailhead Ambassadors and New Hampshire's Granite State Ambassadors, can raise local appreciation for their destination's distinctive assets and serve as proactive information conduits.

Creating a Destination Stewardship Council

Here's a suggested process for adopting a community-endorsed mission and goals that harmonize sustainability, authenticity, economic development, and responsible targeted marketing that avoids overtourism.

A steering committee of motivated volunteers begins by stoking local interest so as to grow into a full, nonpartisan council. Dedicated, enthusiastic local leaders—whether from government, business, or nonprofit groups—are essential to success. Top-down initiatives rarely succeed unless locals can and do take ownership. As the official storytellers for their destinations, DMOs seem to be logical conveners for organizing this collaboration, provided they are allowed to restructure themselves beyond marketing.

A destination stewardship council's first and perhaps most difficult task is to develop a mission or vision statement about tourism, informed in part by polling residents. Here are the steps:

1. For the mission, ask: What do we want? Whom do we want to attract?
2. Then ask: What principles do we want to follow to get there? These principles become consensus reference points, stars to steer by that help keep things on course.
3. Develop a community-supported tourism management plan in accordance with the goals and principles.
4. Adopt one or more projects, with deadlines, preferably engaging the public. This action helps keep the council from degenerating into a desultory discussion group. Whenever possible, use these projects to deepen appreciation for the destination's endemic assets and character, modulating tourism for maximum benefit and minimum harm.

One possible project for the council is certification, especially given that council members will have already met the GSTC's Criterion A1, cited above.

Certification and Overtourism

The practice of certifying tourism businesses for sustainability has become fairly common, but certification for destinations is still in its infancy. To date the GSTC has assessed some thirty destinations, including Cozumel and Sinaloa in Mexico, St. Kitts and Nevis in the Caribbean, Jackson Hole in Wyoming, and the fjords in Norway. That, however, is only a preliminary step. Fewer than a dozen destinations have been fully certified by a GSTC-accredited certification program. Given its multiple stakeholders and institutions, a destination is far more complex than a business. It's no surprise, then, that some destinations have found the certification process unduly bureaucratic and complicated. Nevertheless, it can be worth the hassle.

By 2020, only three major international companies had been approved by the GSTC to certify destinations: EarthCheck, Green Destinations, and RTI Biosphere. All three names imply an environmental focus. Preoccupied with major environmental issues, the sustainable tourism community did not anticipate overtourism any better than governments did, however. Early destination certifications thus focused heavily on the environment, but now crowd control is becoming part of the mix, as GSTC Destination Criterion A8 (formerly A9) suggests:

> A8. Managing visitor volumes: The destination has a system for visitor management which is regularly reviewed. Action is taken to monitor and manage the volume and activities of visitors, and to reduce or increase them as necessary at certain times and in certain locations, working to balance [sic] the needs of the local economy, community, cultural heritage and environment.[45]

Even with its flaws, the destination certification process has great value if done with full community participation and leadership buy-in. It forces a thorough self-evaluation of what often has been a laissez-faire approach unsuited to surging volumes of visitors. All indications are that destination certifications will increase.

Revise and Modernize Metrics

Every year, advances in information technology offer new and better ways to monitor tourism and its impacts. Today, counting arrivals and heads in beds is the equivalent of analysis by abacus. Destination councils would do well to incorporate tech expertise in their operations, focused not only on tourists but on the goals adopted by the community. Metrics can also go far beyond the old GDP model to include tourism costs and nonmonetary benefits, as well as deeper economic analysis: How much money actually stays in the destination? Who gets it? How is it spent? How are bed and other tourism-generated taxes used and to what end? How does that contribute to community and traveler enthusiasm and to destination sustainability?

A Sea Change

Barring planetary catastrophe, tourism is likely to continue its rampant growth for the next few decades, unless some other long-term factor intervenes, such as climate change, a global pandemic, new technological disruption, or perhaps disenchantment with travel itself. So far, the tourism growth trend has already survived wars, epidemics, and the 2008 global financial crisis. It is robust.

This continuing flood tide of tourists requires a massive readjustment of expectations. I revisited Iceland in 2019 and had a fine time as always, seeing old friends, exulting in the wild open scenery, exploring the dynamic geology, and enjoying a couple of the new restaurants that have sprouted up in once-provincial Reykjavik. (Not all tourism impacts are bad.) But tourists were everywhere in places that I can recall once had only a handful. As for that new Arctic Coast Way campaign? If successful, it will send legions of independent travelers along the northern coast, a last remnant of the old Iceland—and yes, economically much in need of tourism to forestall rural out-migration. These places will change—with or without tourism.

After I flew home, I discovered that my expectations had indeed readjusted. For the first time in my adult life, I was not thrilled about the idea of a return visit. Was Iceland "spoiled" for me? Maybe it was, a bit—but perhaps not for any first-time visitor for whom all those tour buses will just be part of the scenery. Oh, I'll probably go back. But something's lost, leaving me with a renewed and poignant appreciation for that remark attributed to Yogi Berra: "Nobody goes there anymore. It's too crowded."

Notes

1. United Nations World Tourism Organization and International Transport Forum (December 2019). *Transport-Related CO_2 Emissions of the Tourism Sector.* https://www.e-unwto.org/doi/pdf/10.18111/9789284416660.

2. Ibid.

3. Rachel Dodds and Richard Butler. (2019). *Overtourism: Issues, Realities, and Solutions.* Berlin/Boston: De Gruyter.

4. David Pilling. (2018). *The Growth Delusion.* New York: Penguin Random House.

5. Pamela Chow. (January 24, 2020). "The Quality Approach." TTG Asia. https://www.ttgasia.com/2020/01/24/the-quality-approach/.

6. Xinhua. (January 3, 2020). "Maldives Attracts 1.67 Mln Tourists in 2019." http://www.xinhuanet.com/english/2020-01/03/c_138676998.htm.

7. Gulf Today. (January 24, 2020). "Greece Expects Tourism Growth This Year Despite a Bumpy 2019." https://www.gulftoday.ae/business/2020/01/24/greece-expects-tourism-growth--this-year-despite-bumpy-2019.

8. Karki Sabina. (January 10, 2020). "Poor Infrastructure May Impair Visit Nepal 2020 Goal." *Khabarhub*. https://english.khabarhub.com/2020/10/67314/.

9. eTurboNews. (January 29, 2020). "Anguilla Tourist Board Hits the Ground Running in 2020." https://www.eturbonews.com/541347/anguilla-tourist-board-hits-the-ground-running-in-2020/.

10. Hans News Service. (January 24, 2020). "Visakhapatnam: Kerala Tourism Sets Ambitious Goals to Enhance Footfalls." https://www.thehansindia.com/andhra-pradesh/visakhapatnam-kerala-tourism-sets-ambitious-goals-to-enhance-footfalls-599493.

11. Financial Mirror (January 15, 2020). "Cyprus Looks Ahead after Breaking More Tourist Records." https://www.financialmirror.com/2020/01/15/cyprus-looks-ahead-after -breaking-more-tourist-records/.

12. Chea Vannak. (January 7, 2020). "Tourist Arrivals Forecasts Continued Growth in 2020." *Khmer Times*. https://www.khmertimeskh.com/50677162/tourist-arrivals-forecasts-continued-growth-in-2020.

13. Seleni Matus. (July 24, 2019). International Institute of Tourism Studies. George Washington University. Personal communication with author.

14. Edward Abbey. (1977). *The Journey Home: The Second Rape of the West*. New York: Dutton. p. 114.

15. Rachel Dodds and Richard Butler. (2019). *Overtourism: Issues, Realities, and Solutions*. Berlin/Boston: De Gruyter. p. 86.

16. Geraldine Enright. (2020). Cliffs of Moher Visitor Experience. Personal communication with author.

17. Henry Zwartz. (October 8, 2018). "Is Tasmania's Iconic Park Being Loved to Death?" ABC News. https://www.abc.net.au/news/2018-10-09/cradle-mountain-tour-operators -keen-for-visitor-cap/10352320.

18. MGMY NextFactor. (November 6, 2018). "Tourism Leaders Are Expanding Their Perspective on Sustainability." http://www.nextfactorinc.com/tourism-leaders-are-expanding-their-perspective-on-sustainability/.

19. Walter Jamieson and Michelle Jamieson. (2019). "Managing Overtourism at the Municipal/Destination Level." Eds. Rachel Dodds and Richard Butler. *Overtourism: Issues, Realities, and Solutions*. Berlin/Boston: De Gruyter.

20. Glenn Mandziuk. (February 26, 2019). Thompson-Okanagan Tourism Association. Personal communication with author.

21. Paul Ouimet. (June 25, 2019). NEXTFactor and DestinationNEXT. Personal communication with author.

22. Ibid.

23. Lucy Kay. (September 4, 2019). Breckenridge Tourism Office. Personal communication with author.

24. Ibid.

25. OTIS. (August 30, 2017). "Re-designed Jamaraat Bridge Will Allow 3 Million

Pilgrims to Safely Reach Mecca." https://www.otis.com/en/us/about/news-and-media/press-releases/re_designed_jamaraat_bridge_will_allow_3_million_pilgrims_to_safely_reach_mecca.aspx.

26. Nick Squires. (June 7, 2019). "Rome Launches Crackdown on Messy Eating, Jumping in Fountains and Other Tourist Misbehavior." *The Telegraph*. https://www.telegraph.co.uk/news/2019/06/07/rome-launches-crackdown-messy-eating-jumping-fountains-tourist/.

27. Jill Talladay. (2020). Care for the Cape and Islands. Personal communication with author.

28. Palau Pledge. (2020). https://palaupledge.com/media/.

29. Walter Jamieson and Michelle Jamieson. (2019) "Managing Overtourism at the Municipal/Destination Level." Eds. Rachel Dodds and Richard Butler. *Overtourism: Issues, Realities, and Solutions*. Berlin/Boston: De Gruyter.

30. CNT Editors. (October 24, 2018). "15 Beloved Places Struggling with Overtourism." *Conde Nest Traveler.* https://www.cntraveler.com/galleries/2015-06-19/barcelona-bhutan-places-that-limit-tourist-numbers.

31. Clare Speak. (March 6, 2019). "Stop Trekking in Flip Flops, Italy's Cinque Terre Begs Tourists" *The Local*. https://www.thelocal.it/20190306/stop-trekking-in-flip-flops-italys-cinque-terre-begs-tourists.

32. Ashley Halpern. (May 23, 2019). "Skip the Golden Circle—Take Iceland's New Arctic Coast Way Instead." *Conde Nest Traveler*. https://www.cntraveler.com/story/iceland-road-trip-arctic-coast-way.

33. MGMY NextFactor. (November 6, 2018). "Tourism Leaders Are Expanding Their Perspective on Sustainability." http://www.nextfactorinc.com/tourism-leaders-are-expanding-their-perspective-on-sustainability/.

34. Ibid.

35. Ashley Halpern. (May 23, 2019). "Skip the Golden Circle—Take Iceland's New Arctic Coast Way Instead." *Conde Nest Traveler*. https://www.cntraveler.com/story/iceland-road-trip-arctic-coast-way.

36. Tourism Council of Bhutan. (2020). https://www.tourism.gov.bt/about-us/minimum-daily-package.

37. Tourism Council of Bhutan. (2020). https://www.bhutan.travel/page/frequently-asked-questions.

38. Sharell Cook. (February 5, 2020). "Travelling in Bhutan: What You Need to Know before You Go." *TripSavvy*. https://www.tripsavvy.com/how-to-travel-to-bhutan-1539176.

39. Oxford Poverty & Human Development Initiative. (2020). https://ophi.org.uk/policy/national-policy/gross-national-happiness-index/.

40. Shefali Anand. (October 15, 2019). "Bhutan Opens Itself to the World-and Tourists." *U.S. News*. https://www.usnews.com/news/best-countries/articles/2019-10-15/bhutan-opens-itself-to-the-world-and-tourists.

41. Global Sustainable Tourism Council .(2020). https://www.gstcouncil.org/.

42. Global Sustainable Tourism Council. (2020). https://www.gstcouncil.org/gstc-criteria/gstc-destination-criteria/.

43. Timothy O'Donoghue. (2020). Riverwind Foundation. Personal communication with author.

44. Geotourism as defined via National Geographic is "tourism that sustains or enhances the geographical character of a place—its environment, culture, geology, aesthetics, heritage, and the well-being of its residents."

45. Global Sustainable Tourism Council. (2020). https://www.gstcouncil.org/gstc-criteria/gstc-destination-criteria/.

Chapter 6.2

Iceland

By Nathan Reigner and María Reynisdóttir

A volcanic island that straddles the Mid-Atlantic Ridge in the middle of the North Atlantic Ocean, Iceland has been a destination for curious tourists since at least the nineteenth century. Capped by glaciers; dotted with volcanoes, geysers, and waterfalls; roamed by sheep and horses; circled by puffins and whales; storied by some of Western Europe's oldest literature; and inhabited by descendants of Viking voyagers, Iceland has become one of the hottest international tourism destinations of the twenty-first century. According to a 2018 survey conducted by the Icelandic Tourist Board, tourists visit Iceland primarily to see and experience its exceptional natural wonders (92 percent of respondents), as well as its unique Nordic culture (74 percent of respondents).[1]

The story of Iceland's grapple with tourism is one of finding balance; in the aftermath of the country's economic recession, tourism was the main sector credited with pulling the country out of its crisis. Now, there are mixed feelings regarding reports of a slowdown, as the economy depends in large part upon international tourism while residents and policy makers are eager for a more sustainable path forward.[2] Iceland is perhaps one of the best examples of how quickly modern tourism can turn on its head, and thus its adaptation measures provide an important benchmark for tourism planners at the national level.

Causes of Growth

Iceland's current tourism boom began in 2008 with the country's banking crisis.[3] The financial collapse both disrupted the domestic economy and made Iceland a less expensive travel destination for foreign tourists. The collapse was followed by the eruption of Eyjafjallajökull volcano in 2010, which grounded European air traffic and focused international attention on Iceland. The attention was, however, not entirely positive. Much of the news coverage portrayed Iceland as a dangerous area suffering from rampant natural disasters.[4] To counter this image, Promote Iceland—the public-private partnership responsible for overseas promotion of Iceland as a destination—launched the "Inspired by Iceland" campaign. This collaborative and creative marketing campaign, combined with favorable exchange rates, increased flight capacity, and attractive stopover programs spurred tourism growth in Iceland at an unprecedented level. Between 2010 and 2018, international arrivals grew from less than half a million to a record 2.3 million, a five-fold increase in only eight years, or on average growth of 23 percent annually, far above both the world average (around 4 percent) and domestic forecasts for the country.[5]

As tourism has increased, it has become clear that the timing and geography of tourism (and its impacts) are unevenly distributed. There has long been a pronounced seasonal peak, with the busiest months (June–August) having approximately twice the number of tourists as the quietest months of January and December.[6] Geographically, tourism in Iceland is unevenly distributed. Virtually all tourists visit the southwest corner of Iceland, home of the island's international airport (Keflavik) and capital (Reykjavik). Beyond the southwest, there is a steep decline in visitation to other regions, which becomes even more pronounced in the winter when driving conditions are difficult, highland attractions are inaccessible, and daylight is limited.[7]

Behavioral and social factors also contribute to the uneven distribution of Iceland's tourism. For example, cruise tourism concentrates large numbers of tourists for short periods of time at port facilities, in often little visited communities. The small town of Ísafjörður in Iceland's remote Westfjords—with a population of twenty-six hundred—received eighty thousand cruise visitors between May and September 2018.[8] According to surveys, inhabitants complain about increased traffic in the town center and pollution, due to cruise ship arrivals.[9]

The featuring of emergent destinations in film and social media has also caused tourist numbers to increase dramatically with little warning or preparation. Fjarðrárgljúfur Canyon was forced to close temporarily in 2019 due to a sharp increase in visitor numbers, due to both its fame as a *Game of Thrones* filming site and those attempting to recreate Justin Bieber's *I'll Show You* music video[10] (a trend four-years strong), which involved tramping on sensitive flora.[11]

Given these factors, it may not be accurate to state, as the widespread international media attention may imply (such as *Skift*,[12] *The Telegraph*,[13] and *National Geographic Traveler*[14]), that Iceland as a whole is experiencing overtourism. Rather, tourism pressure is concentrated in certain places and at specific times, in, for instance, Reykjavik city center and the south coast in the summer.

Impacts of Tourism Growth

Undoubtedly, the rapid growth of tourism has brought many benefits to Iceland. Today, tourism is the leading sector of Iceland's economy, contributing approximately 42 percent of export revenues in 2018, representing approximately US$4.2 billion and 8.6 percent of Iceland's GDP.[15] Statistics Iceland reported that tourism supported around thirty thousand jobs in 2018 (compared to five thousand in 2010) and spurred development of infrastructure and services throughout the country.[16] These developments produce direct benefits for Icelandic residents, who have been able enjoy a wider range of activities and services, thereby improving their daily quality of life. Þórdís Gylfadóttir, Iceland's minister of Tourism, Industry and Innovation, said that "with flights to more destinations at lower prices, we Icelanders . . . can travel abroad more than ever. We also have a much wider variety of services to choose from—restaurants, shops and leisure activities."[17] In addition, tourist interest in Iceland's natural environment has focused international attention on the value of and the need to protect its surrounding nature. It has also helped generate resources to expand and enhance environmental protection programs.[18]

The rapid and dramatic growth of international tourism has, however, also put increasing pressure on Iceland's existing infrastructure, environment, and society. Roads traversed by heavy bus traffic and tourists in rental cars have degraded in some places, making sightseeing more dangerous.[19,20] National parks and nature sites have also been degraded by increased visitation beyond what Iceland's trails and facilities were made to handle. The aforementioned Fjarðrárgljúfur Canyon in the southern region has suffered extensive erosion on its visitor paths along with damage to its vegetation.[21] Similar impacts have also been reported at the Brúarfoss waterfall, less than two hours' away from Reykjavik.[22] Housing prices have increased due to the rapid growth of Airbnb and other short-term, home-based accommodations.[23] Large numbers of foreign workers have entered the Icelandic workforce in response to labor shortages, with one-third of all tourism jobs filled by foreign workers in 2018.[24]

Icelandic residents are well aware of these impacts. According to a 2017 survey, 49.3 percent of Icelanders avoid certain places due to the presence of tourists, and 57 percent are concerned with tourism damaging the quality of Iceland's natural assets.[25]

Furthermore, although tourism has generated economic benefits for Iceland, there is growing concern that the economy may be too heavily reliant on tourism. The Organization for Economic Co-operation and Development warned in its 2019 Iceland economic survey that due to the country's dependency on tourism, a change in visitor demand may result in widespread unemployment and loss of export revenues.[26] It is a relevant fear because international tourism has declined due to, as *Skift* reports, the collapse of budget airline WOW Air in early 2019.[27] This and other factors such as "the strong krona, . . . the Boeing 737 Max fiasco, and rising labor costs" are all contributing to a slowdown of the economy.[28]

Solutions

Although Iceland is facing challenges from the rapid tourism growth that took place between 2008 and 2018, many opportunities for tourists to have exceptional experiences and for communities to benefit from tourism development remain. Furthermore, the issues associated with uneven distribution of tourists throughout the seasons and around the country have been mitigated to some extent in recent years. A marketing concentration on winter activities (for example, northern lights viewing) has improved seasonality. Likewise, there have been increases in bed-nights in all regions as direct international flights to the airport in Akureyri in North Iceland have been introduced. Effective and sustainable management of tourism in Iceland, particularly with respect to the pressures on Icelandic nature and society by tourism, requires acknowledging these distributions and understanding their benefits and negative impacts, while also taking informed and appropriate actions.

Iceland is implementing an ambitious and innovative management program directed at understanding and addressing the issues related to tourism. This program involves an integrated set of coordinated measures, national and local research, policy, regulation, infrastructure, and communication initiatives.

National and Regional Coordination

Coordinating at the national and regional level has been essential for Iceland's response to tourism growth. In 2015, the Ministry of Industries and Innovation, in cooperation with the Icelandic Travel Industry Association, produced the *Road Map for Tourism in Iceland* so as to provide national guidance for coping with the rapid expansion emerging in the tourism sector.[29] The map, created after extensive public and stakeholder consultation, provides seven focal points as a framework for action: coordination, positive visitor experience, reliable data, nature conservation, skills and quality, increased profitability, and distribution of tourists. That same year, a temporary tourism task force—led by government ministers (the minister of Tourism, who is chairman of the board, and ministers of the Environment, Transport, and Finance), municipal policy makers, and leaders from other companies in the tourism sector —was established and given a five-year mandate to implement the *Road Map*. In addition to the coordination efforts of the map and the task force, coordination has also been emphasized at the regional level and within Iceland's national park system. Regional coordination has been centered around seven bottom-up regional destination management plans, each supported by the establishment of destination management organization in that region. There are also plans to improve coordination in the field of nature conservation, with the establishment of a new national park institute to manage the country's national parks and conservation sites, which are currently administered by a variety of independent agencies.

Use of Data

Research and reliable data are cornerstones of Iceland's approach to sustainable tourism management. Among the numerous lines of research and data analysis being conducted are a major national multisector carrying capacity assessment, designed to guide infrastructure investment and policy revision;[30] enhanced and regular surveys of both tourists and Icelandic residents;[31] tourist volume counting;[32] evaluations of labor, housing, and other markets;[33] and the establishment of a tourism data dashboard.

Government Regulations

Coordination has been accompanied by improved policy and regulation. New legislation has been passed to adapt and update government administration, including establishing responsibilities for data gathering and analysis, and regulating short-term rentals and home-sharing (for example, Airbnb). Specific local regulations include a limit on new hotel development in Reykjavik city center to protect its character.[34] Tour buses are now also banned in certain areas of the city center, and instead, the city has created designated pickup spots for tourists.[35] Regulations have been introduced in some of the national parks, such as permit procedures for snorkeling and diving in the Silfra fissure of Thingvellir National Park.[36] Additionally, the need to establish a legal framework for managing tour operations at tourist attractions on public land is being assessed.

Improvements to Infrastructure

Noteworthy infrastructure actions focus on Keflavik airport and nature-based tourism destinations. Keflavik airport is implementing a development master plan to increase its capacity to 14 million passengers per year by 2040, which is 40 percent greater than the 9.8 million passengers it serviced in 2018.[37] A new National Infrastructure Plan, in addition to the already existing Tourist Site Protection Fund, has considerably increased funding for the protection and development of tourist sites.[38]

Communications and Messaging

Iceland is also enhancing its communications channels and messaging to mitigate the negative impacts of tourism. To encourage responsible tourist behavior, visitors to Iceland are asked to take the Icelandic Pledge, which includes a promise to take photos safely and only camp in designated campsites.[39] In addition, SafeTravel.is, a cooperatively funded effort by the Icelandic volunteer search and rescue system, provides travel itinerary registration, activity-specific safety information, condition reports, recommendations, and a smartphone app that allows travelers to update their location with rescuers and call for help if needed.[40]

Conclusion

Throughout the second decade of the twenty-first century, Iceland has been among the world's fastest-growing tourism destinations. Although this explosive expansion has created pressure on the Icelandic environment at certain times (summer) and in certain places (Reykjavik and the south coast), Iceland as a whole does not suffer from overtourism. When considering overtourism in Iceland, it is much more relevant from a policy perspective to be specific about times, places, and activities. Furthermore, it is important to note the benefits rendered by tourism to Iceland's economy, reputation, and daily quality of life for residents.

Although Iceland has historically focused on maximizing the gains from tourism, it is now working hard to mitigate its negative impacts and establish science-based, sustainable management approaches. There are quite a few takeaways from the approach Iceland has undertaken to become a national destination. Destinations should:

- Avail themselves of the full range of available tools;
- Base management actions on sound science and regular monitoring—they should build the capacity and constituency to support these efforts on a continuous basis;
- Establish, in temporary or permanent form, the organizations and networks necessary to fund, coordinate, and implement action;
- Monitor the development of new points of tourism pressure, whether they be sites made popular by social media or social tensions resulting from tourism, and have prepared response plans ready for use;
- Integrate tourism promotion, expectations for tourists' behavior, and locations' capacities with the country's desire to accommodate tourists;
- Finally, distribute both the pressures and benefits from tourism while maintaining a diverse set of experiences and protecting the essential character of the destination through a synergy of transportation, infrastructure investment, and marketing.

Góða ferð! Have a good trip!

Notes

1. Icelandic Tourist Board. (2018). *Tourism in Iceland in Figures.* https://www.ferdamalastofa.is/static/files/ferdamalastofa/talnaefni/tourism-in-iceland-2018_2.pdf.

2. Isaac Carey. (January 16, 2019.) "Why Iceland's Tourism Boom May Finally Be Over." *Skift.* https://skift.com/2019/01/16/why-icelands-tourism-boom-may-finally-be-over/

3. Lavinia Greenlaw. (2017). *Questions of Travel: William Morris in Iceland.* New York: New York Review of Books.

4. Henry Fountain. (April 15, 2010). "Eruption Wasn't That Powerful, but Effects May Linger." *New York Times.* https://www.nytimes.com/2010/04/16/science/16erupt.html

5. Icelandic Tourist Board. (2019). *Numbers of Foreign Tourists.* https://www.ferdamalastofa.is/en/recearch-and-statistics/numbers-of-foreign-visitors; World Tourism Organization. (2019). *International Tourist Arrivals Reach 1.4 Billion Two Years Ahead of Forecast.* http://www2.unwto.org/press-release/2019-01-21/international-tourist-arrivals-reach-14-billion-two-years-ahead-forecasts; Icelandic Tourist Board. (2019). *International Visitors in Iceland.* https://www.ferdamalastofa.is/en/recearch-and-statistics/numbers-of-foreign-visitors.

6. Icelandic Tourist Board. (2019). *Numbers of Foreign Tourists.* https://www.ferdamalastofa.is/en/recearch-and-statistics/numbers-of-foreign-visitors

7. Ibid

8. Cruise Iceland. (2019). *Statistics of Cruise Ship Passengers 2016–2018.* http://www.cruiseiceland.com/about-us/statistics/.

9. Ministry of Industries and Innovation. (2018). *Skýrsla ferðamála-, iðnaðar-, og nýsköpunarráðherra um þolmörk ferðamennsku.* https://www.althingi.is/altext/pdf/148/s/0717.pdf

10. Andie Fontaine. (March 13, 2019). "Icelandic Park Rangers Fend Off Justin Bieber Fans." *Reykjavik Grapevine.* https://grapevine.is/news/2019/03/13/icelandic-park-rangers-fend-off-justin-bieber-fans/

11. Jóhann Páll Ástvaldsson. (January 9, 2019). "Fjaðrárgljúfur Canyon Closed due to Damaged Vegetation." *Iceland Review.* https://www.icelandreview.com/news/fjadrargljufur-canyon-closed-due-to-damaged-vegetation/.

12. Andrew Sheivachman. (2016). "Iceland and the Trials of 21st Century Tourism." *Skift.* https://skift.com/iceland-tourism/.

13. Soo Kim. (December 18, 2017). "Iceland Now Has Seven Times More Tourists Than Local—but Which Country Is Most Overrun with Visitors?" *The Telegraph.* https://www.telegraph.co.uk/travel/destinations/europe/iceland/articles/tourists-outnumber-locals-in-iceland/.

14. Jonathan Tourtellot. (December 21, 2018). "Overtourism: Too Much of a Good Thing." *National Geographic Traveler.* https://www.nationalgeographic.com/travel/features/overtourism-how-to-make-global-tourism-sustainable/

15. Icelandic Tourist Board. (2018). *Tourism in Iceland in Figures.* https://www.ferdamalastofa.is/static/files/ferdamalastofa/talnaefni/tourism-in-iceland-2018_2.pdf

16. Statistics Iceland. (2020). *Short Term Indicators in Tourism.* https://statice.is/statistics/business-sectors/tourism/short-term-indicators-in-tourism/

17. Þórdís Gylfadóttir. (April 24, 2017). "Ræða Þórdísar Kolbrúnar á ráðstefnu um samfélagsleg áhrif ferðaþjónustu á Kosta Ríka, 24. apríl 2017." Government of Iceland. https://www.stjornarradid.is/raduneyti/atvinnuvega-og-nyskopunarraduneytid/ferdamala-idnadar-og-nyskopunarradherra/stok-raeda-ferdamala-idnadar-og-vidskiptaradherra/2017/04/24/Raeda-Thordisar-Kolbrunar-a-radstefnu-um-samfelagsleg-ahrif-ferdathjonustu-a-Kosta-Rika-24.-april-2017/

18. Jukka Kalervo Siltanen. (2017). *Economic Impact of National Parks in Iceland; Case Study of Snæfellsjökull National Park.* Reykjavik: University of Iceland.

19. Phil Sylvester. (October 8, 2019). "Safety Tips for Driving in Iceland: Are the Roads Safe?" *World Nomads.* https://www.worldnomads.com/travel-safety/northern-europe/iceland/tips-for-driving-in-iceland.

20. Vala Hafstad. (June 3, 2019). "Major Off-Road Damage Near Myvatn." *Iceland Monitor.* https://icelandmonitor.mbl.is/news/nature_and_travel/2019/06/03/major_off_road_damage_near_myvatn/.

21. Iceland Magazine staff. (April 3, 2018). "Visitors at Fjarðrárgljúfur Canyon Causing Serious Damage to the Site." *Iceland Magazine.* https://icelandmag.is/article/visitors-fjadrargljufur-canyon-causing-serious-damage-site.

22. Julia Corderoy. (June 8, 2017). "Tourism Boom to Picture-Perfect Iceland Is Ruining Its Delicate Environment." News.com.au. https://www.news.com.au/news/tourism-boom-to-pictureperfect-iceland-is-ruining-its-delicate-environment/news-story/1367b915f3b4f18ddf8157e31b33e55a.

23. Andrew Scheivachman. (September 11, 2019). "The Rise and Fall of Iceland's Tourism Miracle." *Skift.* https://skift.com/2019/09/11/the-rise-and-fall-of-icelands-tourism-miracle/.

24. Statistics Iceland. (2020). *Register Based Employment by Economic Activity.*

https://px.hagstofa.is/pxen/pxweb/en/Samfelag/Samfelag__vinnumarkadur__vinnuaflskraargogn/VIN10020.px/?rxid=5daaeb67-ba0a-4a07-9841-e21fd344be7c

25. Eyrún Jenný Bjarnadóttir, Ásdís A. Arnalds, and Anna Soffía Víkingsdóttir. (2017). "Því meiri samskipti— því meiri jákvæðni. Viðhorf Íslendinga til ferðamanna og ferðaþjónustu." Icelandic Tourism Research Center. http://www.rmf.is/en/reports/index/skyrslur-reports/thvi-meiri-samskipti-thvi-meiri-jakvaedni.

26. Organization for Economic Co-operation and Development. (September 2019). *OECD Economic Surveys: Iceland.* https://www.oecd-ilibrary.org/economics/oecd-economic-surveys-iceland-2019_c362e536-en.

27. Andrew Scheivachman. (September 11, 2019). "The Rise and Fall of Iceland's Tourism Miracle." *Skift.* https://skift.com/2019/09/11/the-rise-and-fall-of-icelands-tourism-miracle/.

28. Ibid

29. Ministry of Industries and Innovation: Tourist Information Center. (2015). *Road Map for Tourism in Iceland.* https://www.stjornstodin.is/

30. EFLA Consulting Engineers. (2018). *Álagsmat umhverfis, innviða og samfélags gagnvart fjölda ferðamanna á Íslandi.* https://www.efla.is/media/utgefid-efni/6732-001-sky-v01-tholmarkarverkefni-lokaeintak_web.pdf

31. Icelandic Tourist Board. (2019). *Tölur og útgáfur.* https://www.ferdamalastofa.is/is

32. Icelandic Tourism Research Center. (2019). *Skýrslur RMF.* https://www.ferdamalastofa.is/is/tolur-og-utgafur/kannanir-og-rannsoknir/sertaekar-kannanir

33. Icelandic Tourist Board. (2019). *Mælaborð ferðaþjónustunnar.* https://www.maelabordferdathjonustunnar.is/

34. Iceland Monitor. (January 5, 2017). "New Gigantic Hotel Opens in Reykjavik's Skeifan Quarter."https://icelandmonitor.mbl.is/news/nature_and_travel/2017/01/05/new_gigantic_hotel_opens_in_reykjavik_s_skeifan_qua/.

35. Travel Reykjavik. (2020). *Bus Stops: Downtown Reykjavik.* https://www.travelreykjavik.com/reykjavik-bus-stops.

36. Pjodgardurinn A Pingvollum. (2020). "Driving." https://www.thingvellir.is/en/plan-your-visit/diving/

37. Isavia. (2018). *Keflavik Airport—Facts and Figures.* https://www.isavia.is/media/1/19-1209-facts-and-figures-2018-1920x1080-20190208.pdf.

38. Stjórnarráðd íslands. (2020). *Landsáætlun um innviði.* https://www.stjornarradid.is/verkefni/umhverfi-og-natturuvernd/natturuvernd/ferdamenn-og-nattura/landsaaetlun-um-innvidi-/

39. Inspired by Iceland. (2019). *The Icelandic Pledge.* https://www.inspiredbyiceland.com/icelandicpledge

40. SafeTravel.is. (2019). https://safetravel.is/.

Chapter 6.3

Lake Tahoe, California and Nevada

By Julie Regan

"The lake burst upon us. A noble sheet of blue water lifted six thousand three hundred feet above the level of the sea, and walled in by a rim of snow-clad mountain peaks. . . . I thought it must surely be the fairest picture the whole earth affords."—Mark Twain, *Roughing It*, 1871

Since the days when Mark Twain camped at Lake Tahoe, the second deepest lake in the United States, it has beckoned visitors to its shores. Situated high in the Sierra Nevada mountain range on the West Coast, its spectacular waters are exceedingly clear, reflecting the cobalt blue sky. At 22 miles long, 12 miles wide, and 1,645 feet deep, Lake Tahoe is gargantuan. It holds 39 trillion gallons of water, has a 72-mile-long shoreline, and sits more than a mile high in the sky at nearly 6,300 feet of elevation.[1] Dubbed "America's year-round playground," Lake Tahoe boasts two peak seasons with snow sports in winter and beach activities in summer, drawing recreational enthusiasts from around the world.

Like other global and regional destinations, Lake Tahoe's tourism industry is thriving. This chapter highlights the overtourism challenges facing the "jewel of the Sierra" and the work underway to face the challenges head on.

Causes and Impacts of Overtourism

The 1960 Winter Olympic Games at Squaw Valley, California, put Lake Tahoe on the map as a snow-sports mecca and unleashed an unchecked building boom that forever changed the ecosystem. Recent studies using mobile device records peg Tahoe's annual visitation at fifteen million, up from previous estimates of five million in 2012.[2] With a year-round population of fifty-five thousand, that equates to roughly 275 tourists for every individual Tahoe resident. Hotel tax revenues are up 140 percent from 2009–2019 in North Lake Tahoe. Sales tax and property tax revenues are up throughout the region. So are traffic congestion and crowding during peak seasons, however.

To put Tahoe's peak visitation in perspective, consider a comparison to Yosemite National Park in California. Yosemite's land mass is more than three times as large the Tahoe Basin's 200,000 acres. At 4.5 million annual visitors, Yosemite has approximately one-third the visitors as Tahoe. With such concentrated visitation in a relatively small area, key questions are emerging. How can Lake Tahoe remain pristine in the face of the ever-growing pressures of human development, tourism, population growth, and overall economic activity? How can we leverage science and technology to protect it? In essence, how can we avoid loving the place to death?

Peak holidays in winter coupled with snow storms frequently result in massive traffic jams with skiers eager to get home but often stranded for hours. Applications such as Waze, HERE WeGo, and other platforms allow visitors easy access to local roads, which creates gridlock on side streets. Local roads are clogged, preventing residents from returning home or holding them captive in their own neighborhoods.

Summers bring even more tourists to enjoy Tahoe, with millions of visitors drawn to the beaches, outdoor concerts, mountain biking, and more. Local governments are enjoying recent tax revenue increases as a result. The increased revenues are driven by several factors: positive regional economic activity, favorable weather for outdoor recreation, the proliferation of vacation home rentals that pay transient occupancy taxes, and local ballot measures to increase tourist taxes. Local governments rely on tax collections from more than five thousand vacation home rental properties in the Tahoe Basin, but struggle with neighborhood compatibility issues such as noise, parking infractions, and trash from short-term rentals in residential neighborhoods.

The Fourth of July holiday is one of Tahoe's peak pressure point occasions. The lake's approximately 350,000-plus Independence Day visitors boost local tax coffers and offer restaurants, shops, and other businesses much-needed revenues, but impacts on quality of life such as trash on the beaches have local residents concerned. According to the League to Save Lake Tahoe, Fourth of July beachgoers leave nearly two thousand pounds of trash behind.[3]

Although local headlines on overtourism mainly highlight neighborhood conflicts, impacts to natural resources are Tahoe Basin officials' top concern. Wildfire is one of the most poignant examples. Approximately 95 percent of wildfires in California are human-caused via tossed cigarettes, unattended campfires, and even arson.[4] An illegal campfire caused the Tahoe region's Angora Wildfire in 2007; nearly 250 homes were incinerated, and the fire caused more than $140 million in property damage. Local families' lives were turned upside down, and the scar of the devastation is still evident on the landscape many years later.[5] Another fire in July 2002 was caused by a tourist who dropped a lit cigarette from a gondola tram car on a sightseeing tour. With millions of tourists crowding Tahoe's forests and beaches every summer, local officials have adopted the mantra, "it's not a matter of if, but when" another wildfire will strike.

Other threats to natural resources from overtourism loom large at Lake Tahoe. Oblivious recreational boaters can transport aquatic invasive species, wreaking havoc on ecosystems. Hiking or biking off-trail can damage sensitive wetlands and disturb wildlife. Careless disposal of trash can harm bears in neighborhoods and the back country.

The jurisdictional landscape in the region is complex, and Lake Tahoe's protection has historically been a shared responsibility. The lake straddles the border of California and Nevada and encompasses six local governments, the US Forest Service, and private property owners.[6] "The only thing more bountiful than love for Lake Tahoe is the diversity of opinion about how it should be managed," said Joanne Marchetta, executive director of the Tahoe Regional Planning Agency.[7]

Solutions

Although there are myriad answers to ensure that this beloved destination isn't loved to death, collaboration is a theme that consistently emerges on how to make tourism more sustainable—collaboration that's guided by science. "We've reached a level of community consensus at Tahoe that the environment is the economy and the economy is the environment," said Steve Teshara, chief executive officer of the Tahoe Chamber of Commerce. "We've found this common ground by relying on credible scientific information to guide our decision-making."[8]

Acknowledging the importance of Tahoe's $5 billion annual tourism-based economy serves as common ground. The symbiotic relationship of the environment and economy has long served as a framework to bridge the divide between polarized positions on development. Collaboration has fostered the recognition that a healthy economy is necessary to preserve Lake Tahoe's irreplaceable environment and that a pristine lake is the region's value proposition in the fiercely competitive destination resort business.

Innovative Partnerships over the Years

In the early days of Lake Tahoe's first tourism boom, unchecked development following the Olympics drew a vociferous public outcry. Scientists raised alarm bells about pollution and called for action. In response, high-level political engagement led to the creation of an interstate compact in 1969 between the states of California and Nevada to protect Lake Tahoe. The result was the formation of the Tahoe Regional Planning Agency (TRPA), the first watershed-based environmental agency of its kind in the United States.[9]

Today, this landmark public policy achievement is heralded as a conservation success story. No other regional government in the United States has the regulatory authority to manage growth at a watershed level in the same manner as TRPA. Limits on growth and development, as well as strong environmental regulations, are the norm at Lake Tahoe. Scientists have studied the lake for more than fifty years; Lake Tahoe has one of the longest-running water quality data sets in the world. These data sets have served a crucial role in aligning interests and providing a rationale to manage growth, development, and tourism policies.

Common ground was hard fought and consensus didn't come easy. Disagreements spawned lawsuits and community discord over many decades—how much if any growth should occur and how to balance environmental protection with private property rights.[10] Since the late 1990s, however, collaborative decision-making has emerged as the preferred governance approach for the Tahoe Basin and is essential to tackling the region's current overtourism challenges.[11] Today, more than seventy-five organizations are part of a partnership in the Lake Tahoe Environmental Improvement Program, which has a series of working groups in areas such as water quality, forest health, invasive species, and most recently, sustainable recreation.

Asking Visitors to Take Care

To combat threats from fire—along with other dangers like the introduction of invasive species into the lake, trail damage from hiking or biking in the wrong locations, and human-wildlife conflicts—Tahoe Basin partners launched a public stewardship campaign in 2015. The "Take Care" campaign relied on extensive market research that found that earnest messages of environmental responsibility and stewardship were often lost on Tahoe's millions of visitors. The campaign therefore employed comedy writers to drive the message home that visitors should behave well at the lake. Since the Take Care campaign began at Lake Tahoe, there have been no new introductions of invasive species nor any damaging wildfires, two of the key components of the Take Care public education program.

Additionally, local organizations are involving visitors through volunteer opportunities. "We organize beach clean ups to bring the community together to protect the lake," said Darcie Collins, chief executive officer of the League to Save Lake Tahoe. "We enlist our visitors to instill a sense of stewardship."[12] Through the takecaretahoe.org website, visitors can also find trail-building volunteer outings, wildlife count excursions, and other outdoor stewardship opportunities. Although trash is still a problem on the beaches, tourism officials hope that as more awareness builds, these impacts will dissipate.

Colorful graphics dominate Lake Tahoe's posters as part of its Take Care campaign. Source: Tahoe Regional Planning Agency.

Shifting Gears: Transportation Strategies

Another key factor driving tourism activity at the lake is its location. The Tahoe Basin is situated less than a half-day's drive from the populated San Francisco Bay area. With such a ripe feeder market, it may seem counterintuitive that tourism officials are promoting more air service to lure longer-stay destination visitors. Research, though, shows that these visitors contribute more to local economies; therefore, Tahoe officials are seeking more air travelers. "For the first time in our history, our organization halted media buys in the San Francisco Bay Area drive-up market in 2018," said Carol Chaplin, executive director of the Lake Tahoe Visitors Authority. "We shifted our spend to markets with solid air service to encourage longer stays which benefit the local economy to a much greater degree."[13]

With the current transportation mode of choice being personal automobiles for Tahoe's millions of visitors, new strategies are needed to relieve peak congestion. Bike and pedestrian trails, at a cost of more than $1 million per mile, are a high priority to give visitors options other than driving to explore the lake. Other strategies being tried include microtransit shuttles (like Uber, but in a van), technological solutions such as real-time parking and travel information, bike-and-scooter-share programs, and smaller-scale parking lots interspersed in neighborhoods bordering tourist areas. The Lime scooter-share service alone has resulted in 100,000 rides annually that would otherwise be taken in private automobiles, clogging roadways.[14] New measures such as parking reservation systems, free public bus service, and improved shuttles are in the works.

Lessons Learned from Lake Tahoe

The body of knowledge on overtourism is resoundingly clear: there is no one-size-fits-all strategy for a tourist destination. At Lake Tahoe, many lessons have been learned about how to balance a healthy environment and economy. Although still a work in progress, key takeaways include the following:

- Practice collaboration. A cohesive strategy must involve government, nongovernmental organizations, tourist resorts, businesses, tourism authorities, communities, and all stakeholders. Alignment around problems and solutions is difficult to achieve but necessary for progress.
- Adopt environmental regulations that guide sustainable growth. Destinations must manage development through strong land use policies and consistent enforcement.
- Use science and technology. Get local scientists to consult on environmental metrics and indicators to monitor the effects of overtourism. Getting community buy-in on the appropriate indicators to track can bring the public and policy makers together.
- Form working groups on specific overtourism issues. Lake Tahoe's sustainable recreation group is developing a strategic plan that organizes the lake into multiple corridors to target solutions around transportation, parking, signage, overcrowding, and roadway safety. Dozens of entities ranging from law enforcement to recreation rental companies are involved.
- Engage the public. Lake Tahoe's Take Care campaign enlists visitors in solutions by having them feel like they contributed to the environment while they enjoyed themselves on vacation. Human behaviors must change to address overtourism problems, and the millions of visitors to a destination must be engaged in the solutions.

- Be sensitive to terminology. The term *overtourism* is often perceived as negative to tourism officials whose missions are to promote the destination. Consider different terms, such *as pressure points* or *bottlenecks*, when discussing overtourism to foster greater collaboration.
- Focus on transportation infrastructure. Microtransit shuttles, bike and pedestrian trails, free bus passes for tourists, and travel platforms with instant information about traffic delays are all necessary to move visitors around bottlenecks and pressure points. Traffic jams are not flattering to any destination, and at Lake Tahoe, air and water quality are impacted as well.
- Exchange ideas with other destinations. Destination managers should seek out opportunities to learn from one another. For example, through shared attendance at a conference on overtourism solutions, the Take Care campaign is now bicoastal—CARE for the Cape and Islands, on Cape Cod in Massachusetts, adopted the public awareness program in 2019 to help visitors harmonize better with their community and environment. (See chapters 5.1 and 6.1.)

The Tahoe Regional Planning Agency and partners remain committed to a collaborative solutions-oriented approach to protect natural assets and ensure that Lake Tahoe remains a vibrant and enjoyable place to live and visit for years to come. Beyond this commitment, it hopes to provide useful lessons learned for other destinations with similar problems so that they can implement similar solutions.

Notes

1. University of California Davis. "Tahoe Environmental Research Center, State of the Lake Report 2017." www.tahoe.ucdavis.edu.stateofthelake.

2. Tahoe Regional Planning Agency. (2017). "Regional Transportation Plan." http://www.trpa.org/regional-plan/regional-transportation-plan/.

3. Tahoe Daily Tribute. (July 6, 2019). "Volunteers Remove 1,875 Pounds of Trash from Lake Tahoe Beaches after July 4th." https://www.tahoedailytribune.com/news/volunteers-remove-1875-pounds-of-trash-from-lake-tahoe-beaches-after-july-4th/; Darcie Collins. (April 2019). Personal interview with author.

4. Staff report based on "Angora: South Lake Tahoe," Sierra Nevada Media Group. (June 24, 2017). *Tahoe Daily Tribune*. https://www.tahoedailytribune.com/news/10-years-ago-today/.

5. L. M. Krieger. (August 12, 2018). *San Jose Mercury News*. https://www.mercurynews.com/2018/08/12/whats-starting-all-these-fires-we-are/.

6. M. Imperial and D. Kauneckis. (2003). "Moving from Conflict to Collaboration: Watershed Governance in Lake Tahoe." *Natural Resources Journal.* 43:4: 1009–55.

7. Joanne S. Marchetta. (April 2019). Personal interview with author.

8. Steve Teshara. (April 2019). Personal interview with author.

9. United States Government Accountability Office. (2007). "An Overview of Structure and Governance of Environment and Natural Resource Compact." https://www.gao.gov/products/GAO-07-519.

10. Tahoe Prosperity Center. (2017). "Measuring for Prosperity, Community and Economic Indicators for the Lake Tahoe Basin." https://tahoeprosperity.org/tahoe-data/.

11. Tahoe Regional Planning Agency. (2018). "Accomplishments Report, Environmental Improvement Program."

12. Darcie Collins. (April 2019). Personal interview with author.

13. Carol Chaplin. (April 2019). Personal interview with author.

14. Paula Peterson. (October 28, 2019). "South Lake Tahoe to Create an Ordinance for Electric Scooters." *South Tahoe Now.* http://southtahoenow.com/story/10/28/2019/south-lake-tahoe-create-ordinance-electric-scooters.

Chapter 6.4

Colorado

By Cathy Ritter

Within the United States, it is possible that the strain of too many people in the same place at the same time is felt more keenly in the West, where solitude and open vistas are cherished. I've met Coloradans who long for the days when they could bike for miles without encountering another soul. To them, the sight of ten, six, or even four cyclists along that same stretch feels like an intrusion.

Disregard for the land is a particular aggravation for people who love their state. I once hiked a Rocky Mountain National Park paved trail lined with signs warning that damage to the adjoining alpine tundra could take ten years to heal. Ahead of me, a grizzled hiker kept shouting at others, "Get back on the trail!" As I caught up, she turned around, completely exasperated with the clueless people around her, to say, "Can't they read the signs?" When no one is reading the signs, tourism is at risk and shouting isn't a strategy.

Colorado resident concerns emerged publicly in 2016 during a series of twenty-three face-to-face listening sessions hosted by the Colorado Tourism Office (CTO), attracting more than one thousand stakeholders. At every stop, concerns about traveler impacts were heard.

Sometimes it was about shrinking water supplies, impacts on land from "social trails," or careless disruption of wildlife habitat or behaviors. But there was also deep frustration with traffic jams and unaccustomed lines at favorite parks and ski resorts. In developing a new statewide strategic plan for the Colorado tourism industry, ultimately known as the Colorado Tourism Roadmap, the CTO tapped into a gathering awareness of a threat to the very nature of our state.

With concerning outcomes of tourism success becoming ever more apparent, the CTO has been charting a new path since 2017 to inspire travelers to seek out less-visited destinations and reduce impacts on the state's natural and cultural resources. As a state-funded tourism agency, it must zealously guard its social license to operate. Should it fail, it may not have the cash to operate.

Protecting the vitality of both tourism and Colorado's natural assets is now widely viewed as a necessary balance for the good of the state. As former Colorado governor John Hickenlooper told *Travel Weekly*, "Investments to drive tourism—it's hard to say whether we're driving tourism or making responsible investments for the health and enjoyment of our population. They go together. That's what's brilliant about it."[1]

Causes of Overtourism

There's no denying that the perspective is different here, especially in the one-third of Colorado that still meets the definition of frontier (those twenty-three counties with fewer than eight people per square mile). But there's also no question that the rapid growth in visitation has made special places feel less spacious—even as it has generated billions in new state and local tax revenues and driven tourism-related economic impact, business earnings, and employment to record highs.

Until recently, Colorado tourism growth was pacing at twice the US average. From 2009 through 2017, Colorado experienced a 41 percent increase in visitors, compared with an average national growth rate of 20 percent. During those same years, Colorado rose from a regional to a national destination, boosting its share of marketable leisure travelers from eighteenth to eighth largest.[2]

Impacts of Overtourism

The mixed blessings of increased tourism are not lost upon locals. A first-time resident sentiment study fielded in May 2018 showed that although 92 percent of Coloradans believed that tourism was important to the state's economy, 85 percent said that it contributes to traffic.[3]

With the full support of the Colorado Tourism Board, Colorado has become the first state to deploy a comprehensive plan to fend off overtourism, primarily because it paid heed to an early warning system: Coloradans themselves.

Solutions

What to do about the problem was a sensitive topic. Some stakeholders were convinced that recognizing visitor pressures could tarnish the image of Colorado's destinations, perhaps even threaten jobs. But as the research phase of the work—with lead planning consultant Nichols Tourism Group—was concluding in late 2016, it became clear that failing to recognize and address stakeholders' real concerns could ultimately risk the future success of Colorado's signature industry.

A Move Away from Growth

One of the first steps was to create a mission statement reinventing the typical definition of tourism success. Rather than strive to attract ever-higher visitor numbers, the aim was switched to increasing visitor spending. The traditional role of destination promotion was broadened to include a new focus on destination development, with a goal of attracting travelers to less-visited places. And to express the newfound commitment to protecting the most precious aspects of the Colorado experience, the CTO pledged to make our efforts sustainable and win hearts along the way.

Of the four new action pillars laddering up to CTO's mission, one staked out entirely new ground: steward. Without really knowing how the organization would fulfill this new commitment, it set out three brand-new objectives:

- Embrace thinking that disperses visitors in productive ways.
- Invite travelers to embrace Coloradans' sustainability ethic while here.
- Create alliances with other stakeholders to magnify the impact of sustainable tourism initiatives.

Dispersing Visitors

Among the CTO's first actions was the May 2017 launch of the Colo-Road Trips microsite, now central to the strategy of dispersing travelers to less-visited places and off-peak experiences.[4] This online collection of about two hundred multiday itineraries is filled with visuals, trip advice, and amazing experiences aimed at taking travelers to places they'd never considered.

CTO's industry partners have a standing invitation to work with us to create new trips. One requirement is that each must include "Sustainability Tips" guiding visitors to volunteer, get educated, or contribute toward causes Coloradans care about. Now a key component of CTO's promotional strategy, Colo-Road Trips itineraries have become one of the most effective social media engagement tools and a powerful call to action for our national campaign.

Creating "Care for Colorado"

The CTO knew that dispersion alone wasn't enough. It seemed that the vast majority of travelers, including several locals, were blissfully unaware of the damage they were causing. Education was the answer, and CTO— as a tourism promoter—had a huge communications platform at its disposal. In addition to the "Come To Life" domestic marketing and public relations program, it operates a top-performing state tourism website in the United States, manages more than one million followers on social channels, and greets around one million visitors a year at ten Colorado welcome centers.

The CTO set its sights on finding a highly credible partner with established messaging that could be customized to support its own messages—and potentially serve as a platform to support a unified message for travelers in states across the United States. Luis Benitez, then-director of the Colorado Outdoor Recreation Industry Office, suggested contacting the Leave No Trace Center for Outdoor Ethics in Boulder.

Founded in the early 1990s by the US Forest Service, Leave No Trace has spun off into a nonprofit organization and has had long-time partnerships with federal public lands agencies, the outdoor industry, and even Subaru. The connection with Leave No Trace executive director Dana Watts was almost instantaneous. After a few months of discussion and a thumbs-up from both of their boards, Leave No Trace and the CTO joined forces through a formal agreement in October 2017.

Their first task was collaborating to create a shared message for Colorado travelers. Based on the iconic "Leave No Trace: Seven Principles," our new Care for Colorado Principles were introduced right at the start of the 2018 summer travel season in a new brochure titled, "Are You Colo-Ready?" With the blessing of Leave No Trace, they kept our message to seven simple declarative statements, like "Stick to Trails," "Keep Wildlife Wild," and "Leave It As You Found It."

They outfitted 350-plus Colorado welcome center volunteers in Leave No Trace/Colorado Tourism Office branded polos and trained them to share the new messaging with travelers. We also equipped all ten centers with water-bottle filling stations. Moreover, they placed Care for Colorado posters on the back of restroom doors for a captive audience of about 660,000 bathroom breakers a year.

As fall rolled around, the alliance introduced the cheeky Care for Colorado video, enlisting a cartoon cast of animals to share an earworm of a song aimed at getting a misbehaving traveler on the right path.[5] The tone of the campaign was kept Colorado style—fun, inviting, and not judgmental. The CTO intentionally avoided a preachy impassioned message that could potentially turn off a less-convinced traveler. In just the first two weeks, the video registered forty-three thousand views and since then has been shared on countless websites and social media feeds.

Knowing that we were putting ourselves in a glass house by exhorting others to care for Colorado, the CTO team took steps to clean up its own act. Since 2018, the CTO has operated its annual Colorado

Care for Colorado principle 3: Leave it as you found it. Source: Colorado Tourism Office.

Governor's Tourism Conference to comply with zero-waste requirements. To date, eight members of the team have been certified by the Global Sustainable Tourism Council.

Building a Coalition

The CTO knew that it could reach even more Colorado travelers through partnerships. In January 2019, the CTO and Leave No Trace announced that the alliance had expanded to include three major tourism-based organizations: the Colorado Hotel and Lodging Association, the Colorado River Outfitters Association, and the Colorado Dude and Guest Ranch Association. By February 2020, membership had swelled to include the Colorado Association of Destination Marketing Organizations, the Colorado Association of Ski Towns, the Colorado Mountain Club, and Bicycle Colorado. The growing Care for Colorado Coalition now has a monthly newsletter with regular distribution of new messaging, including timely social posts and videos on the "etiquette" of hiking, wildlife interactions, and proper disposal of both pet and human waste.

Within this broad alliance, each of these groups is offering its members meaningful methods to share Care for Colorado messaging with its guests. The Dude Ranch Association, for example, created a fun

"Dude Ranch Edition" of the "Are You Colo-Ready?" brochure for ranches to send to guests ahead of their stay.

Across Colorado, a force is building in support of a single, unified goal: safeguarding the health of an industry that drives so much of the state's economy by protecting what makes it viable.

Deploying Timed-Entry Systems and Technology

Smart solutions that may have been unthinkable just a few years ago are winning strong support. Tourism leaders joined land managers in the spring of 2019 to tackle overvisitation at the beloved Hanging Lake near Glenwood Springs. They launched a free shuttle funded by the brand-new reservation system. Resident response on social channels was overwhelmingly positive, even though the new $10 to $12 fee applies to residents as well as tourists.

Well-crafted timed-entry systems could go a long way toward easing long lines and parking congestion at the most popular state and national parks while both enhancing the visitor experience and accommodating resident interests. Technology is now creating pain-free solutions for dispersing travelers. A mountain biking neurophysicist led the development of an app mapping all 750 miles of single-track trails in the legendary home of mountain biking, the Crested Butte–Gunnison Valley. It had a magical unintended consequence: overuse of the two or three most popular trails suddenly dissipated as bikers found their way to single tracks they never knew existed.

The free COTREX (short for Colorado Trail Explorer) app that launched in the fall of 2018 by the Colorado Department of Natural Resources holds similar potential. With navigational guidance to trailheads and more than thirty-eight thousand miles of public trails, COTREX is readymade for hikers, bikers, and off-roaders searching for trails less traveled.

Building a Future Advantage

The initial hunch that educating travelers would also mitigate resident angst was further backed by research. A May 2018 resident sentiment survey showed that 80 percent of residents feel more positive about

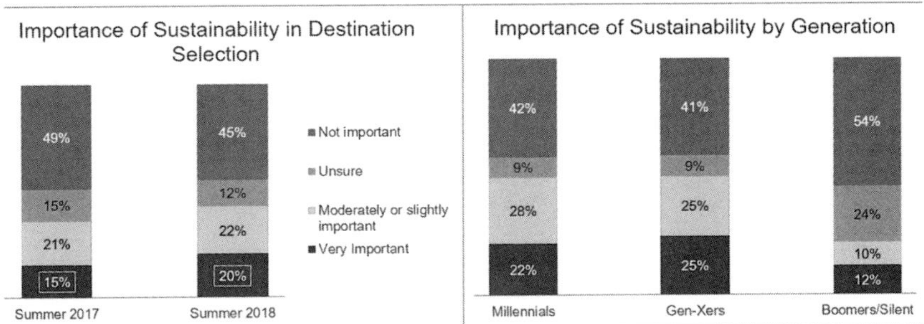

Colorado visitors' interest in sustainability, 2017 and 2018. Source: Colorado Tourism Office.

tourism when a destination educates travelers to reduce their impacts.[6]

Although addressing resident concerns was the primary goal, CTO research has shown that protecting destinations is increasingly important to travelers. Its campaign effectiveness study showed that 27 percent of travelers in the summer of 2019 rated sustainability practices as "very important" in their choice of destinations, up from 20 percent the year before and just 15 percent in summer 2017, with gains among all age groups.[7] It is clear that the work done in Colorado to offset a potential resident backlash against tourism is instead shaping up as a surprising competitive advantage and investment in the future of the state.

Notes

1. Arnie Weissmann. (November 5, 2018). "Exploring and Exploiting Topophilia." *Travel Weekly*. https://www.travelweekly.com/Arnie-Weissmann/Exploring-exploiting-topophilia.

2. Longwoods International. (June 2018). *Colorado Travel Year 2017*. https://industry.colorado.com/longwoods-international.

3. MMGY Global. (July 2018). *Colorado Resident Sentiment towards Tourism.*

4. Colorado Tourism Office. (2020). *Colo-Road Trips*. https://www.colorado.com/colo-road-trips.

5. Colorado Tourism Office. (October 2018). *Care for Colorado—Are You Colo-Ready?* https://www.youtube.com/watch?v=JcY73nUhaQs.

6. MMGY Global. (July 2018). *Colorado Resident Sentiment towards Tourism.*

7. Strategic Marketing and Research Insights. (November 2019). *2019 Summer Advertising Effectiveness.*

Chapter 6.5

New Zealand

By Andrea Insch

Towering mountain ranges, forests and glacier-fed rivers made New Zealand the perfect stand-in for Middle Earth in *The Lord of the Rings* movie series and a cinematic billboard for the country's natural beauty.

Today, jet boats rip down rivers seeking the mythical Isengard, where the wizard Gandalf was imprisoned. "Freedom campers" in rented vans leave trails of waste. Tens of thousands of helicopter trips annually deposit visitors, some in flip-flops, on New Zealand glaciers that were once the realm of expert climbers. One tour group had to be rescued after trying to walk barefoot to Mount Ngauruhoe, in apparent homage to J.R.R. Tolkien's Mount Doom.[1]

That is how a May 2018 *Wall Street Journal* article, datelined Queenstown, New Zealand, opens and juxtaposes the world-renowned scenic landscapes of New Zealand, with the lesser-known, but increasingly negative impacts of overtourism in some of the country's tourist hotspots. When so many people are dependent on tourism, it is difficult to take a hard look at the sustainability of the sector and the impacts it has had.

Growth of Tourism and Emergence of Overtourism

Since 1990, New Zealand has seen a four-fold growth in international tourist arrivals, from a base of just under 1 million (976,010) in 1990 to 3.9 million in 2019.[2] Although the annual arrivals growth rate was, on average, 5 percent from 1990 to 2019, it exceeded 10 percent in 1994, 2000, 2004, and 2016.[3] Total international tourism spend increased from just under NZ$6 billion (US$9.6 billion) in 1999 to NZ$17.2 billion (US$27.5 billion) in 2019, a growth of 2.9 fold.[4] The annual rate of growth in total international tourist expenditure was, on average, 5.6 percent between 1999 to 2019.[5]

Tourism is a major part of the New Zealand economy and the leading export sector, directly contributing in NZ$16.2 billion (US$10.1 billion) value added, or 5.8 percent of gross domestic product (GDP) in 2019. Supporting industries added NZ$11.2 billion (US$7 billion), constituting another 4 percent of GDP.[6] The value of New Zealand's domestic tourism expenditure is even larger, totaling NZ$23.7 billion (US$14.8 billion) in expenditure in 2019.[7] Overall, the tourism industry plays a major role in the livelihoods of many New Zealanders, employing 229,566 people in 2019, which constitutes 8.4 percent of the total work force.[8]

By 2025, international tourist numbers are expected to reach 5.1 million, growing at 4 percent annually between 2018 and 2025. Australia is forecast to remain New Zealand's largest source market, but the government projects that the increase in international arrivals "will be driven in the short-term by strong growth in the US market, and over the longer-term by growth in Asian markets, especially China."[9] Tourist spending is forecast to be NZ$203 (US$126) per person per day by 2025, and overall total international tourist expenditure is forecast to reach just under $15 billion in 2025. It is projected that total tourism expenditure will grow at an annual rate of 4.3 percent or by a total of 34 percent from 2018 to 2025. Because New Zealand is a long-haul destination for international travelers, the average time spent within New Zealand is substantial. The average length of stay is expected to increase from eighteen days in 2018 to nineteen days by 2025.[10]

Fundamental to the rapid growth of New Zealand's tourism has been *100% Pure New Zealand*, the successful marketing campaign first launched in 1999 by Tourism New Zealand (TNZ), the country's destination marketing organization. The campaign, which targeted international markets, welcomed 3.1 million international visitors in 2015, up from 1.6 million in 1999. Between 2015 and 2019, international arrivals again rose significantly, by another 857,000 visitors, propelled by the *Lord of the Rings* film-inspired tourism, an increasingly mobile global middle class, and a growing number of airlines servicing New Zealand.[11]

Domestic tourism has also grown, although precise statistics are not available. One indicator of the volume of domestic tourism, the number of guests staying in commercial accommodation including hotels, motels, backpackers, and "holiday parks" (campgrounds), grew by 25 percent between September 2009 and September 2019.[12] Domestic travel has been fueled by decreasing costs, more domestic flight capacity, and more mobile retired New Zealanders.[13]

The country's accelerated growth in both international and domestic tourism has led to reports of overtourism in a number of locations at different times of the year. It has been most noticeable in sites featured in the *Lord of the Rings* trilogy.[14] Soon after the release of the first movie in 2001, a guidebook was published that provided the location of eighty-eight film sites, with maps, GPS coordinates, and touring information.[15] This guidebook, together with the Department of Conservation's release of GPS coordinates for "Rings locations," stimulated "the growth of both self-driven and organised tourism in relation to the film sites."[16] Further fueling the campaign, Tourism New Zealand rebranded its original 100% Pure campaign as "100% Middle-earth, 100% Pure New Zealand," and Air New Zealand proclaimed itself as the "Airline to Middle-earth," decorating four planes with *Lord of the Rings* imagery. Although there are no precise figures of how much New Zealand invested in promoting the film trilogy, one study concluded that "for a country the size of New Zealand, the resources committed to leveraging the effect of the *Lord of the Rings* are unprecedented."[17]

Impacts of Overtourism

Overtourism in New Zealand is challenging the quality of life in some of the country's smaller communities and popular visitor attractions.

To assess its impacts, it is important to consider the perceptions of residents, business operators, parks officials, visitors, and other key stakeholders. The following are some of the most apparent impacts of overtourism.

Quality of Life in Queenstown

As noted above, Queenstown, a popular resort town of approximately 15,650 residents located in the South Island, is one of the places most acutely experiencing the pressure of overtourism propelled by the *Lord of the Rings* phenomena.[18] Tourism is extremely important to the Queenstown–Lakes District region, which recorded the second highest expenditure in New Zealand by international and domestic tourists in 2019. Total visitor spending in the region has more than doubled in a decade, from NZ$1.3 billion (US$816 million) in 2009 to NZ$3.1 billion (US$1.9 billion) in 2019, and tourism arrivals in the region grew 1.5 fold during the decade, from 2.86 million visitor nights in 2009 to 4.64 million visitor nights in 2019.[19] Queenstown's tourism market is a mix of high-value spend visitors, including business and luxury travelers, average-spend holiday makers, and lower-spend visitors such as backpackers and freedom campers (discussed below).

Queenstown's tourism boom is challenging residents' quality of life as transport infrastructure, essential services, and accommodations, both in the town center and the outlying areas, are stretched. The government has proposed to begin scheduled jet flights at the nearby Wānaka airport to further grow visitor numbers; local residents and other stakeholders are strongly opposed, however, in part because the plan would expand the noise boundaries in Queenstown to permit the increased air traffic. "It seems to me that the limits of tolerance for growth at any cost have been reached in this community," health advocate Marion Poore told a public hearing called to discuss the proposed airport expansion.[20]

Today, annual international tourist arrivals are nearly double the size of New Zealand's population, and a backlash appears to be growing. According to the November 2019 "Mood of the Nation" survey, 78 percent of Queenstown residents believe that international tourism puts too much pressure on New Zealand in a variety of ways. The survey results reveal that 76 percent of Queenstown respondents perceive that international tourism results in increased traffic congestion on holiday routes, 61 percent perceive it causes a higher number of road accidents, 61 percent perceive that tourism makes accommodation too expensive for New Zealand residents, and 56 percent perceive that it results in damage to New Zealand's natural environment. These results from Queenstown are significantly higher than for residents in the surrounding Otago region and for New Zealand as a whole.[21]

Other Communities and Sites

Unlike the country's urban centers, including Auckland, Wellington, and Christchurch, which are better equipped to host large numbers of visitors, smaller towns, villages, recreation spots, and camping sites suffer from insufficient and inappropriate infrastructure to manage the mounting tourism influx, especially during the high season (December, January, and February). Sites, towns, and attractions impacted include the Church of the Good Shepherd at Lake Tekapo, Akaora, Aoraki Mount Cook National Park, Fiordland (including Milford Sound), Waiheke Island, Westland glaciers, Routeburn track, and Tongariro National Park.

Among the most visible signs of overtourism are traffic congestion and accidents, littering, damage to sites of historical significance, and degradation of the natural environment. Where government agencies are seen as not responding adequately to local problems arising from overtourism, residents have sometimes engaged in direct action. Take, for instance, the quintessentially New Zealand practice of "freedom camping"—putting up tents or parking camper vans in areas not designated for camping—which has been growing exponentially. As one tourism website states, "Unfortunately, free camping is having an increasingly negative effect on New Zealand's clean, green environment due to the increasing numbers of freedom campers—some of whom create litter problems, dispose of human waste inadequately, and discharge grey water outside of dump stations. Free campers tend not to be popular with local residents."[22] In the West Coast settlement of Kakapotahi, for instance, residents blocked access to a freedom camping site after local authorities failed to enforce the site's fifteen vehicles limit. According to one of the protesters, some fifty vehicles were found camping at the site, so "we're doing what should be done, but isn't."[23]

Attitudes of Residents

The "Mood of the Nation" survey of New Zealanders indicated that 52 percent of respondents nationwide perceived that predicted tourism growth is "too high" and 40 percent perceived that international tourism places "too much pressure" on the country. New Zealand residents also responded that the country lacks the infrastructure to support the growing number of international visitors and that tourism is having a negative impact on the environment.[24]

The Environment

Recent government evaluations of the impacts of international tourism in New Zealand since 1990 have identified a range of physical and ecological effects including soil erosion, vegetation destruction, damage to geological features, and disruption to wildlife.[25] In particular, tourist interactions with wild animals, including rare and endangered populations, has gone "largely unmanaged."[26] Apart from the site-specific biophysical impacts, tourism growth has placed pressure on New Zealand's environment more broadly, particularly the effects of climate change (for example, sea-level rise, glacier retreat) due to carbon emissions.[27] New Zealand's inbound and outbound tourism is dependent on long-haul flights and thus has a substantial carbon footprint. As a 2019 study argued, without the introduction of new aviation travel technologies, "carbon limits will ultimately mean that air travel cannot remain the cheap, convenient, fast and regular form of 'everyday' travel that it is today."[28]

Solutions

One of the major challenges in understanding the complex problem of overtourism and designing solutions to address it is the nature of the tourism industry itself. In a spatiotemporal analysis of the environmental impacts of tourism in New Zealand, the industry is described as "an open system that is loosely coordinated, fragmented and hard to clearly define. It comprises a multitude of agencies and actors who interact at crosscutting levels to produce social, economic and environmental outcomes that should be in the collective best-interest."[29]

In addition, long-haul destinations like New Zealand have limited ability to influence the tourism market, where international travelers are often influenced by global factors such as economic downturns, political instabilities, pandemics, and blockbuster films that are beyond control of the receiving country. Thus, designing solutions to address overtourism in New Zealand, as part of the global tourism system, involves adopting systems thinking and a long-term view of the role of tourism in the economy and the lives of New Zealanders. There is evidence that government and industry bodies, as well as some business operators, are taking a wider and longer view aimed at strengthening the sustainability of tourism in New Zealand. At present, however, many initiatives are short term, aimed at addressing the limitations in tourism infrastructure to cope with the growth in tourist numbers. Below are some of the reforms that are in place or under discussion.

New Tourism Tax

In July 2019, New Zealand introduced an international visitor conservation and tourism levy (IVL) of NZ$35 (US$22) per visitor to increase funding needed for critical infrastructure and conservation projects. Some tourists, including those from the largest source market, Australia, are exempt from the IVL, which is expected to generate NZ$80 (US$50) million per year.[30] Bruce Parkes, Department of Conversation's director general of policy and visitors, stated that the IVL will "make a significant difference" in protection of flora and fauna. As he elaborated, "Boardwalking over a wetland to protect the biodiversity would be the sort of investment that would achieve both a better visitor experience and better biodiversity outcomes."[31] Funds from the levy have been allocated to a range of projects, including creating three new habitat sites for the long-term sustainable management of the kākāpō, a threatened native species of bird.[32]

Role of Tour Operators

Many of New Zealand's tourism operators, who often work on the front lines of overtourism, are agreeing to implement measures to enhance the industry's sustainability. Tourism Industry Aotearoa, a nongovernmental body representing all sectors of the industry, led the development of the Tourism Sustainability Commitment (TSC) whose overarching aim is to achieve eight industry goals related to economic, visitor, host community, and environmental sustainability.

More than one thousand tour operators have signed up to support the TSC since its launch in November 2017. The goal is to have business members achieve fourteen sustainability targets by 2025. They include educating visitors about New Zealand's cultural and behavioral expectations, actively engaging with the communities in which they operate, contributing to ecological restoration initiatives, and paying a fair wage to all staff.[33] For many operators, meeting these commitments will require a clear departure from their usual business practices.[34]

Attracting High-Value Visitors

After decades of focus on attracting greater numbers of tourists with relatively modest daily spend rates and with Australia as the most important source market, the New Zealand government is undergoing a policy shift. Many government agencies at the national, regional, and local levels are attempting to target high-value tourism segments, including from the United States, who will spend more in the local economy. In addition, TNZ has reduced its budget for destination-wide marketing to focus more on regional tourism product development. This policy shift aims to encourage regional dispersal of visitors and decrease the effects of seasonality on the most overtouristed attractions and destinations.

One example, led by TNZ, is the promotion of Wellington, Tasman, Marlborough, and Nelson to so-called free independent travelers from China.[35] The "Heart of the Long White Cloud" campaign, which features respected Chinese cultural influencers, is reported to have succeeded in attracting Chinese visitors to these regions in the off-peak seasons.[36] Domestic tourists and repeat Australian visitors will also be targets of promotional activities by TNZ and regional tourism organizations with the aim of dispersing visitors beyond major destinations, such as Auckland, Christchurch, Queenstown–Lakes District, and Wellington.[37]

An initiative that has cross-sectoral support is the "Tiaki Promise—Care for New Zealand," a pledge that encourages all visitors (and residents) in New Zealand to commit to behave in ways that protect the environment, respect culture, and keep everyone safe. The promise states, "While travelling in New Zealand I will: care for land, sea and nature, treading lightly and leaving no trace; travel safely, showing care and consideration for all; and respect culture, travelling with an open heart and mind."[38] Key industry players, including TNZ, Air New Zealand, Department of Conservation, Tourism Industry Aotearoa, Local Government New Zealand, New Zealand Māori Tourism, and Tourism Holdings Ltd., launched the pledge in 2018. The Tiaki Promise relies on government authorities, tourism operators, and other tourism entities, as well as residents, to make visitors aware of behavioral expectations and to hold individuals accountable by self-policing and calling out inappropriate behavior. International visitors on board Air New Zealand flights learn about the Tiaki Promise through an in-flight entertainment channel.[39]

Going Forward: Focus on Sustainability

Efforts to address overtourism in New Zealand are facing a fork in the road. There is growing evidence that the country's "existing high growth model is not sustainable under the current management approach."[40] Thus, addressing overtourism and its underlying causes and consequences needs to be tackled with new thinking that puts long-term sustainability as the basis for all decision-making and action. By implication, doing so will require farsighted choices and rethinking the assumption of the long-standing management model of continuous growth in tourist numbers.

As noted, some steps have already been taken to increase investment in infrastructure to protect fragile ecosystems and endangered species, to more effectively disperse visitors to less popular destinations, to improve the industry's capacity to deliver high quality visitor experiences, and to raise visitor support for sustainable tourism practices. There is also evidence of greater interagency coordination at all levels among those responsible for conservation, economic productivity, business growth, and tourism.

It is doubtful that this level of investment and cooperation is sufficient to ensure the long-term sustainable development of New Zealand's tourism industry, however. Further actions are vital including obtaining better insights into the different visitor segments, particularly domestic tourists and their preferences and travel choices, which will help encourage greater regional dispersal and circumvent overcrowding at iconic sites. Understanding the value of each visitor (within each visitor segment) is also required to take actions designed to dampen growth in tourist numbers so as to maintain the country's conservation values, the industry's social license to operate, and the quality of the visitor experience.[41] Some of the specific recommendations being urged are greater consistency and compliance across New Zealand, including certification of responsible camping. At present, there are guidelines for responsible freedom camping and fines for illegal activities such as dumping of waste, but no official certification program.[42] In addition, there is a need for more low-impact, regenerative wildlife tourism projects, such as Whale Watch Kaikoura, a Māori-owned company dedicated to sustainable tourism practices that regulates access to certain species and reinvest profits in activities to support their conservation.[43] To redress carbon emissions generated through air travel, the New Zealand government is considering several options, including a carbon tax on aviation fuels, agreed-upon environmental standards in bilateral air service agreements, and emissions-based passenger taxes.[44]

Notes

1. Rachel Pannett. (May 22, 2018). "Anger over Tourists Swarming Vacation Hot Spots Sparks Global Backlash." *Wall Street Journal.* https://www.wsj.com/articles/anger-over-tourists-swarming-vacation-hot-spots-sparks-global-backlash-1527000130.

2. Stats NZ. (December 9, 2019). *Tourism Satellite Account: 2019.* https://www.stats.govt.nz/information-releases/tourism-satellite-account-2019.

3. Ministry of Business, Innovation & Employment. (May 2017). *New Zealand Tourism Forecasts 2017–23.* https://www.mbie.govt.nz/immigration-and-tourism/tourism-research-and-data/international-tourism-forecasts/previous-international-tourism

-forecasts/2017-2023/; Ministry of Business, Innovation & Employment. (May 2019). *New Zealand Tourism Forecasts 2019–25*. https://www.mbie.govt.nz/assets/a8bba25fdf/new-zealand-tourism-forecasts-2019-2025.pdf.

4. FigureNZ. (December 10, 2019). *International Tourism Expenditure Compared with Selected Primary Exports: Year Ended March 1999–2019, NZD Billions*. https://figure.nz/chart/EdaXEBTxUQsijkEQ.

5. Stats NZ. (2019). *Tourism Satellite Account: 2019*. https://www.stats.govt.nz/information-releases/tourism-satellite-account-2019.

6. Ibid.

7. Ibid.

8. Ibid.

9. Ministry of Business, Innovation & Employment. (May 2019). *New Zealand Tourism Forecasts 2019–25*. https://www.mbie.govt.nz/assets/a8bba25fdf/new-zealand-tourism-forecasts-2019-2025.pdf.

10. Ibid.

11. FigureNZ. (February 17, 2020). *International Visitor Arrivals to New Zealand Year:1922–2019, Millions*. https://figure.nz/chart/3wdeCHBIP84nSEp7.

12. Stats NZ. (November 13, 2019). *Accommodation Survey: September 2019*. https://www.stats.govt.nz/information-releases/accommodation-survey-september-2019.

13. James Higham, Stephen Espiner, and Sabine Perry. (January 2019). "The Environmental Impacts of Tourism in Aotearoa New Zealand: A Spatio-Temporal Analysis." *Parliamentary Commissioner for the Environment*. https://www.pce.parliament.nz/media/196977/the-environmental-impacts-of-tourism-in-aotearoa-new-zealand-a-spatio-temporal-analysis.pdf.

14. Ibid.; Anne Buchmann. (2007). *In the Footsteps of the Fellowship—Understanding the Expectations and Experiences of Lord of the Rings Tourists on Guided Tours in New Zealand*. PhD thesis. Lincoln University. p. 21.

15. Ian Brodie. (2002). *The Lord of the Rings Location Guidebook*. Auckland: Harper Collins.

16. Anne Buchmann. (2007). *In the Footsteps of the Fellowship—Understanding the Expectations and Experiences of Lord of the Rings Tourists on Guided Tours in New Zealand*. PhD thesis. Lincoln University. p. 21.

17. Sue Beeton. (2005). *Film-Induced Tourism*. Clevedon, UK: Channel View Publications. p. 81.

18. Stats NZ. (October 21, 2019). *Subnational Population Estimates: At 30 June 2019 (Provisional)*. https://www.stats.govt.nz/information-releases/subnational-population-estimates-at-30-june-2019-provisional.

19. Ministry of Business, Innovation & Employment. (March 5, 2020). *Annual Tourism Spend Grouped by TA, RTO, Country of Origin and Product Category*. https://www.mbie.govt.nz/immigration-and-tourism/tourism-research-and-data/tourism-data-releases/monthly-regional-tourism-estimates/latest-update/annual-tourism-spend-grouped-by-ta-rto-country-of-origin-and-product-category/; Stats NZ. (November 13, 2019). *Accommodation Survey: September 2019*. https://www.stats.govt.nz/information-releases/accommodation-survey-september-2019.

20. Debbie Jamieson. (August 26, 2019). Queenstown and Wānaka Airport Expansion Strategy Grounded Again. *Stuff*. https://www.stuff.co.nz/business/ 115275863/queenstown-and-wnaka-airport-expansion-strategy-grounded-again.

21. Tourism Industry Aotearoa. (2019). *Mood of the Nation—November 2019*. https://tia.org.nz/resources-and-tools/insight/mood-of-the-nation/.

22. New Zealand Tourism Guide. (2020). https://www.tourism.net.nz/new-zealand/about-new-zealand/free-camping.html.

23. Jenna Sherman. (February 14, 2019). "Residents Enforcing Freedom Camping Limit." *Otago Daily Times.* Online source, available at: https://www.odt.co.nz/regions/north-otago/residents-enforcing-freedom-camping-limit.

24. Tourism Industry Aotearoa. (2019). *Mood of the Nation—November 2019*. https://tia.org.nz/resources-and-tools/insight/mood-of-the-nation/.

25. Ibid.; Parliamentary Commissioner for the Environment. (December 19, 2019). *Pristine, Popular . . . Imperilled? The Environmental Consequences of Projected Tourism Growth*. https://www.pce.parliament.nz/publications/pristine-popular-imperilled-the-environmental-consequences-of-projected-tourism-growth; Ministry of Business, Innovation & Employment and Department of Conservation. (May 2019). *New Zealand-Aotearoa Government Tourism Strategy*. https://www.mbie.govt.nz/dmsdocument/5482-2019-new-zealand-aotearoa-government-tourism-strategy-pdf.

26. James Higham, Stephen Espiner, and Sabine Perry. (January 2019). "The Environmental Impacts of Tourism in Aotearoa New Zealand: A Spatio-Temporal Analysis." *Parliamentary Commissioner for the Environment.* https://www.pce.parliament.nz/media/196977/the-environmental-impacts-of-tourism-in-aotearoa-new-zealand-a-spatio-temporal-analysis.pdf.

27. Ibid.

28. Ibid.

29. Ibid.

30. Kelvin Davis and Eugenie Sage. (September 27, 2018). "Visitor Levy to Raise $80 Million to Future Proof Tourism." New Zealand Government. https://www.beehive.govt.nz/release/visitor-levy-raise-80-million-future-proof-tourism.

31. Amanda Cropp. (August 11, 2019). "A Much Debated Visitor Levy Is Raising Millions, but How Should We Spend It?" *Stuff.* https://www.stuff.co.nz/business/114794528/a-much-debated-visitor-levy-is-raising-millions-but-how-should-we-spend-it. 11 August 2019.

32. Ministry of Business, Innovation & Employment. (2020). *Projects Funded by the IVL*. https://www.mbie.govt.nz/immigration-and-tourism/tourism/tourism-funding/international-visitor-conservation-and-tourism-levy/projects-funded-by-the-ivl/.

33. Tourism Industry Association. (2020). *Tourism Sustainability Commitment, 2020*. https://sustainabletourism.nz/about-us/about-sustainable-tourism-commitment/components/.

34. Rotorua Daily Post. (February 25, 2019). "Tourism Businesses 'Committed to Sustainability' to Be Pointed Out at TRENZ." *NZ Herald.* https://www.nzherald.co.nz/rotorua-daily-post/business/news/article.cfm?c_id=1503434&objectid=12207148.

35. Stuff. (June 7, 2017). "Chinese 'Influencers' Promote Wellington, Tasman, Nelson and Marlborough in New Campaign." https://www.stuff.co.nz/travel/93383593/chinese-influencers-promote-wellington-tasman-nelson-and-marlborough-in-newcampaign.

36. David Smol, Jenn Bestwick, and Kevin Malloy. (September 30, 2019). *Optimising Tourism New Zealand's Future Role and Contribution to New Zealand.* New Zealand: Ministry of Business, Innovation and Employment.

37. Ibid.

38. Liz Carlson. (November 26, 2018). "Our Tiaki Promise Protects New Zealand for Future Generations." *Stuff.* https://www.stuff.co.nz/travel/kiwi-traveller/108812349/our-tiaki-promise-protects-new-zealand-for-future-generations; *Wilderness.* (November 20, 2019). "Behaviour Pledge for Te Araroa Trail." https://www.wildernessmag.co.nz/behaviour-pledge-for-te-araroa-trail/.

39. Air New Zealand. (May 1, 2019). "Air New Zealand Launches 'Tiaki—Care for New Zealand' Inflight Entertainment Channel." https://www.airnewzealand.com/press-release-2019-airnz-launches-tiaki-care-for-new-zealand-inflight-entertainment-channel.

40. James Higham, Stephen Espiner, and Sabine Perry. (January, 2019). "The Environmental Impacts of Tourism in Aotearoa New Zealand: A Spatio-Temporal Analysis." *Parliamentary Commissioner for the Environment.* https://www.pce.parliament.nz/media/196977/the-environmental-impacts-of-tourism-in-aotearoa-new-zealand-a-spatio-temporal-analysis.pdf.

41. David Smol, Jenn Bestwick, and Kevin Malloy. (September 30, 2019). *Optimising Tourism New Zealand's Future Role and Contribution to New Zealand.* New Zealand Ministry of Business, Innovation and Employment.

42. Patrick Timm. (20180. "What Does Freedom Camping in New Zealand Mean? Responsible Camping and How to Do It!" The Vertical Adevnturer. https://www.theverticaladventurer.com/single-post/What-Does-Freedom-Camping-in-New-Zealand-Mean-Responsible-Camping-and-How-to-Do-It.

43. Whale Watch. (2020). *Conservation Policy/Tiaki Promise.* https://www.whalewatch.co.nz/our-place/conservation-policy/.

44. James Higham, Stephen Espiner, and Sabine Perry. (January 2019). "The Environmental Impacts of Tourism in Aotearoa New Zealand: A Spatio-Temporal Analysis." *Parliamentary Commissioner for the Environment.* https://www.pce.parliament.nz/media/196977/the-environmental-impacts-of-tourism-in-aotearoa-new-zealand-a-spatio-temporal-analysis.pdf.

Chapter 6.6

Trolltunga, Norway

By Christina Beckmann and Kaitie Worobec

During the Ice Age, a wedge of rock appeared in western Norway, jutting out high over Lake Ringedalsvatnet. Known as Trolltunga in Norwegian, or "Troll's Tongue," this strip of gneiss is what remains of a mountain when glacial ice froze in its crevices and eventually broke it into chunks, leaving behind a jutting promontory. Standing atop this rock formation over a 2,296-foot (700-meter) sheer drop-off, and the corresponding photo opportunity, is the reward for adventurous travelers who hike for hours to reach the cliff face.

Up until 2010, fewer than eight hundred people per year ascended the mostly rock trail to the cliff's edge. "Back in 2010, we rarely saw more than 10 people at a time on the trail," commented Åse Marie Evjen, manager at Trolltunga AS, the nonprofit organization responsible for managing the area.[1] In a 2019 article for *The Telegraph* entitled "Liked to Death—Five Places Instagram Put on the Map, Then Ruined," Greg Dickinson noted the difference a decade can make: "[In 2009,] there were no queues, no hashtags. 4G did not exist yet and, even if it did, you wouldn't be able to access it in this remote part of the world. It was just a beautiful view, overlooking a lake."[2]

Causes

Trolltunga's popularity can be attributed to the unique photo opportunity provided by the rock ledge in combination with the incredible backdrop of the Norwegian fjord landscape.[3] In 2010, images of Trolltunga were used to promote tourism in the Hardanger region as part of an effort to revitalize the area's economy. Between 2012 to 2014, Trolltunga received the following highlights: a review on Tripadvisor with the title "You want this picture on your [Facebook] wall," a mention by *HuffPost* as "The scariest Instagram spot on earth," and top ranking on *BuzzFeed*'s list of "The 17 most stunning places in the world to take a selfie."[4] In less than a decade, this wild, exposed cliff transformed from a relatively unknown place into one of Norway's most famous tourist landmarks.

Visitation to Trolltunga reached its peak in 2018 with 87,000 visitors, compared to fewer than 1,000 visitors in 2010. An Instagram search in August 2019 for #trolltunga returned 153,086 posts, an increase of more than 34,000 since January 2019 (118,837 posts).[5] Although that doesn't mean that an additional 34,249 people visited since January, it does indicate the rapid circulation of images of Trolltunga and the powerful influence of social media in raising the profile of destinations.

Trolltunga is not an easy place to get to. From the main trailhead outside the village of Skjeggedal, the round-trip hike is 17.4 miles (28 kilometers) with an ascent of more than 2,600 feet (800 meters).[6] The hike generally takes eight to twelve hours to complete, and according to local guide Erlend Indrearne, at least one or two visitors in every group have to turn around before reaching the viewpoint.[7] Many are inexperienced hikers who don't understand the intensity of high mountain weather in Norway and ignore warnings not to undertake the hike in cold and stormy conditions.

Impacts of Overtourism

Overtourism has been a strain on Trolltunga's sensitive landscape, threatens visitor safety, and has had a limited economic impact on the surrounding communities. Although the impacts caused by one hiker looking for a selfie may seem negligible, the cumulative impacts from tens of thousands of visitors are significant.

Environmental Impacts

Exponential growth in visitor numbers, combined with the short summer season for trail work and limited staff and volunteer resources, has resulted in erosion of the trail. According to Haaken Michael Christensen, senior advisor to adventure tourism with Innovation Norway, "What we saw is when the volume of people rose quickly, trails were expanded in width, so two or three parallel trails were running alongside each other."[8] In addition, used toilet paper, abandoned tents, food wrappers, and plastic bottles have all been found littered on the trail to Trolltunga.[9] There are currently no toilet facilities or garbage cans along the trail.

Safety

Despite the narrow rock ledge and strong wind, some visitors jump in the air or do yoga poses or even hand stands—all in hopes of capturing the perfect shot for social media.[10] One young woman died in 2015 after falling more than 650 feet (200 meters) from Trolltunga.[11] In addition, some hikers must wait for over an hour to have their photo session on the cliff due to the congestion, preventing them from getting back before dark. There were forty search and rescue missions between July and September 2016, but only a very small proportion of emergency calls are injury-related; the vast majority come from cold, lost, and tired hikers.[12] This situation puts a strain on local emergency services, including Red Cross volunteer mountain rangers in the nearby towns of Tyssedal and Odda. Many rescues are by helicopter, and hikers are not responsible for the costs accrued.

Limited Economic Impact

The influx of visitors has provided a boost to local businesses in surrounding towns, but the exponential increase in visitation has not been proportional to the economic impact. Visitors do not pay an entrance fee to hike Trolltunga, and the only revenue generated at the site is through parking fees. Furthermore, the vast majority of hikers choose to go on their own, so there are limited revenue opportunities for guide companies.[13]

Solutions

Trolltunga is not a national park or World Heritage Site, although it is surrounded by protected areas. This location makes for a unique tourism management structure in which a nonprofit organization established by the local municipalities in the Odda district of Vestland County (formerly Hordaland County), called Trolltunga AS, has, in collaboration with Vestland County Council and others, taken the lead. The organization's aim is to ensure that "every hiker has a safe, spectacular outdoor adventure at Trolltunga" while safeguarding the fragile mountain environment.[14] Trolltunga AS's main responsibilities include constructing hiking trails, updating visitor information boards, maintaining toilets and emergency shelters, managing the parking lots and organizing the shuttle service to and from the trailhead, providing mountain ranger services, and maintaining the 3.7-mile (6-kilometer) Skjeggedalsvegen road.[15] It also publishes updated information about trail conditions and weather on its website, with clear warnings about when to avoid the hike or to proceed only with an expert guide.[16]

After Trolltunga became a victim of its own popularity, Trolltunga AS recognized that it was necessary to better regulate the hike and ensure visitor safety. The group's main objectives were to define the trail as clearly as possible and to raise awareness among hikers to ensure they have a safer, more enjoyable hiking experience.

The Norwegian concept of *fjellvettreglene*, better known as Norway's mountain code, also played into Trolltunga AS's management approach. *Fjellvettreglene* encourages a respectful relationship with nature and inspires careful planning for hiking trips, including bringing the necessary equipment, seeking shelter if necessary, and feeling no shame in turning around.[17] As hiking guide Erlend Indrearne explained, "*Fjellvettreglene* taught us nature doesn't care about our egos. We should show as much respect and take as much caution as possible. For people who aren't experienced hikers, [the Trolltunga hike] is considered extreme. Not many tourists realize that."[18]

Addressing the situation at Trolltunga and shifting visitor behavior involved the implementation of several management measures by Trolltunga AS in collaboration with Vestland County Council and Visit Norway (the national tourism board). Having a dedicated team in place at Trolltunga AS allows for the coordinated management of the area as a whole, including maintaining the trail and surrounding landscape, having consistent contact with tour operators and guides, maintaining relationships with cabin owners in the area, and liaising with local businesses to ensure that the community benefits from tourism. Trolltunga AS also collaborates with Visit Hardangerfjord and Miljødirektoratet (Norwegian Environment Agency).[19]

Regulation of Hiking Season

Trolltunga AS put new guidelines in place recommending hiking with a guide between October and May, due to the challenging conditions and sudden weather changes.[20] It markets this period as an opportunity to hike to Trolltunga without the crowds—but with a guide. Between June and September, experienced hikers are permitted to go on their own, but Trolltunga AS still promotes and encourages guided hiking.

Visitor Education

Trolltunga AS erected improved informational signs at the beginning of the trail, reinforcing messages such as having suitable clothing and footwear and bringing garbage back down to the parking lot. In addition, Vestland County Council advised Visit Norway on key messages for their *Safety First* video, which features a guide who explains the importance of suitable clothing, sticking to the trail, and using local guide services.[21] The video encourages hikers to stay on the trail and implement other measures to limit erosion and damage to the wider sensitive landscape. The tourist information center, hotels, and taxis also aligned the information they distributed to communicate a consistent message to visitors.

Mountain Rangers

In 2017, Trolltunga AS tested the use of on-site rangers who stayed in the mountains around the clock during the summer season. Their role was to assist people who were cold and tired and to talk to those who seemed unfit and poorly prepared, encouraging them if necessary to turn back.[22] The mountain ranger initiative was expanded in 2018 for an extended period beyond the summer months.

The mountain rangers also provide detailed information about the conditions for the hike, including the weather forecast and what to pack, on the Trolltunga Trail Information Facebook page.[23] The presence of the mountain rangers also means that fewer people litter, and hikers are now given garbage bags to encourage them to take their waste back to the trailhead.

#BeSafie Marketing Campaign

In 2016, Visit Norway developed a national campaign to address the dangerous behaviors exhibited by visitors who were unprepared to hike and willing to risk their safety for a perfect selfie.[24] The campaign was named #BeSafie. One element included a series of installations depicting large images of cliff edges and wild animals, allowing people to take dramatic selfies in a safe environment. The campaign also outlined the nine guidelines of the Norwegian Mountain Code, promoting safe trekking.

Destination Marketing

Trolltunga AS and Visit Norway have made efforts to extend visitors' length of stay and encourage broader exploration of the surrounding area, strengthening the economic impact from visitation for local businesses. Trolltunga AS's goal is for visitors to stay a minimum of two nights and three days and experience the local culinary traditions as well as other natural attractions near Odda such as glacier trekking and two nearby national parks. These alternative tourism products beyond the Trolltunga hike are promoted on their website and on new signage along the hiking trail.[25]

ASCENT Project

Vestland County Council also participated in the ASCENT project, an international initiative examining locations in five Northern European Upland sites (Ireland, Northern Ireland, Finland, Iceland,

and Norway) where increased visitation has led to challenges for sensitive landscapes.[26] Although the Vestland County Council's plans and jurisdiction cover a broader geography beyond Trolltunga, including some nearby national parks, this site is at the forefront of the aims of this project. In 2019, the council released a Visitor Management Strategy for Trolltunga under the ASCENT project.[27] One goal of this report was to describe the site's progress in aligning with the criteria for National Tourist Trails, which require monitoring and improvement around infrastructure, visitor safety, education, and conservation. The project has helped the local council and other organizations better understand how to protect and manage the area by learning from others with similar challenges.

Trail Maintenance

The dramatic increase in visitors to Trolltunga has meant that Trolltunga AS and Vestland County Council have had to divert considerable funds and human resources to maintaining the trail.[28] A considerable increase in the parking fee has been key to financing all the trail work, facilities at the trailhead, and the mountain guards.[29] Information on how the parking fee revenue is allocated is displayed next to the parking ticket machine, and visitors are responding favorably despite the relatively high price.

Transportation

Another success was the implementation of tiered pricing for the three parking lots. The closer the parking lot is to the trail, the higher the parking fee. Shuttle buses operate from the town center at a lower cost, and a coordinated schedule allows shuttle buses to operate despite a very narrow access road where two buses cannot pass each other.

Conclusion

As a result of these measures, particularly enhanced visitor education and the increased presence of the mountain rangers, the number of annual search and rescue missions declined from forty-two in 2016 to twenty in 2019.[30] Tourists are noticeably better equipped and have a better understanding of the demands of the hike. Increased trail maintenance has improved the environmental impacts of tourism and also enabled hikers to get to Trolltunga and back in less time and with less risk of injury.

An indirect benefit of the efforts to curb overtourism has been the shift in media attention. Whereas earlier media coverage on Trolltunga focused on photo opportunities at Trolltunga, the ASCENT project in particular has succeeded in increasing attention to issues of safety and sustainably managing natural areas,[31] as witnessed by a 2017 article titled, "Why Norway Is Teaching Travellers to Travel."[32]

The rapid growth in visitor numbers to Trolltunga has created both challenges and opportunities. Trolltunga AS, Visit Norway, and Vestland County Council have worked hard to improve hiker safety and manage the area responsibly, and it is clear that the measures implemented have had a positive impact. The challenge that remains is to preserve the natural landscape for future generations while encouraging visitors to engage with this spectacular place beyond a photo shot.

Notes

1. Marta Rongved Dixon. (July 25, 2019). Vestland County Council. Personal communication with Kaitie Worobec.

2. Greg Dickinson. (January 23, 2019). "Liked to Death—Five Places That Instagram Put on the Map, Then Ruined." *The Telegraph*. https://www.telegraph.co.uk/travel/news/most-instagrammed-destinations/

3. Michelangelo Stanislav Magasic. (October 2018). "Connected Tourists: From Sightseeing to Sitesharing." PhD thesis, Curtin University. https://espace.curtin.edu.au/bitstream/handle/20.500.11937/75446/Magasic%20M%202018.pdf?sequence.

4. Ibid.

5. Instagram. "#trolltunga." https://www.instagram.com/explore/tags/trolltunga/.

6. Fjord Norway. (2020). "Trolltunga." https://www.fjordnorway.com/things-to-do/trolltunga-p958013.

7. Shannon Dell. (October 16, 2017). "Why Norway Is Teaching Travelers to Travel." BBC. http://www.bbc.com/travel/story/20171015-why-norway-is-teaching-travellers-to-travel.

8. Stephanie Pearson. (September 3, 2019). "Norway's Bold Plan to Tackle Overtourism." *Outside*. https://www.outsideonline.com/2401446/norway-adventure-travel-overtourism.

9. Ibid.

10. Greg Dickinson. (January 23, 2019). "Liked to Death—Five Places That Insta-

gram Put on the Map, Then Ruined." *The Telegraph.* https://www.telegraph.co.uk/travel/news/most-instagrammed-destinations/.

11. Ove Sjøstrøm, Marianne Nilsen, Cathrine Krane Hansen, and Sigrid Haaland. (September 5, 2015). "Kvinne i 20-årene døde etter fall fra Trolltunga." *Aftenposten.* https://www.aftenposten.no/norge/i/xbXp/kvinne-i-20-aarene-doede-etter-fall-fra-trolltunga.

12. Marta Rongved Dixon. (July 25, 2019). Hordaland County Council. Personal communication with Kaitie Worobec.

13. Åse Marie Evjen. (February 28, 2020). Trolltunga AS. Personal communication with Kaitie Worobec.

14. Trolltunga AS. (2020). https://trolltunga.com/.

15. Ibid.

16. Åse Marie Evjen. (February 28, 2020). Trolltunga AS. Personal communication with Kaitie Worobec.

17. Shannon Dell. (October 16, 2017). "Why Norway Is Teaching Travelers to Travel." BBC. http://www.bbc.com/travel/story/20171015-why-norway-is-teaching-travellers-to-travel.

18. Ibid.

19. Åse Marie Evjen. (February 19, 2020). Trolltunga AS. Personal communication with Kaitie Worobec.

20. Trolltunga. (2020). https://trolltunga.com/plan-your-trip/information-about-trolltunga/.

21. ASCENT Project. (2019). *ASCENT to Summit: A Journey to Sustainable Management in European Uplands & Natural Environments—Conference Report 2019.* https://www.ascent-project.eu/wp-content/uploads/2019/07/ASCENT-to-Summit-Conference-Report-2019.pdf.

22. Marta Rongved Dixon. (July 25, 2019). Vestland County Council. Personal communication with Kaitie Worobec.

23. Facebook. "Trolltunga Trail Information." https://www.facebook.com/Trolltunga.trail.information/

24. Molly Jones. (June 21, 2016). "Visit Norway Encourages Safe Selfies Through #BeSafie Campaign." *Norwegian American.* https://www.norwegianamerican.com/visit-norway-encourages-safe-selfies-through-besafie-campaign/.

25. Trolltunga AS. (2020). *Surroundings.* https://trolltunga.com/surroundings/.

26. ASCENT Project. (2020). *Trolltunga.* https://www.ascent-project.eu/trolltunga/.

27. Astrid Rongen. (September 7, 2019). "Besoksstrategi for Trolltunga." *Norconsult.* https://www.ascent-project.eu/wp-content/uploads/2019/11/DESIGN_HCC-T3.1-Visitor-strategy-for-Trolltunga_NO-09.august-2019-.pdf.

28. Marta Rongved Dixon. (July 25, 2019). Vestland County Council. Personal communication with Kaitie Worobec.

29. Ibid.

30. Åse Marie Evjen. (February 28, 2020). Trolltunga AS. Personal communication with Kaitie Worobec.

31. Ibid.

32. Shannon Dell. (October 16, 2017). "Why Norway Is Teaching Travelers to Travel." BBC. http://www.bbc.com/travel/story/20171015-why-norway-is-teaching-travellers-to-travel.

Chapter 7

From Overtourism to No Tourism: Finding a New Normal

By Martha Honey

For the first two decades of the twenty-first century, the travel industry was on a roll, and in December 2019, tourism's continued growth appeared almost unstoppable. As the world's largest service industry and largest employer, tourism had outpaced the global gross domestic product for nine straight years and was growing year after year even faster than the UN World Tourism Organization had forecast. Within this context, overtourism had emerged as the concern de jure facing many destinations and was being grappled with by governments, international agencies, civic organizations, and many in the tourism industry.

In early 2020, however, tourism met its match as COVID-19 defied and reversed all the trends in tourism. Within a few short weeks, virtually all the destinations examined in this book went from overtourism to no tourism. As Jonathan Tourtellot reflected from self-isolation in his northern Virginia mountainside home, "This is one hell of a way to cure overtourism. Not at all what those of us working on the problem had in mind. The coronavirus has turned the destination-tourism relationship on its head, from 'over' to 'under' in the blink of an eye and the bark of a dry cough."[1]

COVID-19 brought tourism—and the world as we knew it—to a

standstill, yet this standstill will not last forever. Indeed, at the time of this writing in 2020, many parks and beaches have reopened and are beginning, once again, to experience the stresses of overcrowding.

As the world begins to emerge from the pandemic nightmare, the "new normal" is likely to once again include overtourism. There are, however, also signs that the pandemic pause has stimulated a new commitment to safe, healthy, and sustainable travel, including preferences for small group, outdoors, and crowd-free activities that could permanently transform the travel industry. Amid the horrors of COVID-19 are glimmers of hope that tourism will not simply return to its old habits, but will emerge with a stronger determination to embrace principles such as equity, conservation, community, collaboration, and good governance. This final chapter summarizes the lessons learned from overtourism and explores how those lessons, as well as ones arising from the COVID-19 pandemic, can make tourism healthier and more sustainable in the postpandemic world.

National Parks: Overtourism Solutions

Of the different types of destinations examined in this book, national parks emerge at the top, and in the United States, the National Park Service (NPS) emerges as one of the best positioned to effectively address overtourism, including in the postpandemic era. Like other types of destinations, the impacts of overtourism have been uneven, with, as Richard Bangs put it in chapter 3.1, "marquee parks," including Acadia, Yosemite, Zion, Glacier, Yellowstone, Grand Canyon, and Rocky Mountain, most likely to experience "greenlock," or gridlock in natural surroundings.

In 2019, total visitors to the US national parks topped 327 million, up from 287 million visits in 1999, an increase of 40 million in two decades.[2] Although the reasons for steep jumps in visitor arrivals to specific parks are not always clear, they include close proximity to urban areas, increased travel by Asia's new middle class, and expanding social media, particularly the explosive growth of selfies, bucket lists, and Instagram posts that promote specific parks. Park visitation across the national system has been boosted as well by the decade-long period of economic prosperity in the United States and the growing number of retired baby boomers with the time and resources to travel. Many were crafting do-it-yourself vacations, often by road, which also contributed to greater visitor flows to the parks.

As visitor numbers grew, so did the management challenges. In fact, the NPS undertook its first experiment in dealing with overtourism—in the form of too many uncontrolled rafters on the Colorado River—in the early 1970s, some four decades before the term had entered our public discourse. The park service's multifaceted solution for the Colorado River included an allotment system to limit the number of rafters and new regulations for guiding, camping, cooking, and waste removal to ensure that the natural environment remained pristine. It worked, and this allotment system has been in place, with periodic tweaks and updates, ever since.

Today, the NPS, now more than a century old, has a well-developed process for dealing with overtourism at both the national level and within individual parks. In seeking solutions to overtourism, Bangs explained that the NPS uses a planning framework known as visitor use management (VUM) that encompasses three broad categories of components: education, engineering, and enforcement (see chapter 3.1). Most parks employ a mix of approaches based on these three categories. The first, educational programs and tools, are designed to provide travelers with trip-planning information via websites, park newspapers, signage, and digital platforms. Montana's Glacier National Park, for example, has a no-nonsense "Crowds" page with a pristine Instagram photo side by side with a "reality" photo of an overlook filled with visitors. It includes a list of where and when to expect crowds and offers tips for avoiding overtourism, including visiting less crowded nearby parks. The second VUM approach, engineering techniques, includes new, improved, or redesigned roads, visitor centers, trails, parking lots, park entrances, and shuttle services. Yellowstone, for instance, is reducing the bottleneck at its main entrance by adding a fast lane, modeled after E-ZPass, for those with park passes. The third, enforcement solutions, typically involves tools such as reservation systems, managed access, permitting, and visitor caps, all designed to curb congestion.

One lesson the NPS has drawn from its experiences in dealing with overtourism is that although too much visitation can certainly degrade resources and visitor experiences, setting a fixed visitor capacity number—what is typically referred to as carrying capacity—is not a panacea. Instead, an analysis by the government's Visitor Use Management Council concludes that "managing use levels to a visitor capacity is one of many strategies for dealing with visitor use management issues. Changing visitor behavior, modifying where and when use occurs, or building facilities that can accommodate heavy use are [all] critical strategies for protecting desired conditions."[3]

In addition to its clear management systems, the park service's ability to effectively address overtourism has been enhanced by two other realities: a largely guaranteed funding source and wide public support. About 90 percent of the NPS annual budget comes from Congress, and although at times it is inadequate, this funding has given the US national parks more economic stability than other types of destinations; a World Heritage Site (WHS), for example, may receive only limited support from UNESCO and is often heavily dependent on raising resources through tourism. The other reality is that US national parks, even in the face of overtourism, have remained enormously popular with the public, and user satisfaction is very high. In its 2019 visitor survey, for instance, some 98 percent of nearly sixty thousand respondents rated the quality of facilities, services, and recreational opportunities in national parks as "very good" or "good," whereas only 2 percent said they were "average."[4] This positive outlook came despite, by the end of 2019, record numbers of visitors—more than three hundred million—for the fifth straight year in a row.

National Parks: Postpandemic

In reflecting on the national parks' experience in dealing with overtourism, Donald Leadbetter, NPS tourism program manager, said that the NPS, with its "tool kit that has lots of different tools within it," has been "leading the way [in the United States] in finding ways to address overtourism."[5] These innovative efforts, however, abruptly stopped in March 2020 with the arrival of the intertwined catastrophes, the coronavirus pandemic and the economic collapse. Beginning in mid- to late March, parks across the United States began closing, and, in turn, wildlife—black bears, deer, bobcats, and pronghorn antelope—began congregating in areas of parks normally teeming with visitors.[6] "With threats like vehicle collisions or being harassed for a selfie largely eliminated, the coyotes are trotting through Camp Curry hunting mice, and the black bears [are] wandering around a mostly empty Yosemite Village," said Beth Pratt, National Wildlife Federation's regional executive director for California.[7] This Garden of Eden moment proved short-lived, however. In late spring and early summer, as the national lockdown in the United States began to gradually lift, parks and public lands at the local, state, and national levels began to reopen.

It quickly became clear that US parks offer many qualities that the sheltered-in-place public seeks in the brave new world of postpandemic travel. Health and safety are top of mind. In addition, surveys and media reports revealed that the newly liberated public prioritizes US over international travel; drive-to destinations over cruises, long-haul flights, or indoors museums, performances, and events; outings in small groups with family and friends; travel experiences that are conducive to social distancing; and outdoor recreation with fresh air and pristine natural surroundings.[8] According to a June 2020 survey from Destination Analysts, "Road trips, staying at a beach resort, and visiting national and other parks are among the highest rated relaxing travel experiences."[9] The parks' wide popularity comes with a certain irony: although the US national parks succeeded in addressing overtourism quite effectively prepandemic, now these same parks, along with beaches, are the destinations most likely to see a rapid postpandemic return of overtourism.

Fortunately, the NPS quickly recognized the renewed threats of overtourism. Even without a return of international travelers, who in recent years have made up about 20 percent of US national park visitors, the NPS braced for a flood of domestic visitors arriving by road. "In managing the post-pandemic era," the NPS tourism chief Leadbetter explained, "We're using the same tools and assessment processes of Education, Engineering, Enforcement" used prepandemic to address overtourism. "The work of the last few years in developing these framework tools and guidelines has better enabled us to move quickly and adaptively now. Before it was just dealing with overtourism. Now it's using these tools to also protect public health. The framework has not changed, but now we have one more management objective."[10] In addition, in June 2020, the US Senate, in a rare bipartisan move, passed a major land conservation bill giving the national parks $9.5 billion in additional funding through 2025.[11]

Although there is a general management framework, the NPS also offers, said Leadbetter, "a lot of leeway for park by park decision making because situations are so different."[12] In early June 2020, for instance, Colorado's Rocky Mountain National Park adopted for the first time what Leadbetter described as a "controversial" but temporary reservation system designed to ensure social distancing. It includes a requirement that cars must prepurchase $25 entry passes good for only one day, timed entry in two-hour blocks, and an in-park shuttle that operates at around 20 percent capacity. Overall, the park decided initially to permit entry equal to just 60 percent of the park's maximum parking capacity each day, and some areas and activities remained closed to the public.[13]

By contrast, Yellowstone National Park, which sits primarily in Wyoming with smaller pieces in Montana and Idaho, developed a different strategy under the banner "Be safe. Be flexible. Monitor. Adjust." It does not include advance reservations. Rather, beginning in May 2020, Yellowstone issued a three-phase "COVID-19 Reopening Plan" beginning with opening essential services, including roads, restrooms, trails, and medical clinics, and concluding with opening "as health conditions allow" all services, including hotels and full-service restaurants, ranger programs tours, and commercial bus tours.[14]

The reopening plan was linked closely to the communities around the park, and it grew out of "the opinions and recommendations provided by surrounding county health officers, leadership officials, and state governors. These professionals understand the transboundary nature of visitors traveling to Yellowstone and are best positioned to understand local and regional health concerns."[15] As park visitation ramped up, however, some gateway communities on the edge of Yellowstone and other national parks voiced real concerns that the flood of visitors would also bring the novel coronavirus to rural communities where infections had been low. These communities often lack sufficient masks, gloves, and other safety equipment or medical facilities to handle an upsurge.[16]

Many of the practices that the NPS is putting in place to help prevent a resurgence in COVID-19 are also sensible steps that will help keep overtourism at bay: fewer visitors, more distancing, improved hygiene, and thoughtful planning that puts stakeholder voices at the forefront. In addition, the lockdown itself, when the parks were closed and the animals roamed free, offers lessons for more long-term sustainable practices. As the *Los Angeles Times* wrote in May 2020, "National parks could consider rotating more visitor-free resting times for nature, or limiting entry, to keep the environment from being loved to death."[17] This strategy could also help to keep overtourism in check.

Beaches and Coastlines: Overtourism Solutions

For decades, beaches, islands, and coastlines have been the most rapidly bought up and built up pieces of tourism real estate. This upsurge has led to a range of problems, including marginalization of established coastal communities; competition for fresh water, land, and other resources; rising real estate prices and cost of living; mounting traffic, trash, and pollution; and overcrowding, which often morphs into overtourism. Such problems, in turn, lead to a decline in the quality of life for locals and in the quality of the vacation experience for visitors.

Overtourism in beach destinations, perhaps more acutely than other types of destinations, is intertwined with another major calamity, climate change. As Andrea Sachs noted in chapter 5.1, "Seaside towns are struggling with sand erosion, the loss of coral reefs, a plunge in marine life, algae blooms, and beached animals. Although climate change is a primary cause, overtourism is no friend to the fragile ecosystem either. In fact, among overtouristed areas, the combined forces of these perils seem to be hitting beachfront destinations the hardest." The convergence of climate change and overtourism has enhanced the complexities of seeking solutions.

To date, there are few international, national, or regional templates or sets of procedures for visitor management on beaches such as have been developed for national parks and protected areas. Instead, there were, through 2019, a range of experiments to address overtourism at specific beach sites; if the method appeared to work, it might be duplicated elsewhere. Some of the most extreme methods used in recent years have been beach closures to stop coral and beach erosion and let the ecosystem rest and recover. As Sachs explained in chapter 5.1, full closure has been used most notably, beginning in 2018, at two famous and overly popular beach destinations in Asia: Maya Bay in Thailand and Boracay Island in the Philippines. The once secluded and pristine Maya Bay, which began attracting five thousand visitors a day after appearing in Leonardo DiCaprio's film *The Beach*, is now to remain closed until 2021. Beyond that, it may cap visitors at twelve hundred a day. Boracay Island was closed for six months, during which time illegal buildings were demolished, a new waste infrastructure was installed, and a long list of "can no longer do" restrictions were put in place. Since reopening, Boracay has become an exemplar of sustainable beach management. Because total beach closure is unpopular with both tourists and local communities, however, these extreme measures are likely to be reserved only for the most poorly managed beaches on the brink of environmental collapse.

More common are programs to limit the number of beachgoers and better manage their behavior. For instance, as both Sachs and Frank Haas describe (see chapters 5.1 and 5.4) , Hanauma Bay Nature Preserve, outside Waikiki in Hawaii, has implemented a program to curb the crowds by halting entrance once the three-hundred-spot parking lot is full. Visitors must also watch a film about the bay's ecology before proceeding to the beach. The preserve's program to control entrance numbers and educate visitors has been hailed a success and is being viewed as a model for other beaches in Hawaii.

On Scotland's Isle of Skye, which Sachs noted was being overrun by "volume tourists"—short-stay, selfie-snapping visitors jamming, in particular, the North Point at sunset—authorities have set out to change visitor behavior. They began to promote other trails and scenic lookouts, encouraging travelers to visit during off-peak months, to slow down, and to stay longer, all under the directive, "Stay Longer. See Less. Experience More."

A number of coastal, beach, and island destinations have used improved stewardship to build a "culture" that controls overtourism. This technique takes many forms. For instance, on Catalina Island, just off the coast of Southern California, where both space and fresh water are limited, the chamber of commerce has created "Care for Catalina," a widely distributed guide targeting the million-plus visitors a year. As Sachs put it, the guide "preaches the holy commandments advanced by many overtouristed destinations"—avoid peak travel times and get off the beaten track—as well as good behavior tips on water and fuel conservation, plastic reduction, trail usage, wildlife protection, and cell phone etiquette.

On the other side of the United States, on Cape Cod in Massachusetts, another stewardship program, CARE for the Cape and Islands, engages tourists, businesses, local organizations, and residents in raising funds and awareness for environmental stewardship programs. In oceanfront destinations around the world are other examples of good stewardship programs, often started and led by tourism businesses themselves. In Jamaica, for instance, boutique hotelier Chris Blackwell has created the Oracabessa Foundation, which runs a successful fish sanctuary and supports a range of community projects financed through guest donations. On a grander scale, in the Dominican Republic, the large international resort corporation Grupo Punta Cana runs a foundation that supports a range of social and environmental programs, including coral replanting, a private ecological park, a huge recycling facility, and a high school that offers training and jobs to local students. Grupo Punta Cana has also overseen construction of one of the first environmentally sensitive airports, "green" golf courses, and low-rise hotels designed to control density and protect the environment.

A final category of reforms to address overtourism in beaches, coastlines, and islands is better management by tourism organizations and government agencies. In recent years, there have been a range of initiatives to broaden the role of destination marketing organizations (DMOs) from their traditional laser focus on increasing tourism arrivals (marketing) to a more holistic focus that balances the needs of visitors and residents through sustainable master planning for the destination (management). As Jonathan Tourtellot elaborated in chapter 6.1, "Ever so slowly, the DMOs' pure focus on marketing is beginning to change."

In Hawaii, for instance, the Hawai'i Tourism Authority (HTA), the island's DMO, is striving to expand its mandate beyond marketing. Since 2019, the HTA has defined its four areas of focus as natural resources, Hawaiian culture, community, and brand marketing and has tied its success not simply to increased arrivals but to several key performance indicators. As HTA director of communications Marisa Yamane explained to Sachs, "HTA has shifted its priorities away from increasing visitor arrivals and is focusing its efforts on enhancing the quality of life for residents and the visitor experience." She added, "The measures of success are visitor satisfaction, resident sentiment, per person per day spend, and total visitor expenditures."[18] This approach works to address overtourism, as well as other tourism-related problems, through the all-important dual framework of good management and sustainable practices.

Beaches and Coastlines: Postpandemic

As with national parks, the coronavirus ended the overtourism crisis on beaches and islands around the world almost overnight. With the crowds gone, wildlife quickly returned for a brief period. Endangered species of sea turtles, for instance, quickly reclaimed beaches for nesting. Along Brazil's northeast coast, about one hundred hawksbill turtles were spotted hatching on beaches after locals were forbidden to visit, while in Juno Beach, Florida, scores of undisturbed loggerhead nests were visible on the beaches, and the turtles were spotted mating closer to shore in waters once dominated by boaters.[19] Around Miami, no people and boats also meant less trash, including plastics, less noise, and fewer fuel leaks. Biscayne Bay, at a crucial tipping point of losing its sea grass ecosystem, was suddenly beaming with life, including pods of dolphins and two rare, highly endangered smalltooth sawfish. As videographer Mike Ruiz, who captured these sea creatures on camera, marveled, "The ocean was crystal clear, neon blue; you could see right through the bottom of the ocean. I've never seen the water that blue in the years that we've been filming in South Florida."[20]

Keeping a lockdown on beaches proved difficult, though, especially as temperatures rose and summer approached. In late April, California, a state that had ordered one of the earliest lockdowns in the United States, opened a handful of beaches along its southern coast—with deeply disturbing results. Tens of thousands of people packed the beaches over the next weekend, abandoning physical distancing and other protocols and reigniting fears that large crowds in public spaces could reverse progress on containing COVID-19. "This virus doesn't go home because it's a beautiful, sunny day along our coast," warned California governor Gavin Newsom.[21]

There were similar worrying scenes in other US states, including Florida, Texas, and South Carolina, where there were little to no restrictions on beachgoing.[22] And even where beaches were opening with a range of restrictions, most tourism businesses remained shuttered, dealing devastating blows to tourism-dependent beach communities—and adding more pressure to open up in time for the summer holiday season.

Going forward, beaches in the United States, like national parks, will likely become even more popular travel choices, but unlike its parks system, the United States has no well-tested national-level framework for managing beaches and controlling crowds. Although undoubtedly states and beach communities will draw upon some of the techniques used, prepandemic, to control crowding and sustainably manage beaches, the reality is that overtourism, now complicated by public health and safety issues, is likely to become an ongoing challenge for beaches, coastlines, and islands.

World Heritage Sites: Overtourism Solutions

In 2019, on the eve of the coronavirus pandemic, UNESCO's portfolio of World Heritage Sites included 1,121 natural and cultural properties in 167 countries around the world. WHS properties are defined as "places on Earth that are of 'Outstanding Universal Value' (OUV)" that should be "protected for future generations to appreciate and enjoy."[23] WHS designation has proved to be a double-edged sword, however: it brings, as stated in chapter 4.1, "significant international attention and assistance" and often the growth of tourism. Without sufficient controls and management, though, overtourism can result, degrading the site and overwhelming the host communities. And because these sites are places deemed to be of global value, WHS designation brings with it a special urgency to deal with overtourism.

During the pandemic, the focus was on loss of tourism and loss of revenue. There was little reflection on the overtourism problems that had, through 2019, plagued many cultural and natural heritage sites.

Prior to the pandemic, many sites—from Venice and Barcelona to the Serengeti in Tanzania, Galapagos Islands in Ecuador, Machu Picchu in Peru, and Luang Prabang in Laos—had been grappling with how to rein in overtourism. As in other types of destinations, it is typically a potpourri of factors that have brought these unique World Heritage places to the brink of overtourism.

Three of the most important factors across all World Heritage Sites relate to the structure and functioning of UNESCO's World Heritage Committee and program itself. One is that most sites have ineffective or insufficient management plans and, typically, no tourism management plan. Although the UNESCO World Heritage Centre website lists "unchecked tourist development" as among the leading problems threatening World Heritage Sites, UNESCO has not required that sites have tourism management plans; according to a 2017 study, only 5 percent of natural World Heritage Sites have publicly-available current tourism plans that were not part of larger general management plans.[24] As Dan Peltier wrote in *Skift*, "UNESCO sites—of all places—should be the destinations taking leadership roles in tourism management planning as they're often the first place many tourists think about when deciding what to do on a trip. But data show that's far from reality for a variety of political and economic reasons."[25]

To effectively address overtourism, the World Heritage Convention's operational guidelines need to be reformed to require that sites produce and maintain up-to-date tourism management plans based on data such as visitor trends, arrival numbers, and tickets sold. In addition, there is a need to develop a common methodology and template for data collection that is applied across all sites. As discussed in chapter 4.1, World Heritage Sites that are experiencing overtourism should put in place measures to limit impacts of visitation by encouraging visitation in nonpeak seasons, adjusting entrance fees, regulating the supply of accommodation and parking, adding visitor centers, limiting souvenir shops, requiring advance bookings, and charging visitors a higher price for short-term visits.

The current management shortcomings are compounded by a second internal issue: that the World Heritage Committee's definition of overtourism is narrow and its remedial procedures, including the World Heritage in Danger List, are slow, limiting the committee's' ability to respond quickly and holistically to some places that are widely viewed as overtourism hot spots.

In assessing what puts a site's OUVs at risk, the committee focuses on damage to built or natural structures rather than excessive arrival numbers or softer societal issues such as deteriorating quality of residents' lifestyle or the visitor experience. As Peltier further noted, UNESCO's World Heritage Committee doesn't view "the risk from crowds as the most problematic" to OUVs, and therefore its "latest Endangered List leaves out many sites plagued by overtourism."[26] As noted in chapter 4.1, "It would appear to be beneficial if the committee created an additional category and legal tools to evaluate and address the impacts of overtourism."

The third serious internal issue that contributes to overtourism is financial. Like the NPS, UNESCO has a lofty mission to conserve and protect its special places for the pleasure and enjoyment of humankind into perpetuity. Whereas the NPS receives most of its budget from the federal government, however, World Heritage Sites receive only modest support from UNESCO and, often, minimal funding from the countries in which they are located— particularly for sites in poorer developing countries. Instead, sites rely on tourism, along with international development assistance, to provide funds for protecting and managing these sites. In addition, the World Heritage Centre's budget and staff have both faced cuts in recent years. These cuts come at a time when the number of World Heritage Sites around the globe is growing, making protection and preservation of the sites, as well as equipping them to receive tourists, more difficult and complex.

One positive development in recent years has been UNESCO's efforts to better integrate sustainable tourism principles and practices into management of World Heritage Sites. In 2010, the World Heritage Committee began an overhaul of its Sustainable Tourism Programme to create "a new paradigm" that takes "a holistic and strategic approach to World Heritage properties and destinations that will include bottom-up as well as top-down measures to ensure sustainability that reflects not only high-level goals but also local needs and the ability to attain these goals."[27] The World Heritage Centre is also discussing the possibility of creating a voluntary eco-certification program or set of uniform sustainability guidelines for World Heritage Sites, which would build on existing management and assessment tools.

These important efforts to incorporate sustainability standards and practices in all World Heritage Sites have, however, been limited by a lack of sufficient funds. The Sustainable Tourism Programme is financed through extrabudgetary government and private-sector grants rather than through UNESCO's annual budget. There is never enough, and this fact, plus UNESCO's cumbersome bureaucracy, has meant, some experts say, that sustainability remains more aspirational than practical. In the assessment of one Malaysian heritage expert, "Currently UNESCO has no clear guidelines or effective methods to control the commercialisation of world heritage sites, and its talk on sustainability is more a verbal exercise than enforceable."[28]

Despite these significant shortcomings, UNESCO's World Heritage Sites remain a crowning achievement of the international community. As noted in chapter 4.1's conclusion, "Many of these thousand-plus iconic, irreplaceable treasures would not have survived without the WHS designation and the support the UNESCO brand has generated, including from a rising tide of tourism."

World Heritage Sites: Postpandemic

The COVID-19 lockdown proved devastating to many World Heritage Sites that are dependent on revenue from tourism. By late April 2020, with the coronavirus sweeping around the world, sites had been closed in 90 percent of countries, causing a "dramatic downturn in tourism."[29] According to one account at that time, "Many of China's heritage sites and tourism areas have been closed down. This leaves the fate of the normally massive 127.1 billion dollars' worth of tourist dollars within China in an uncertain state."[30]

World Heritage Sites hold an international reputation that parallels that of the NPS in the United States as being among our most cherished public institutions. Their postpandemic popularity will be defined by several factors, including how the country or region where the WHS is found has handled the pandemic, the accessibility of the site, and the rigor of its management protocols including for social distancing, health, and sanitation. Urban sites, where oversized crowds had become common, are predicted to be less likely to attract visitors than those in accessible natural settings. And given the public's fear of flying, combined with the suspension of flights and national lockdowns of many airports, sites that are far flung from sending markets or in countries with continuing COVID-19 cases are predicted to experience steep declines in tourism for many months. They include Machu Picchu in Peru, Galapagos Islands in Ecuador, the Serengeti and Masai Mara National Parks in Tanzania and Kenya, Angkor Wat in Cambodia, Luang Prabang in Laos, and Mount Everest in Nepal, all sites that are long-haul destinations for the majority of their traditional visitors.

The ancient Inca citadel of Machu Picchu, designated as both a cultural and natural WHS, was, prepandemic, Peru's most important tourism attraction, and tourism revenue was helping to push the country towards middle-class status. With a five-fold increase in international visitors between 2000 and 2019, Machu Picchu was also in the forefront of the worldwide battle against overtourism. In the early 2000s, a multistakeholder legislation initiative began to better manage tourism trekking along the Inca Trail leading to the citadel. Beginning in 2017, the focus has been on controlling visitation inside Machu Picchu itself, mainly through regulating the number of visitors to just under six thousand a day and the length of stay to just four hours. The plan was fully implemented in 2019, to mixed reviews by local guides who thought that the plan was more focused on maximizing economic benefits than on ensuring environmental and cultural protection or a high-quality visitor experience (see chapter 4.2).

When COVID-19 ended this experiment and Machu Picchu was officially closed in March 2020, the impacts were double-edged. Although the ancient citadel was given a welcome respite, the tourism-dependent mountain towns around the site suffered major economic and health hardships. Although Peru had the second highest number of COVID-19 cases in the region, officials came under stiff pressure to reopen Machu Picchu in July for domestic as well as international tourism. To ensure social distancing and other protocols, only 675 visitors—one-fourth the number in 2019—were to be admitted per day, guided tours would be limited to seven visitors, and everyone would be obligated wear a mask. Although long-haul international travelers are not expected back anytime soon—probably not before a vaccine is found—officials had banked instead on domestic tourism, offering free entry to children, public employees, and the elderly.[31] The reopening plan, however, was postponed after a surge of new cases in the country.[32] At other World Heritage Sites, especially in Europe and China, predictions are also that, for the foreseeable future, visitors would be mainly domestic travelers.

Historic Cities: Overtourism Solutions

When overtourism first erupted into the public consciousness, it was primarily via historic cities in Europe. Recently, however, technology and geography have coalesced to make Europe in general an epicenter of overtourism. Its compact size, concentration of historic cities, relative affluence, prevalence of budget flights, and good railway connections combined to make vacation travel in Europe increasingly easy, routine, and affordable. Alongside traditional two-week tours of the continent's historic highlights, shorter, cheaper, more frequent, and frenetic city breaks have mushroomed, catering to hordes of young adults.

These trends, alongside the explosion of the home- and car-sharing economy, have upended urban accommodation and transport and transformed once tranquil, stable neighborhoods into transient housing zones. Long-time locals have seen the stock of rental properties decline, housing prices climb, and their quality of life degraded. As cities become overwhelmed with short-term rentals, growing numbers of residents are selling and leaving, causing urban populations in cities like Barcelona and Venice to drop significantly.

In portside cities, the tourism problem is compounded by the exponential growth of cruise tourism as several times a day thousands of cruise passengers disembark from ever larger vessels. They pour into historic city centers, generating noise, pollution, and congestion while spending far less than land-based travelers. Around ports and in historic city centers, cheap souvenir shops and fast-food restaurants have proliferated, displacing long-established, higher-quality shops catering to residents and stayover travelers.

"As destinations across Europe become increasingly submerged with tourists," wrote Francesca Street in chapter 2.1, "concerns over the livability and endurance of historic cities such as Venice, Dubrovnik, and Amsterdam are growing ever louder. With protests intensifying and cities cracking down on visitor behavior, the region has become the proverbial canary in the coal mine for historic cities in the Americas, Asia, and elsewhere that are just beginning to experience the early symptoms of overtourism—and to experiment with potential reforms."

The early seeds of reform have also been sown in Europe's historic cities, scene of so many vocal, creative, and sustained antitourism protests. On occasion, extreme overcrowding has led to strikes and closures of popular destinations. In 2019, for instance, staff at the Louvre in Paris staged several strikes, protesting that the museum was "suffocating." The Louvre, which holds the dubious distinction as the most visited museum in the world, had been receiving over ten million tourists a year.

European cities have also spawned a range of activist civic groups and engaged academics who have critiqued the overtourism crisis; lobbied for change; offered alternative models of small, sustainable, and off-the-beaten-path tours; and demanded a seat at the table as city governments began grappling with how to rein in overtourism. In Barcelona, as explained in chapter 2.2, "critical voices began to emerge warning that the growth in visitor numbers was overwhelming public spaces, degrading residents' quality of life, creating tensions between residents and visitors, and raising housing and real estate prices." Despite the city's dependence on tourism, by 2017 the tide had turned: residents declared tourism as "the worst" problem facing the city, and half of visitors said that Barcelona was too crowded for them to enjoy their visit.

Faced with both resident and visitor resentment, Barcelona crafted a dynamic, integrated strategy to manage overtourism, and it has become a model for other cities. In 2015, Barcelona made history by electing its first woman mayor, Ada Colau, who was also the first high-profile politician in Europe to run on a tame tourism ticket. As noted in chapter 2.2, the Colau government in Barcelona "laid out a collective approach to develop the tools required to manage tourism in a sustainable and responsible way. Barcelona's new policies tried to overcome the traditional sectorial approach to tourism and to create

Selfie seekers at the Mona Lisa in the Louvre, Paris, France. Source: Tarik (Unsplash).

an innovative and holistic framework that goes beyond the narrative of 'success-by-growing.'"

In Barcelona as well as in other cities, two central precepts guided the government reforms. One, as Street wrote, was to "flip from destination marketing to destination management, with a focus on increasing length of stay and spending per visitor, rather than increasing visitor numbers." The other was to move from a siloed approach with tourism managed only by tourism agencies to a dynamic holistic approach that engaged local businesses, civic groups, academics, and multiple government departments.

Under these two new precepts of destination management and holistic public policy planning, European cities have sought to tackle overtourism with a variety of tools and programs. They have involved visitor reeducation and dispersal, new technologies, and regulation and taxes. For instance, Amsterdam's Enjoy and Respect civic group has launched a media campaign to transform the behavior of the young British and Dutch tourists who flood the city on weekends for bar hopping and all-night partying. The city government has imposed penalties for bad behavior, banned tourist stores in the historic center, limited access to the famed Red Light District, and sought to redirect tourist flows by moving the iconic "I Amsterdam" letters away from the overcrowded Museumplein square to less congested sections of the city. The long-term goal, wrote Street in chapter 2.1, is to "reorient the city's public image" in an effort "to ensure that Amsterdam is a livable and workable city for local residents, as well as a quality destination for visitors."

Across Europe, a number of new apps and websites, like Avoid-Crowds.com, are aggregating tourist data to help track crowds in key European cities and offer alternative routes and itineraries. Meanwhile, a growing number of initiatives, including Venezia Autentica, BestVeniceGuides, and Unrating Vienna, are promoting sustainable tourism through small group tours that avoid the crowds and highlight lesser known attractions.

Venice has also experimented with temporary regulations to manage pedestrian and water traffic and separate travelers from locals, including designating certain areas for residents only and those visitors with special Venezia Autentica passes. Venice, Edinburgh, Barcelona, and other cities are enacting tourist taxes to help discourage short-term visitors and to raise additional funds for coping with overtourism. Amsterdam, for instance, has imposed a tourist tax that specifically targets cruise visitors in an attempt to address its overtourism issues.[33]

Some of the most concerted efforts at controlling overtourism in European cities have focused on two other areas: regulating short-term rentals and, in port cities, controlling cruise tourism. In Madrid, for instance, legislation to require separate entrances for visitors and full-time residents aims to prevent many city center apartments from being operated as vacation rentals. In Valencia, private vacation rentals are not permitted above the first floor so as to keep ocean views for locals. And in Barcelona, where the number of vacation flats had more than quadrupled between 2011 and 2014, a network of residents and inspectors, along with an online "flat detector" tool, identified and closed down nearly five thousand unlicensed short-term rentals between 2016 and 2018.

An alternative to Airbnb is Fairbnb.coop, launched in Europe in 2019 as "a cooperative accommodation booking platform that promotes and funds local initiatives and projects." Fairbnb.coop permits only one home rental per host and donates part of the booking fee to local community projects.[34] Although the initiative is still in its infancy, in the spring of 2020 it was operating in six European cities with a pledge to donate 50 percent of its booking fees to coronavirus relief funds in its destinations.[35]

At the same time, a growing number of US cities have moved to regulate short-term rentals. Charleston, South Carolina, which has been repeatedly voted the "Best US City" by *Travel + Leisure* magazine, has adopted strict rules to police vacation home rentals using compliance software designed to identify, monitor, and fine unpermitted rental properties. According to Dan Riccio, in 2018, Charleston's city council passed an ordinance requiring that short-term rentals be licensed, pay a 4 percent bed-night tax, and be owner-occupied with the owner present when rented (see chapter 2.3). Rentals of entire houses are prohibited. Officers with the new Department of Livability and Tourism, created to focus full time on short-term rental compliance, determined that more than half of the vacation home rentals advertised online were unlicensed.

Controlling cruise tourism has been a top priority in a number of port cities, including Dubrovnik, Venice, and Barcelona, which are all home to important World Heritage Sites. Of the three, Dubrovnik has been most successful in implementing some reforms. In recent years, Dubrovnik has seen roughly two million visitors a year, split quite evenly between stayover tourists and cruise ship passengers. These numbers were putting severe strains on its aging infrastructure and on the patience of its residents. So beginning in 2018, the city's mayor personally negotiated with the head of CLIA, the international cruise line association, to permit docking of only two ships a day, with a total of five thousand passengers. Under the banner "Respect for the City," the government has taken other measures to reduce overcrowding, including staggering ships' departure and arrival times, reducing by 80 percent the number of souvenir stands in the historic city, and cutting by 20 percent the number of restaurant tables and chairs. As Mato Frankovic, mayor of Dubrovnik, explained, "We are ready to lose some money, but we will have a better quality of life for citizens and tourists."[36]

Venice, by contrast, has been involved for many years in complex and still unresolved efforts to redirect and limit cruise tourism. For years, UNESCO's World Heritage Committee has, via critical reports and onsite missions, warned that if Venice did not make reforms, including stopping cruise ships from using the Venetian Lagoon, the city would be put on the World Heritage Sites in Danger list. In late 2019, following the city's worst-ever flooding, UNESCO declared Venice as "highly vulnerable," threatened by both climate change and cruise tourism.[37] Despite these warnings, the municipal government has maneuvered, through a series of studies and plans, to successfully keep Venice off the endangered list. As Street wrote, although international pressure from UNESCO "has clearly brought global attention to Venice's plight, it has not, as yet, succeeded in compelling the government to implement comprehensive long-term solutions."

In summarizing the efforts by historic cities to rein in overtourism, Street concluded, "Led by local governments, reform strategies must also be informed by consultations with city residents, academics, tourism professionals, and other stakeholders, promoted by civic campaigns and social media, and designed to protect the historic character of the city, its ongoing livability, and its future development." And, as Street noted, "Some historic European cities are experimenting with suites of solutions, but it is still too soon to judge if these initiatives will be enough to curb and control overtourism effectively." Indeed, although many of these reform efforts had shown short-term success, their long-term viability in curbing overtourism was cut short in early 2020 by the coronavirus pandemic.

Historic Cities: Postpandemic

In early 2020, the world watched in horror as Italy became the early epicenter of the COVID-19 in Europe. By late February, Italy's borders had closed, and tourism had stopped. Then, by early June as Italy began cautiously reopening, protesters were back in the streets in Venice—now with proper social distancing and masks—to oppose a new cruise dock that would bring tourists through one of the city's last remaining livable neighborhoods. Their larger purpose, according to the *New York Times*, was to "seize a once-in-a-lifetime opportunity to show that another, less tourist-addled future was possible." As the protest signs read, "Nothing Changes If You Don't Change Anything."[38]

The coronavirus has laid bare the underlying weaknesses of the societies it has ravaged, including in Italy "an addiction to tourism that has priced many residents out of historic centers and crowded out creativity, entrepreneurialism, and authentic Italian life."[39] An open letter to the Italian government, cosigned by museum directors and professors in Venice plus celebrities like Mick Jaggar and Francis Ford Coppola, presents ten commandments for a new Venice, including stricter regulations on tourism flows and the Airbnb market and support for the long-term rental market. As the rector of one university said, "In this moment, there is a temporary window."[40]

Jane da Mosto, who heads the nonprofit group We Are Here Venice, agrees that the pandemic can be a turning point. She envisions a new Venice emerging in the postpandemic world that is committed to the principles of sustainable tourism. As da Mosto put it, "The problem for Venice isn't the lack of tourists, it's the lack of permanent residents. And with more residents, the city will reflect more the Venetian culture and the wonderful lifestyle that this extraordinary city offers and future visitors to the city will be able to enjoy Venice more."[41]

There were similar sentiments from overtourism weary residents in Amsterdam. "It's like the city is ours again," was a common sentiment among Amsterdammers. "The cause of this crisis is very sad, but for us it's a blessing in disguise," said a retired economist, returning from an unusually quiet trip to the market, adding, "Tourism here has become too much. We are sick of it, just sick."[42]

Of Europe's historic cities, it is perhaps likely that the walled city of Dubrovnik along the Croatian coast could most swiftly and effectively adjust its tourism to a new normal in the postpandemic era. Although COVID-19 halted cruise tourism and brought the city's campaign to curb overtourism to an abrupt halt, the reforms begun in 2019 did hold promise. And in early 2020, the Croatian government moved swiftly and decisively to address the coronavirus pandemic. Even though Croatia is located near the main European epicenters of COVID-19, its infection and death rates were among the lowest in Europe.

Despite these achievements, however, Dubrovnik also has, as one writer put it, several "Achilles heels" that make a quick rebound of tourism unlikely.[43] Most importantly, 90 percent of tourists reach Dubrovnik by air, and most recent visitors have been from Britain and the United States, along with South Korea and Australia, all long-haul markets that will not return quickly. As one article in Dubrovnik summed up the situation, "2020 looks like being a year to forget for the tourism industry in Dubrovnik, or in a more positive vein, a year to plan, create and contemplate a brighter and different future. Rather like a boxer taking an 8-count to catch his breath Dubrovnik can use this time to not only move away from mass and commercial tourism but even start moving in a greener more sustainable future."[44]

Europe as a whole began opening to tourism unevenly, based on epidemiological evidence that the spread of the virus had been significantly decreased and was likely to remain stable as tourism increased. In May 2020, the European Commission released guidelines for "the safe and gradual restoration of tourism activities." The Baltic states (Latvia, Lithuania, and Estonia) quickly announced they were creating a "travel bubble," allowing citizens to travel freely within the region.[45] As in the United States, Europeans emerging from the pandemic are favoring either "staycations" at or near home or drive-to vacations, either within their own country or to nearby destinations. Two of the preferred drive-to vacation destinations are the Croatian and Montenegro coastlines, which had had almost no COVID-19 cases.

But these places do not include Dubrovnik, which is located at the country's southern tip and is therefore not an easy drive from Germany and other important sending markets in Europe.

In fact, tourists are expected to return more slowly to Dubrovnik, Venice, and other historic cities than to other destinations. International travelers continue to be barred by ongoing travel restrictions, and many Europeans fear that social distancing will prove difficult in urban areas. Tourism, which accounts for about 10 percent of Europe's gross domestic product, is projected to lose about $75 billion as a result of the global health crisis. It is not expected to fully recover in Italy and other hard-hit destinations until 2023.

In the meantime, most visitors to historic cities will be nationals. Although the economic downturn is extremely difficult for these tourism-dependent cities, the crisis has also created opportunities to find ways to better serve the local population. In Venice, for instance, the city moved to address its long-term housing shortage caused in part by the explosion of Airbnb vacation rentals by signing, in April 2020, a protocol to convert empty units into longer-term student housing.

In addition, with the halt of cruise ships during the pandemic, local gondoliers again reclaimed the canals, delivering groceries to older Venetians who could not safely leave their homes. These manually powered boats, together with the absence of cruise ships, dramatically improved Venice's air and water quality, raising hopes, said Da Mosto of We are Here Venice, "that smaller, more sustainable boats will be more prevalent even after the pandemic is over." He adds, "There's a huge potential for traditional rowing to be part of everyday life, and not as in going back to the past but actually in looking toward the future."[46]

The Future for Sustainable Tourism

Despite the unspeakable hardships wrought by the coronavirus pandemic and the economic crises, many tourism destinations around the world are envisioning a slim silver lining in the opportunity to strengthen sustainable forms of tourism. In examining the future for sustainable tourism, there are lessons to be drawn from both overtourism and the pandemic. On the face of it, these two phenomena are polar opposites, yet they are forever linked in time: overtourism collided historically with the coronavirus pandemic, which, almost overnight, led to no tourism.

The lockdown interlude gave a glimpse of what many of our most loved places could look like, if free of overtourism. The brief closures of overtouristed destinations, from beaches and national parks to historic cities and World Heritage Sites, demonstrated what balanced ecosystems should—and could—be. This break is leading to more focus on both the value of planned closures to provide destinations with time to heal and the need to cap numbers through reservation and allotment systems, limits on parking, and other techniques. Going forward, it is likely that these measures will become more commonplace.

A central lesson from the pandemic, as from overtourism, is the critical importance of good government and effective management. The countries that most successfully beat back the coronavirus pandemic, including Iceland,[47] New Zealand, Costa Rica,[48] and Croatia,[49] did so through early, aggressive, and consistent policies grounded in good science; clear and informative public messaging; civic engagement;

good public health systems; and effective government leadership. They shared a common playbook that included swift surveillance, quarantine and closures, testing and tracing, social distancing, and face masks.

To take one example, New Zealand had, by early June 2020, announced that it had eradicated COVID-19: the country had no active cases and no new cases and so "could return to a form of pre-pandemic normal."[50] Retail, hospitality, and tourism across the country were back without social distancing, masks, or other restrictions. The country's borders, however, remained closed, with plans to first open up a so-called trans-Tasman COVID-safe travel zone with Australia, its most important tourism market. The keys to success, tweeted New Zealand's former prime minister Helen Thomas, are "clear leadership & an engaged public [which] have produced this result. Principles of inclusion, resilience & sustainability should now guide recovery in NZ & globally."[51] Not surprisingly, New Zealand and other countries that have controlled the coronavirus pandemic are also leaders in sustainable tourism, based on strong policies and practices.

By contrast, a success story outlier is Hong Kong, which defeated COVID-19 through an effective citizen-led campaign rather than good government. Hong Kong is widely viewed as a "failed state," and its deeply unpopular, pro-Chinese chief executive fumbled the official response to the pandemic. But, as Zeynep Tufekci wrote in *The Atlantic*, "the secret sauce of Hong Kong's response was its people and, crucially, the movement that engulfed the city in 2019. . . . The organizational capacity and the civic infrastructure built by the protest movement played a central role in Hong Kong's grassroots response."[52] This popular movement, as chronicled in *The Atlantic*, used social media, mobilization of medical professionals, street demonstrations, public education, border closures, masks, and other techniques to swiftly beat back the coronavirus. By May 2020 life had largely returned to normal, with museums, libraries, schools, and offices reopening.[53] The inspiring case study of Hong Kong demonstrates that, in all likelihood, citizen engagement is even more critical than effective government in combating the coronavirus pandemic.

Other best practices that emerged from the coronavirus pandemic and economic collapse also hold potential for strengthening sustainable tourism. For instance, the pandemic brought out a renewed commitment to "small is beautiful," including small businesses. Around the world, governments sought to shore up small businesses with financial assistance to keep people employed and businesses from permanently closing. Although in the United States the government programs were poorly administered and far less generous than in Europe, Canada, and elsewhere, the crisis unleashed a wide range of community initiatives.

To help save small businesses, communities across the United States started "buy local" campaigns, "take out Tuesday" initiatives, grant and loan programs, crowdfunding drives, and virtual sidewalk sales. As a local business owner in Rhinebeck, New York, wrote after receiving a small grant from a new civic organization, Rhinebeck Responds, "Rhinebeck Responds is helping to keep Rhinebeck's charming village and its small businesses intact through this pandemic. We need to help each other now more than ever and I am overwhelmingly grateful for the generosity shown during this time."[54] Although the long-term prognosis for many small businesses remains uncertain, one positive and, it is hoped, long-term result is increased community collaboration and caring for what is special about small towns and local neighborhoods, which, in turn, helps strengthen the important sustainable tourism concepts of sense of place and destination stewardship.

For rural towns like Rhinebeck, whose economies partially rely on tourism, the postpandemic travel preferences also favor small, local, and outdoors. Surveys showed that cabin-fevered Americans were eager to get out, but at least in the near term, they will travel closer to home, in small groups of family and friends, by car, and for short vacations. As discussed, parks and nature are top on the list. Also preferred are small rural towns surrounded by natural attractions where social distancing and hygiene protocols are respected.[55] According to a May 2020 survey, 52 percent of respondents said that their first trip would be to visit friends and family, 76 percent plan to travel by car, and 86 percent said that their travel would be domestic.[56] Wellness experiences and active vacations, including hiking, biking, and walking, are all priorities for the newly liberated public. This package of preferences may help curb the reemergence of mass tourism in favor of more sustainable attractions, activities, and destinations.

Although the COVID-19 pandemic lockdown may ultimately

strengthen travel that is domestic, small group, by road, outdoor, and crowd-free, other tourism sectors are emerging far more battered and bruised. The airlines in general, along with international leisure and business travel, are way down. In May 2020, for instance, passenger traffic on US airlines was down 95 percent,[57] and a mere 3 percent of Americans surveyed said they planned to travel to an international destination in the next six months.[58] In the near term, airlines are making reforms to lessen human interactions and promote social distancing, with a positive focus on health and safety. As one industry analysis put it, "Creating a contactless journey is key to restoring passenger confidence and accelerating recovery."[59]

Although air travel will certainly rebound over time, it may well never fully return to its robust prepandemic levels. It is likely that going forward, more business, including conferences and meetings, will take place virtually, thereby reducing the aviation industry's carbon footprint. As Rebecca Heilweil puts it in her analysis of the postpandemic future of air travel, "The Covid-19 pandemic has also reminded everyone to consider the massive carbon footprint created by the airline industry." She goes on to conclude, "Perhaps, trips once deemed worthy of a flight, especially business-related ones, could just happen over Zoom."[60]

Although the environment will win with less air travel, those who stand to lose the most from this drop in air traffic will be tourism-dependent countries in Africa, Latin America, and Asia that rely on long-haul international travelers. The future looks bleak not only for the people in these countries, but also for their World Heritage Sites that are heavily dependent on tourism revenue.

The postpandemic fate of two other travel sectors appears to hold more promise for strengthening sustainable tourism. Efforts to control and reform home-sharing via Airbnb and other platforms may receive a boost in the postpandemic period. By late May 2020, Airbnb had lost $1.5 billion in bookings since the coronavirus began and laid off nineteen hundred employees, or about 25 percent of its workforce.[61] As a *New York Times* analysis put it, "The future of home sharing depends largely on whether travelers see rentals as private, often cheaper, alternatives to hotels, or a source of exposure to strangers' germs. The vagaries of cancellation policies among rentals will also have an impact."

Airbnb is upgrading its information about listings to include whether hosts are practicing stringent cleaning guidelines and a minimum twenty-four-hour or maximum seventy-two-hour waiting period between bookings.[62]

The future of home-sharing also depends on whether cities, in particular, decide to use the postpandemic period to push for more substantial reforms and controls. For example, in Lisbon, Portugal, many essential workers have been forced to live outside the city because so many properties had been converted into vacation home rentals. One of the most ambitious moves came from Lisbon's mayor, Fernando Medina, who announced in July 2020 that the city was "prioritizing affordable housing" and would convert all short-term rentals into "safe rent" homes for hospital staff, transport workers, teachers, and others who were providing essential services. He added, "Now we want to bring the people who are Lisbon's lifeblood back to the centre of the city as we make it greener, more sustainable, and ultimately, a better place to both live and visit." Medina called the move a "bold strategy that offers landlords long-term stable incomes and gives us the chance to recreate a more vibrant, healthier and equitable city." He is also working with mayors from other cities around the world "to determine what a green and just recovery from Covid-19 pandemic looks like."[63] Lisbon's innovative shift is helping reinvigorate the city's housing stock away from mass tourism and back to equitable living. It just might become a model for other cities, too.

Of all the tourism sectors, it is without doubt large cruise lines that proved most vulnerable to the coronavirus pandemic and are likely to have the longest recovery. The three megacruise corporations, Carnival, Royal Caribbean, and Norwegian, which control 70 percent of the global cruise market, ended 2019 with $6 billion in profits and high hopes for smooth sailing in 2020. But, as Stand.earth and five other environmental groups put it in an open letter to the heads of the three megalines, "the new coronavirus had other plans."[64]

Cruise ships quickly became one of the most dangerous places to be during the pandemic. Although the risks were clear by early February 2020, the industry chose to keep cruising while the US government proved reluctant to shut it down. It was not until March 13 that cruising was officially stopped, following US Centers for Disease Control and Prevention warnings of the increased risk of COVID-19 infection on cruise ships.

As the environmental group's open letter explained, the world looked on in horror as "ships were refused port entries, passengers were quarantined in their rooms for weeks while the virus spread like wildfire, and dozens of people ultimately lost their lives to the Covid-19 virus they contracted on board."[65] According to the *Miami Herald*, "The cruise industry—which downplayed the dangers to consumers and kept sending out ships despite outbreaks on board and warnings from public health officials—has largely stayed silent about the toll." Based on its own investigation, the *Miami Herald* concluded in late April 2020 that fifty-four ships, or roughly one-fifth of the global ocean cruise fleet, had had COVID-19 cases.[66]

As the pandemic began to recede, the cruise lines pledged that they were making "significant" modifications in their health, hygiene, and safety practices, including eliminating buffets, conducting passenger health screenings, and contingency planning for when infection occurs. In late 2020, the three major US lines had not fully resumed operations, and the ratings agency Moody's predicted that the cruise lines would not resume operations until 2021.[67]

The main challenge was winning public confidence. Even though cruise passengers are a deeply loyal subset of travelers[68] who vacation repeatedly via cruises, travel agents reported in June 2020 that the majority of their cruise bookings were for the summer of 2021, more than a year away.[69] According to a Harris Poll that ranked the amount of time respondents said that they would wait to undertake travel-related activities after the COVID-19 curve was flattened, "taking a cruise" ranked longest. The poll concluded, "The cruise industry is expected to have the longest road to recovery," with a majority of respondents saying that "it will take a year or more before they will take a cruise."[70]

Although the public remained wary of getting back on board, civic groups in North America, Europe, Australia, and elsewhere stepped up their public challenges to the cruise industry and calls for far broader reforms than the cruise lines were considering. On June 5, 2020, the environmental coalition hosted a "Global Day of Action to Clean Up Cruising" via an online rally with community members from several dozen ports around the world. As their open letter stated, "The failure of your companies to respond immediately and appropriately to the scale of the unfolding pandemic had another unforeseen consequence. It dragged your industry into the harsh light of public scrutiny, and many are shocked by the ugly truths they're seeing for the first time."[71]

Indeed, as the summer of 2020 progressed, the global coalition of cruise critics grew to include civic groups in more ports of call and a widening set of demands for reform, ranging from the predictable issues of public health and air and water pollution to deeper concerns regarding the economic and cultural impacts of cruise tourism on host communities, wage and labor conditions for crew, and cruise ships' impacts on climate change. It represented the most full-throated and broad-based critique to date of the cruise industry. Although the outcome remains uncertain, the pandemic had inadvertently raised public concerns about the health, safety, and growing array of other cruise practices and impacts, which bodes well for efforts to strengthen sustainable tourism.

The pandemic has spawned even broader efforts to alter the course of tourism for the better. The Future of Tourism Coalition, launched in mid-2020, is led by the Center for Responsible Travel and five other organizations and advised by the Global Sustainable Tourism Council. Together they share a global mission "to place destination needs at the center of tourism's new future." The coalition's platform calls for recentering the tourism industry around a set of thirteen guiding principles that are "vital for long term deep-rooted growth." The strategy is participatory, bringing together governments, the private sector, academia, nonprofit organizations, and others to help shepherd solutions.[72] As this initiative signals and as the chapters of this book chronicle, we see emerging from the back-to-back calamities of overtourism and the COVID-19 pandemic both resilience and resolve to make lemonade out of lemons—a way to discern lessons from these twin catastrophes and demonstrate that there is, indeed, a future for sustainable tourism.

Notes

1. Jonathan Tourtellot. (March 23, 2020). "Corona-crisis: A Destination Management Opportunity." Destination Stewardship Center. https://destinationcenter.org/2020/03/corona-crisis-a-destination-management-opportunity/.

2. Neal Herbert. (February 27, 2020). "National Park Visitation Tops 327 Million

in 2019." Press release. National Park Service. https://www.nps.gov/orgs/1207/2019-visitation-numbers.htm.

3. InterAgency Visitor Use Management Council. (July 2016). *Visitor Use Management Framework A Guide to Providing Sustainable Outdoor Recreation.* Edition One. pp. 83–94. https://visitorusemanagement.nps.gov/Content/documents/highres_VUM%20Framework_Edition%201_IVUMC.pdf.

4. Pacific Consulting Group and National Park Service. (2019). "National Park Service 2019 Visitor Survey Card Data Report." https://irma.nps.gov/Datastore/Reference/Profile/2268309.

5. Donald Leadbetter. (October 28, 2019). National Park Service. Personal communication with editors.

6. Jeremy Miller. (May 21, 2020). ""We've never seen this': wildlife thrives in closed US national parks." Society of Environmental Journalists. https://www.sej.org/headlines/weve-never-seen-wildlife-thrives-closed-us-national-parks.

7. Katia Hetter. (April 20, 2020)."With many national parks closed, their animal residents are getting a break (from Us)." CNN Travel. https://www.cnn.com/travel/article/national-park-week-healing-trnd-wellness/index.html.

8. Donald Leadbetter. (May 28, 2020). National Park Service. Personal communication with author.

9. Destination Analysts. (June 1, 2020). " https://www.destinationanalysts.com/insights-updates/.

10. Donald Leadbetter. (May 28, 2020). National Park Service. Personal communication with author.

11. Carl Huise. (June 17, 2020). "Senate Passes Major Public Lands Bill." *New York Times.* https://www.nytimes.com/2020/06/17/us/senate-national-parks-funding-bill.html.

12. Donald Leadbetter. (May 28, 2020). National Park Service. Personal communication with author.

13. "Alerts in Effect." Rocky Mountain National Park Colorado. National Park Service. https://www.nps.gov/romo/planyourvisit/fees.htm#timedentryfaq.

14. National Parks Service. (May 2020). "COVID-19 Reopening Plan. Yellowstone National Park." https://www.nps.gov/yell/learn/news/upload/Yellowstone-COVID-19-Reopening-Plan-2020.pdf.

15. Ibid.

16. Nick Mancall-Bitel. (June 25, 2020). "As Tourists Flock to National Parks, Nearby Restaurants Brace for the Coronavirus." Eater Travel. https://www.eater.com/2020/6/25/21302173/national-parks-reopening-coronavirus-covid-19-restaurants-tourists-campers-travel.

17. Los Angeles Times Editorial Board. (May 6, 2020). "Editorial: Coronavirus Is Teaching Us Lessons on How to Coexist with Nature." *Los Angeles Times.* https://www.latimes.com/opinion/story/2020-05-06/editorial-coronavirus-is-teaching-us-lessons-on-how-to-coexist-with-nature.

18. Marisa Yamane. (September 3, 2019). Director of Communications. Hawaii Tourism Authority. Email communication with author.

19. Lizzy Rosenberg. (April 1, 2020). "Without the Usual Crowds, Sea Turtles Have Started Hatching on Brazil's Empty Beaches." *The Guardian.* https://www.theguardian.com/world/2020/mar/29/newborn-endangered-sea-turtles-throng-brazils

-deserted-beaches; "Experts Say COVID-19 Closures Could Have Positive Impact on Marine Life." (April 19, 2020). WVLT News. https://www.wvlt.tv/content/news/Experts-say-people-staying-home-due-to-COVID-19-could-have-positive-impact-on-marine-life--569770441.html.

20. Louis Aguirre. (April 13, 2020). "Closed Beaches Due to Coronavirus Has South Florida's Oceans Looking Clear and Beautiful." Local 10 News. https://www.local10.com/news/local/2020/04/13/closed-beaches-waterways-due-to-coronavirus-has-south-floridas-oceans-looking-clear-and-beautiful/.

21. Sam Levin and Vivian Ho. (April 27, 2020). "Thousands of People Pack California Beaches Despite Coronavirus Concerns." *The Guardian.* https://www.theguardian.com/us-news/2020/apr/27/california-beaches-coronavirus-orange-county.

22. Sam Levin and Vivian Ho. (April 27, 2020). "Thousands of People Pack California Beaches Despite Coronavirus Concerns." *The Guardian.* https://www.theguardian.com/us-news/2020/apr/27/california-beaches-coronavirus-orange-county.

23. UNESCO. (Accessed March 2020). "What is World Heritage?" https://whc.unesco.org/en/faq/19.

24. Susanne Becken and Cassandra Wardle. (January 2017). *Tourism Planning in Natural World Heritage Sites.* Griffith Institute for Tourism; UNESCO. https://www.griffith.edu.au/__data/assets/pdf_file/0034/18889/UNESCO-WHA-Report13Finalfinal-1.pdf.

25. Dan Peltier. (April 19, 2017). "Nearly Half of UNESCO Sites Don't Have Plans to Manage Overtourism Challenges." *Skift.* https://skift.com/2017/04/19/nearly-half-of-unesco-sites-dont-have-plans-to-manage-overtourism-challenges/.

26. Dan Peltier. (July 31, 2018). "UN Agency's Latest Endangered List Leaves Out Many Sites Plagued by Overtourism." *Skift.* https://skift.com/2018/07/31/un-agencys-latest-endangered-list-leaves-out-many-sites-plagued-by-overtourism/

27. "5C. World Heritage Convention and Sustainable Development." (June 24–July 6, 2012). UNESCO. Convention Concerning the Protection of the World Cultural and Natural Heritage. World Heritage Committee. Thirty-sixth session. 36 COM 5C. Saint Petersburg, Russian Federation. 5 p. https://whc.unesco.org/archive/2012/whc12-36com-5C-en.pdf.

28. Laignee Barron. (August 30, 2017). "'Unesco-cide': Does World Heritage Status Do Cities More Harm than Good?" *The Guardian.* https://www.theguardian.com/cities/2017/aug/30/unescocide-world-heritage-status-hurt-help-tourism.

29. Robert Collins. (May 1, 2020). "UNESCO Hosts Meeting of Culture Ministers to Discuss Covid-19 Pandemic." *News.* Museums Association. https://www.museumsassociation.org/museums-journal/news/01052020-unesco-hosts-covid-19-meeting-culture-ministers.

30. Colin Wilson. (April 29, 2020). "Avoiding Heritage: Cultural Heritage in the Age of Pandemic." Thoughts on World Heritage. https://medium.com/thoughts-on-world-heritage/avoiding-heritage-cultural-heritage-in-the-age-of-pandemic-7e05a943d176.

31. Carlos Mandujano. (June 15, 2020). "Machu Picchu to Sharply Limit Visits after July Reopening in Peru." Agence France-Presse. https://www.thejakartapost.com/travel/2020/06/15/machu-picchu-to-sharply-limit-visits-after-july-reopening-in-peru.html.

32. Traveling and Living in Peru. (June 22,2020). "Machu Picchu Reopening Post-

poned by Cusco Officials." https://www.livinginperu.com/machu-picchu-reopening-postponed-by-cusco-officials-peru/.

33. Kaye Holland.(July 26, 2019). "Cruise Ship Routes to Dubrovnik Come under Scrutiny." *The Telegraph*. https://www.telegraph.co.uk/travel/cruises/news/cruise-ship-routes-to-dubrovnik-come-under-scrutiny/.

34. Fairbnb.coop. https://fairbnb.coop/.

35. Ibid.

36. Helen Coffey. (October 2, 2018). "Dubrovnik to Cap the Number of Cruise Ships Allowed to Dock Each Day." *The Independent*. https://www.independent.co.uk/travel/news-and-advice/dubrovnik-cruise-ship-cap-croatia-overtourism-two-dock-a8565166.html.

37. UNESCO. (November 13, 2019). "UNESCO Closely Follows Tides and Flooding in Venice World Heritage Site." WHC.UNESCO.org. https://whc.unesco.org/en/news/2055/.

38. Jason Horowitz. (June 3, 2020). "Venice Glimpses a Future with Fewer Tourists, and Likes What It Sees." *New York Times*. https://www.nytimes.com/2020/06/03/world/europe/coronavirus-venice-tourists.html.

39. Jason Horowitz. (June 3, 2020). "Venice Glimpses a Future with Fewer Tourists, and Likes What It Sees." *New York Times*. https://www.nytimes.com/2020/06/03/world/europe/coronavirus-venice-tourists.html?searchResultPosition=12.

40. Ibid.

41. Barbie Latza Nadeau. (May 16, 2020). "Deserted Venice Contemplates a Future without Tourist Hordes after Covid-19." CNN World. https://www.cnn.com/world/live-news/coronavirus-pandemic-05-16-20-intl/h_0a9ed8e97c3bffc63a2d591694a36551.

42. Tim Igor Snijders. (May 6, 2020). "'The City Is Ours Again': How the Pandemic Relieved Amsterdam of Overtourism." *Washington Post*. https://www.washingtonpost.com/travel/2020/05/06/city-is-ours-again-how-pandemic-relieved-amsterdam-overtourism/.

43. Mark Thomas. (May 31, 2020). "What Will Tourism Look Like in Dubrovnik in 2020? Three Factors That Make For Depressing Reading." *The Dubrovnik Times*. https://www.thedubrovniktimes.com/news/dubrovnik/item/9122-what-will-tourism-look-like-in-dubrovnik-in-2020-three-factors-that-make-for-depressing-reading.

44. Ibid.

45. Kate Whiting. (May 14, 2020). "These Countries Are Making 'Travel Bubbles' for Post-Lockdown Tourism." World Economic Forum. https://www.weforum.org/agenda/2020/05/tourism-coronavirus-travel-bubble-lockdown/.

46. Janna Brancolini. (May 18, 2020). "European Hot Spots Like Venice, Italy, Are Reopening, but without Many Tourists." *Los Angeles Times*. https://www.latimes.com/world-nation/story/2020-05-18/europe-venice-italy-reopens-minus-tourists-coronavirus.

47. Elizabeth Kolbert. (June 2, 2020). "How Iceland Beat the Coronavirus." The New Yorker. https://www.newyorker.com/magazine/2020/06/08/how-iceland-beat-the-coronavirus2

48. Douglas Broom. (May 7, 2020). "This Latin American Country Is Keeping COVID-19 Firmly under Control. How?" World Economic Forum. https://www.weforum.org/agenda/2020/05/costa-rica-winning-battle-coronavirus-covid-19/.

49. Mark Thomas. (April 7, 2020). Croatia Has Strictest Response to Coro-

navirus in the World States Oxford University." *The Dubrovnik Times.* https://www.thedubrovniktimes.com/news/croatia/item/8669-croatia-has-strictest-response-to-coronavirus-in-the-world-states-oxford-university.

50. Damien Cave. (June 6, 2020). "With No Virus, New Zealand Lifts Lockdown." *New York Times.* https://www.nytimes.com/2020/06/08/world/australia/new-zealand-coronavirus-ardern.html.

51. Ibid.

52. Zeynep Tufekci. (May 12, 2020). "How Hong Kong Did It: With the Government Flailing, the City's Citizens Decided to Organize Their Own Coronavirus Response." *The Atlantic.* https://www.theatlantic.com/technology/archive/2020/05/how-hong-kong-beating-coronavirus/611524/.

53. Ibid.

54. 55. Melaine Rottkamp, Vice President, Dutchess Tourism. (June 2020). Several interviews with author.

56. Longwoods International. (May 5, 2020). "Americans Look Forward to Travel, Especially to Reunite with Family and Friends." COVID-19 Travel Sentiment Study-Wave 8. https://longwoods-intl.com/news-press-release/covid-19-travel-sentiment-study-wave-8.

57. "The Future of Travel: How the Industry Will Change after the Pandemic." (May 5, 2020). *New York Times.* https://www.nytimes.com/interactive/2020/05/06/travel/coronavirus-travel-questions.html.

58. Longwoods International. (May 5, 2020). "Americans Look Forward to Travel, Especially to Reunite with Family and Friends." COVID-19 Travel Sentiment Study-Wave 8. https://longwoods-intl.com/news-press-release/covid-19-travel-sentiment-study-wave-8.

59. Collins Aerospace. (June 2020). Air Travel Post-Covid-19re-Imaginingair Travel for a Post-Pandemic World. White Paper. https://www.collinsaerospace.com/-/media/project/collinsaerospace/collinsaerospace-website/product-assets/marketing/c/contactless-passenger-journey/re-imagining-air-travel-for-a-post-pandemic-world.pdf.

60. Rebecca Heilweil. (May 27, 2020). "The Pandemic Could Change Air Travel Forever." Vox. Recode. https://www.vox.com/recode/2020/5/27/21263647/pandemic-flight-future-airplanes-airports.

61. Ashlea Halpern. (May 20, 2020). "How Airbnb Could Change After the Pandemic—for the Better." *Conde Nast Traveler.* https://www.cntraveler.com/story/how-airbnb-could-change-after-the-pandemic-for-the-better?verso=true.

62. "The Future of Travel: How the Industry Will Change after the Pandemic." (May 5, 2020). *The New York Times.* https://www.nytimes.com/interactive/2020/05/06/travel/coronavirus-travel-questions.html.

63. Fernando Medina. (July 4, 2020). "After Coronavirus, Lisbon Is Replacing Some Airbnbs and Turning Holiday Rentals into Homes for Key Workers." *The Independent.* https://www.independent.co.uk/voices/coronavirus-lisbon-portugal-airbnb-homes-key-workers-a9601246.html.

64. "An Open Letter to Cruise Giants." (June 7, 2020). Advertisement. *Miami Herald.* https://www.stand.earth/sites/stand/files/openlettertocruise-giants-miamiheraldad.pdf.

65. Ibid.

66. Taylor Dolven, Sarah Blaskey, Nicholas Nehamas, and Alex Harris. (April 23, 2020). "Cruise Ships Sailed on Despite the Coronavirus. Thousands of People Paid The Price." *Miami Herald.* https://www.miamiherald.com/news/business/tourism-cruises/article241640166.html.

67. Will Feuer. (June 20, 2020). "Cruise Lines Voluntarily Suspend All Trips Out of U.S. Ports until Sept. 15, Trade Group Says." CNBC. https://www.cnbc.com/2020/06/19/cruise-lines-voluntarily-suspend-all-trips-out-of-us-ports-until-sept-15-trade-group-says.html.

68. Martha Honey, editor. (2019). *Cruise Tourism in the Caribbean: Selling Sunshine.* Oxford, UK: Routledge Press.

69. Cruise Advisor. (June 16, 2020). "Cruise Adviser Survey: Agents See Bookings Spike." https://cruise-adviser.com/cruise-adviser-survey-agents-see-bookings-spike/.

70. Suzanne Rowan Kelleher. (March 22, 2020). "Poll: Americans Are Wary about Traveling after COVID-19 Curve Flattens." The Harris Poll. https://theharrispoll.com/poll-americans-are-wary-about-traveling-after-covid-19-curve-flattens/.

71. "An Open Letter to Cruise Giants." (June 7, 2020). Advertisement. *Miami Herald.* https://www.stand.earth/sites/stand/files/openlettertocruise-giants-miamiheraldad.pdf.

72. Future of Tourism Coalition. (2020). https://www.futureoftourism.org/.

Contributors

Martha Honey is the cofounder and director emeritus of the Center for Responsible Travel (CREST). She led CREST as its executive director for sixteen years before transitioning to her project-based role of director emeritus in 2019. Honey has written and lectured widely on ecotourism, impact tourism, cruise and resort tourism, coastal and marine tourism, climate change, and certification issues. Her books include *Coastal Tourism, Sustainability, and Climate Change in the Caribbean*, Vols. 1 and 2, *and Marine Tourism, Climate Change, and Resilience in the Caribbean*, Vols. 1 and 2 (Business Expert Press, 2017); *Ecotourism and Sustainable Development: Who Owns Paradise?* (Island Press, 1999 and 2008); and *Ecotourism and Certification: Setting Standards in Practice* (Island Press, 2002). She is the executive producer of CREST's 2016 film *Caribbean 'Green' Travel: Your Choices Make a Difference*, released in May 2016. Most recently, she is an editor and author of a new study on cruise tourism, published in Spanish as *Por el Mar de las Antillas: 50 Años de Turismo de Cruceros en el Caribe* and in English as *Cruise Tourism in the Caribbean: Selling Sunshine*. Previously, Honey worked for twenty years as a journalist based in East Africa and Central America. She holds a PhD in African history from the University of Dar-es-Salaam, Tanzania.

Kelsey Frenkiel is a program manager at CREST, where she coordinates research and advocacy to promote responsible travel policies and practices globally. Her particular area of focus is in tourism and wildlife conservation. Frenkiel is also a freelance travel writer and researcher, and her work has appeared in *National Geographic Traveler*, *The Washingtonian*, and other outlets. She holds a bachelor's degree in anthropology from the College of William and Mary and a master of science degree in primate conservation from Oxford Brookes University, where she studied the intersection between tourism and slow loris conservation in Java, Indonesia.

Albert Arias-Sans is the head of the Strategic Tourism Plan for Tourism Barcelona 2020.

Arias-Sans is a trained geographer and holds a master's degree in urban management from Erasmus University, Rotterdam. He has worked as a researcher, consultant, and lecturer in urban and tourism issues since 2004 and is currently an associate lecturer at the Universitat de Barcelona, as well as a PhD candidate at the Universitat Rovira i Virgili.

Roberta Atzori is an assistant professor at California State University, Monterey Bay, where she teaches and conducts research within the Sustainable Hospitality Management program. Her research interests include sustainable tourism, ecotourism, and climate change mitigation and adaptation in tourism destinations. Atzori has been a CREST academic affiliate since 2017. She earned her PhD in hospitality management from the Rosen College of Hospitality Management, University of Central Florida.

Richard Bangs has often been called the father of modern adventure travel and the pioneer in travel that makes a difference, travel with a purpose. He has spent thirty years as an explorer and communicator and along the way led first descents of thirty-five rivers around the globe, including the Yangtze in China and the Zambezi in southern Africa. He is coauthor, with Pasquale Scaturro, of the 2005 book *Mystery of the Nile*. Bangs's 1999 book, *The Lost River: A Memoir of Life, Death, and transformation on Wild Water*, won both the National Outdoor Book Award in the literature category and the Lowell Thomas Award for best travel book. His PBS series, *Adventures with Purpose*, has won a series of broadcast awards and honors. He is on the board of CREST, cofounded www.MTSobek.com, was on the founding executive team of www.Expedia.com, and cofounded the travel storytelling platform www.steller.co.

James R. Barborak is codirector of the Center for Protected Area Management at Colorado State University. A World Commission on Protected Areas member, Barborak has worked in more than twenty-five countries on protected area management; conservation finance, policy, and governance; capacity building; and development of opportunities for ecotourism and outdoor recreation. He has also been actively involved in efforts to increase benefits to local communities and Indigenous people living in and around protected areas. His bachelor's and master's degrees in natural resources are from Ohio State University, and he took additional coursework at Yale University's School of Forestry and Environmental Studies.

Elizabeth Becker is the author of *Overbooked: The Exploding Business of Travel and Tourism*, which was named an Amazon Book of the Year in 2013 and was considered "required reading" by Arthur Frommer. An award-winning journalist, Becker has been a war correspondent in Cambodia for the *Washington Post*, a senior foreign editor of National Public Radio, and an international economics correspondent for the *New York Times*. She has reported from Asia, Europe, the Middle East, Africa, South America, and Australia.

Christina Beckmann is a senior director for strategy and impact at the Adventure Travel Trade Association. She has twenty years of consulting and research experience working at the intersection of tourism, environment, economic development, and entrepreneurship. A frequent collaborator and speaker, Beckmann's writing can be found in numerous travel trade, academic, and consumer publications. She holds a BA in communication from Cornell University; an MBA in entrepreneurship from American University; and an MA in communication, culture, and technology from Georgetown University. Originally from Alaska, Beckmann now lives in San Francisco, California, with her family.

Thiago do Beraldo-Souza has worked for the Chico Mendes Institute for Biodiversity Conservation (ICMBio) since 2002. Thiago is currently the coordinator of ecotourism for ICMBio and is in charge of the institutional agendas for ecotourism planning, economic impact analysis, community-based tourism, and long-distance trails. In 2016, he earned a PhD in interdisciplinary ecology with a focus on tourism and recreation management from the University of Florida. His research there included the development of a methodology to measure the economic impacts of tourism in protected areas of Brazil. Thiago is a member of IUCN's World Commission on Protected Areas and serves as knowledge development coordinator of its Tourism and Protected Areas Working Group.

David Blanton is the director of Serengeti Watch, a project of Earth Island Institute, a US-based nonprofit conservation organization. Serengeti Watch raises funds for community conservation, advocates for conservation, and promotes responsible tourism through its program Friends of Serengeti. Blanton served in Uganda as a Peace Corps volunteer and developed a social studies curriculum for Kenya's Utalii College. He is the founder of the International Galapagos Tour Operators Association, which raises funds from tour companies and travelers for conservation. He has a bachelor's degree in science and anthropology from Indiana University and a master's degree in communications from Fairfield University.

Robyn Bushell is a professor in heritage and tourism studies at the School of Social Science and the Institute for Culture and Society at Western Sydney University in Australia. Working at the interface of heritage, tourism planning, and community well-being, mostly in Southeast Asia, her research focuses on policy, planning, and capacity building. She collaborates with many UN agencies, particularly the United Nations Educational, Scientific and Cultural Organization (UNESCO), the World Health Organization (WHO), and the United Nations World Tourism Organization (UNWTO), as well as the International Union for Conservation of Nature (IUCN), the Association of Southeast Asian Nations (ASEAN), and aid agencies on strategic visitor management in world heritage destinations. Through this work, she aims to contribute to the following sustainable development goals: good health and well-being, reduced inequalities, and more. She has a PhD in community planning from the University of Sydney.

Frank Haas is the president of Marketing Management, a consultancy focused on travel and tourism in Hawai'i and internationally. He was previously vice president and director of marketing for the

Hawai'i Tourism Authority and has worked on three of the authority's strategic plans. Haas has held executive positions in tourism and hospitality representing airlines, destinations, attractions, and tour operators. Internationally, he has worked on projects in Asia, North Africa, and the Middle East. He is a former national chair of the American Marketing Association. Haas holds a bachelor's degree and an MBA in advertising, both from Northwestern University.

Carter A. Hunt is an associate professor of recreation, park, and tourism management and anthropology at Penn State University. His research focuses on the relationship between tourism development and biodiversity conservation. He has conducted fieldwork in Colombia, Costa Rica, Ecuador, Guatemala, Nicaragua, Peru, and Tanzania. Hunt is currently a Fulbright scholar working in coordination with the Charles Darwin Foundation to assess impacts of tourism on the Galapagos Islands' social and environmental systems. He holds a bachelor of arts degree from the University of Kentucky and master of science and PhD degrees from Texas A&M University and is a former postdoctoral fellow at Stanford University.

Andrea Insch is an associate professor at the University of Otago, Dunedin, New Zealand. Before undertaking her doctorate at Griffith University in Brisbane, Australia, she worked at Queensland's Department of State Development. In 2005, Insch moved to New Zealand to join the marketing department at the Otago Business School. Her interdisciplinary studies involved connecting marketing, urban studies, and tourism. She is on the editorial advisory board of the *Journal of Management History*, a member of the editorial board of *British Food Journal,* and is the book review editor and regional editor (Australia and New Zealand) for *Place Branding and Public Diplomacy*.

Birendra KC is an assistant professor in the Department of Hospitality and Tourism Management at the University of North Texas. He is also a visiting and affiliate professor of international sustainable tourism for the Tropical Agricultural Research and Higher Education Center in Costa Rica. KC's research focus is on nature-based tourism, policy, and planning in sustainable tourism, community-based tourism, and protected area management. He received his BS in forestry from Tribhuvan University, MS in forestry from the University of Kentucky, and PhD in parks, recreation, and tourism management from North Carolina State University.

Laura Kasa is a consultant for Kasa Consulting. As the executive director for Save Our Shores, she collaborated with local government, businesses, and community groups to lead the effort of banning single-use plastics in forty-two jurisdictions. She left the organization in 2015 and later started her own consulting firm. Her consulting project for California State University, Monterey Bay helped coordinate ecotourism conferences in Monterey in 2017 and 2019. Kasa holds a bachelor's degree in Spanish language from Villanova University and a master's degree in environmental policy from Columbia University.

Aileen Lamb is tourism and creative economy manager with Scottish Enterprise and has worked in Scottish tourism for more than twenty-five years. At Scottish Enterprise, Scotland's economic development agency, she has developed the organization's approach to destination development, informed government policy on tourism management, and jointly created the award-winning Tourism Destination Leaders Programme. Her recent work has been with national and international partners to develop a SMART (technology and data-led) approach to Scotland's future tourism delivery. Lamb is a founding board member of Women in Tourism and a board member of National Museums of Scotland Enterprises. She is an alumnus of Glasgow Caledonian University.

Juarez Michelotti has worked for the Social Service of Commerce in São Paulo, Brazil, since 2009. He helped lead the planning team for the organization's Bertioga private nature preserve along Brazil's southeastern coast and now coordinates its management. He has designed and taught short courses related to trails, universal design, protected area management, and infrastructure and environmental interpretation. Previously, he headed the Volta Velha Outdoor Education Center. Michelotti, a forester, earned his bachelor of science degree from the Federal University of Paraná, and he has a specialization in biodiversity from Joinville University. He is currently a master's student in Colorado State University's Conservation Leadership Program.

Louise Norton is the owner of Apu Lodge in Ollantaytambo, Peru. She lived in Peru between 2003 and 2017, working with the porters on the Inca Trail to improve their working conditions, as well as running her inn in Ollantaytambo. In Peru, she also worked with the Alma Children's Education Foundation and was vice president of the Association of Hotels and Restaurants of Ollantaytambo. Previously, she worked as a researcher for *The Guardian* journalist Madeleine Bunting and at Cambridge University. She has a bachelor of arts degree from Cambridge University and master of science in responsible tourism management from Leeds Beckett University, United Kingdom.

Aina Pedret is a project officer at the Strategic Plan for Tourism 2020 Barcelona in Spain and was responsible for the Tourism Mobility Strategy for the plan. She is a trained geographer, and since 2007, she has worked as a consultant, researcher, and teacher in urban management and planning, specializing in the fields of mobility and tourism.

Paulo Eduardo Pereira-Faria is an environmental analyst and specialist in recreation and ecotourism at the Chico Mendes Institute for Biodiversity Conservation in Brazil. He has coordinated, implemented, and collaborated with trail projects, visitation planning, and development of ecotourism and recreation opportunities in more than forty Brazilian protected areas. He has also coordinated training in tourism planning, visitation monitoring, and trail management and participated in technical teams to prepare general management plans for Brazilian parks and reserves. He has a bachelor's degree in biological sciences from the Federal University of Santa Catarina, Brazil, and participated in Colorado State University's protected areas management course.

Julie Regan is the chief of external affairs and deputy director for the Tahoe Regional Planning Agency, where she has served as an executive since 2003. She has worked in the fields of professional communications and government affairs since the 1990s in both the public and private sectors. Regan is currently pursuing a PhD in environmental science at the University of Nevada, Reno. Her previous experience ranges from publishing *Treasures by the Sea* magazine to running the marketing and real estate arm of ResortQuest International in Bethany Beach, Delaware. She also worked in the water, electric, and natural gas utility industries. Regan has a bachelor of arts degree in communications from the University of Delaware and a master's in journalism from Temple University in Philadelphia. She is accredited by the Public Relations Society of America and is the cochair of the nationally focused Network for Landscape Conservation.

Nathan Reigner is an applied social scientist who focuses on tourism and visitor use of parks, protected areas, and natural and cultural heritage areas. He is a research affiliate of the Icelandic Tourism Research Center and was the 2019 Icelandic Ministry of Foreign Affairs Arctic Fulbright Scholar. Reigner's work helps special and sensitive places plan and manage for sustainable tourism and recreational use. Nathan has worked throughout the US National Park system, as well as in Iceland, Greenland, the Faroe Islands, Russia, and Oman. He holds a PhD in natural resource management from the University of Vermont.

María Reynisdóttir is a specialist in the Department of Tourism at Iceland's Ministry of Industries and Innovation. Prior to joining the ministry in 2015, she worked in the field of tourism marketing, first at Visit Reykjavik and later at leading travel agency, Hey Iceland. In her job at the ministry, Reynisdóttir is involved in projects concerning tourism policy, carrying capacity, infrastructure development, and destination management. She also serves on the Organisation for Economic Co-operation and Development's tourism committee on behalf of Iceland. Reynisdóttir holds a master of science degree in tourism planning and development from the University of Surrey, United Kingdom.

Dan Riccio is the director of the Department of Livability and Tourism for the City of Charleston, South Carolina. He began his career with the city as a Charleston police officer in 1988 and retired from the police department at the rank of lieutenant in 2010. He currently oversees code enforcement operations pertaining to residential and commercial property standards, tourism management and enforcement, short-term rental enforcement, and special events management for the Department of Livability and Tourism. Riccio earned his bachelor of science degree in business management from Limestone College and master's in human resources management from Webster University.

Cathy Ritter is director of the Colorado Tourism Office. She leads a $20 million initiative to maximize the potential of traveler spending in a state where the stakes for managing tourism success are high. In 2016, she initiated the development of the Colorado Tourism Roadmap. She also led an initiative to establish eight new Colorado travel regions, and she directs a national marketing campaign that consistently ranks among the top 10 percent in the United States for return on investment. Ritter serves on the executive committee of the U.S. Travel Association and is the incoming chair of the National Council of State Tourism Directors. In 2019, she was named as the Colorado Hotel and Lodging Association's Industry Partner of the Year. Previously, she served four years as Illinois state travel director.

Andrea Sachs has been an award-winning reporter with the *Washington Post* since 1997. Since 2000, she has been covering the travel industry for the Travel section and has also written pieces for the *Post Magazine*, *Style*, and *KidsPost*. In her free time, she works with rescue animals, fostering dogs and cats, and serving as a flight volunteer in Puerto Rico and Colombia. Sachs received her bachelor's degree in art history at Tufts University and a master's degree in journalism from Northwestern University.

Natalia Sánchez Castro is a project officer at the Strategic Plan for Tourism 2020 Barcelona in Spain. She was responsible for the Territorial Strategy for the plan. She is a trained geographer, and since 2014, she has worked as a consultant specializing in the fields of sustainability, strategic planning, urban management, tourism, and local economic development.

Francesca Street is a London-based digital journalist at CNN Travel. At CNN, she investigates topics including the growth of facial recognition at airports and overtourism. In 2019, Street won a Travel Media Award and MHP's 30 under 30 Young Journalists Gold Award in the Culture, Entertainment and Lifestyle category. She was also a finalist in the 2019 Business Travel Journalism Awards. Francesca's writing has been featured in *The Telegraph*, *Herald-Scotland*, and *The Independent*. In 2017, she was a runner-up in *The Telegraph*'s Cassandra Jardine Prize. Street graduated from the University of Edinburgh in 2016 with an master's degree (Hons.) in English literature.

Jonathan B. Tourtellot is the chief executive officer of the Destination Stewardship Center (DSC) and a National Geographic Fellow Emeritus. During a thirty-one-year career as a senior editor at the National Geographic Society, he founded and directed the society's former Center for Sustainable Destinations, now the independent DSC. While at National Geographic, he introduced the concept of geotourism, based on sustaining or enhancing the geographical character of a place, and instituted the landmark Destination Stewardship surveys reported annually in *National Geographic Traveler*, 2004–2010. Tourtellot is the primary author of the Geotourism Charter, a set of stewardship principles adopted by the Organization of American States. He writes about destination stewardship, sustainable tourism, World Heritage, and climate change.

Arnie Weissmann is editor in chief of *Travel Weekly*. He was the founder of the travel industry's first destination information service, Weissmann Travel Reports; authored a popular geography textbook; and was group publisher of critical hotel and destination guides in the United States and United Kingdom. He is a regular contributor to the PBS program *The Travel Detective* and frequently speaks at events in and outside the travel industry. He is on the board of the nonprofit organization Tourism Cares and has been a guest lecturer at Cornell University School of Hotel Administration. Weissmann holds a BA in creative writing from the University of Illinois.

Kaitie Worobec is a tourism consultant specializing in sustainable destination management and development. She has a particular interest in using a data-driven approach to help destination stakeholders better understand and measure the broader impact of tourism. Worobec has worked with organizations such as Airbnb, the United Nations' World Tourism Council, and the Adventure Travel Trade Association, as well as several destination marketing organizations in Canada and domestic and international tour operators. She holds a bachelor of commerce degree from the University of Alberta and a master of science degree in responsible tourism management from Leeds Beckett University in the United Kingdom.

About CREST

The Center for Responsible Travel (CREST) is a unique policy-oriented research organization dedicated to increasing the positive global impact of responsible tourism. CREST assists governments, policy makers, tourism businesses, nonprofit organizations, and international agencies with finding solutions to critical issues confronting tourism, the world's largest service industry. CREST provides interdisciplinary analysis and innovative solutions through field projects, research, consultancies, and outreach, recognizing tourism's potential as a tool for poverty alleviation and biodiversity conservation.

Center for Responsible Travel

Acknowledgments

As we were finishing up the final sections of this book on overtourism in early 2020, the global novel coronavirus pandemic hit, and tourism, which had been on an upward trajectory for years, was brought to a standstill. Overtourism, the hot-button issue of concern through 2019, was replaced—almost overnight—by no tourism. These twin, polar opposite calamities of overtourism and pandemic-induced no tourism are now and forever historically tied together.

Tourism will return in time, and in some destinations where adequate management and community engagement are not in place, so, too, will overtourism. The journey of recovery offers opportunities to build on what we've learned from both overtourism and the pandemic. As is laid out here, lessons from overtourism—both the factors that cause it and the remedies that serve to control it—can help build back tourism with resilience and sustainability. Indeed, the concluding chapter in this book reflects on recovery from the pandemic that offers a "new normal," tilting toward small group, outdoors, and crowd-free types of travel. This approach appears to bode well for curbing overtourism and strengthening and broadening the definition of sustainable travel with a stronger focus on health and safety.

Overtourism: Lessons for a Better Future begins with a broad overview from the point of view of the travel industry before diving deeper into five different types of destinations: historic cities, national parks, World Heritage Sites, beaches and coastlines, and countries and regions. The lead essays for each chapter are written by experienced travel writers and editors who have been covering overtourism for years. The tone is set in the foreword by Elizabeth Becker, author of the seminal 2013 work *Overbooked*, which chronicled the arrival of the age of overtourism. As a former *New York Times* foreign correspondent, Becker also foreshadowed the central role that journalists, including the travel media, would play in documenting the impacts of overtourism around the world. This volume is enriched by the deep on-the-ground expertise, thoughtful analysis, and engaging writing styles of these journalists as they offer insights into how overtourism has affected each type of destination and what measures are being attempted to reign it in.

The lead essays are complemented by a rich panorama of case studies examining how overtourism is playing out in well-loved travel places around the world, from Machu Picchu to Mount Everest, Barcelona to Banff, and Big Sur to the Serengeti. The authors include academics, government officials, consultants, community organizers, nonprofit leaders, and tourism professionals who live or work in the destinations they write about. We are grateful to all these contributors for the high quality of their submissions and for the time and care they gave to the final spit and polish of their work.

Undertaking an edited volume like this one is challenging and has required significant behind-the-scenes assistance from the staff and research interns at the Center for Responsible Travel (CREST). We would especially like to thank three of our CREST colleagues who have generously and expertly contributed to this book: Samantha Bray, managing director, and Ellen Rugh, program manager, who contributed through research, manuscript review, and copyediting assistance; and Rebekah Stewart, director of communications, who provided promotional support for the book. Our bank of exceptional research interns included Emily Ganem, Christopher Gillespie, Ariel Klein, Cassie McCabe, Madison Mitchell, Megan Reese, and Rashi Tolani. We thank them for their competent and diligent assistance on tasks large and small, including research, copyediting, and reviewing essays and case studies, and photo selection and permissions. We also give special thanks and appreciation to our colleague and friend Jonathan B. Tourtellot for his keen, soup-to-nuts interest in this book and his ongoing assistance with its content and framing.

Finally, it is a pleasure and an honor to have Island Press as our publisher. We are grateful to senior editor Emily Turner and president David Miller for initially accepting the book proposal and enthusiastically promoting it to their team. And it has been an ongoing pleasure to have Emily as our editor. Her wise counsel, sound advice, gentle nudges, and competent editing have improved the quality of this work and shepherded it over the finish line. We thank you for your support and hope you are as proud as we are of this volume. We extend our gratitude to you all.

Martha Honey and Kelsey Frenkiel

Indagare®

How you travel matters

THE CURTIS & EDITH MUNSON FOUNDATION

Index

Page numbers followed by "f" indicate figures.

Island Press | Board of Directors